THE PHYSICS BOOK

Books by Clifford A. Pickover

The Alien IQ Test
Archimedes to Hawking
A Beginner's Guide to Immortality
Black Holes: A Traveler's Guide
Calculus and Pizza
Chaos and Fractals
Chaos in Wonderland
Computers, Pattern, Chaos, and Beauty
Computers and the Imagination
Cryptorunes: Codes and Secret Writing
Dreaming the Future
Egg Drop Soup
Future Health
Fractal Horizons: The Future Use of Fractals
Frontiers of Scientific Visualization
The Girl Who Gave Birth to Rabbits
The Heaven Virus
Jews in Hyperspace
Keys to Infinity
Liquid Earth
The Lobotomy Club
The Loom of God
The Mathematics of Oz
The Math Book
Mazes for the Mind: Computers and the Unexpected
Mind-Bending Visual Puzzles (calendars and card sets)
The Möbius Strip
The Paradox of God and the Science of Omniscience
A Passion for Mathematics
The Pattern Book: Fractals, Art, and Nature
The Science of Aliens
Sex, Drugs, Einstein, and Elves
Spider Legs (with Piers Anthony)
Spiral Symmetry (with Istvan Hargittai)
Strange Brains and Genius
Sushi Never Sleeps
The Stars of Heaven
Surfing through Hyperspace
Time: A Traveler's Guide
Visions of the Future
Visualizing Biological Information
Wonders of Numbers

THE PHYSICS BOOK

From the Big Bang to Quantum Resurrection, 250 Milestones in the History of Physics

Clifford A. Pickover

An Imprint of Sterling Publishing
387 Park Avenue South
New York, NY 10016

ISBN 978-1-4027-7861-2

Library of Congress Cataloging-in-Publication Data

Pickover, Clifford A.
The physics book : from the Big Bang to Quantum Resurrection,
250 milestones in the history of physics / Clifford A. Pickover.
p. cm.
Includes bibliographical references and index.
ISBN 978-1-4027-7861-2 (alk. paper)
1. Physics--History. I. Title.
QC7.P49 2011
530.09--dc22

2010051365

Distributed in Canada by Sterling Publishing
c/o Canadian Manda Group, 165 Dufferin Street
Toronto, Ontario, Canada M6K 3H6
Distributed in the United Kingdom by GMC Distribution Services
Castle Place, 166 High Street, Lewes, East Sussex, England BN7 1XU
Distributed in Australia by Capricorn Link (Australia) Pty. Ltd.
P.O. Box 704, Windsor, NSW 2756, Australia

For information about custom editions, special sales, and premium and corporate purchases, please contact
Sterling Special Sales at 800-805-5489 or specialsales@sterlingpublishing.com.

Manufactured in China

6 8 10 9 7

www.sterlingpublishing.com

"We all use physics every day. When we look in a mirror, or put on a pair of glasses, we are using the physics of optics. When we set our alarm clocks, we track time; when we follow a map, we navigate geometric space. Our mobile phones connect us via invisible electromagnetic threads to satellites orbiting overhead. But physics is not all about technology. . . . Even the blood flowing through arteries follows laws of physics, the science of our physical world."

—Joanne Baker, *50 Physics Ideas You Really Need to Know*

"The great equations of modern physics are a permanent part of scientific knowledge, which may outlast even the beautiful cathedrals of earlier ages."

—Steven Weinberg, in Graham Farmelo's *It Must Be Beautiful*

Contents

Introduction

The Scope of Physics

> *"As the island of knowledge grows, the surface that makes contact with mystery expands. When major theories are overturned, what we thought was certain knowledge gives way, and knowledge touches upon mystery differently. This newly uncovered mystery may be humbling and unsettling, but it is the cost of truth. Creative scientists, philosophers, and poets thrive at this shoreline."*
>
> —W. Mark Richardson, "A Skeptic's Sense of Wonder," *Science*

The American Physical Society, today's leading professional organization of physicists, was founded in 1899, when 36 physicists gathered at Columbia University with a mission to advance and disseminate knowledge of physics. According to the society:

> Physics is crucial to understanding the world around us, the world inside us, and the world beyond us. It is the most basic and fundamental science. Physics challenges our imaginations with concepts like relativity and string theory, and it leads to great discoveries, like computers and lasers, that change our lives. Physics encompasses the study of the universe from the largest galaxies to the smallest subatomic particles. Moreover, it's the basis of many other sciences, including chemistry, oceanography, seismology, and astronomy.

Indeed, today physicists roam far and wide, studying an awesome variety of topics and fundamental laws in order to understand the behavior of nature, the universe, and the very fabric of reality. Physicists ponder multiple dimensions, parallel universes, and the possibilities of wormholes connecting different regions of space and time. As suggested by the American Physical Society, the discoveries of physicists often lead to new technologies and even change our philosophies and the way we look at the world. For example, for many scientists, the Heisenberg Uncertainty Principle asserts that the physical universe literally does not exist in a determinist form, but is rather a mysterious collection of probabilities. Advances in the understanding of electromagnetism led to the invention of the radio, television, and computers. Understanding of thermodynamics led to the invention of the car.

As will become apparent as you peruse this book, the precise scope of physics has not been fixed through the ages, nor is it easily delimited. I have taken a rather wide view and have included topics that touch on engineering and applied physics, advances in our understanding of astronomical objects, and even a few topics that are quite philosophical. Despite this large scope, most areas of physics have in common a strong reliance on mathematical tools to aid scientists in their understandings, experiments, and predictions about the natural world.

Albert Einstein once suggested that the most incomprehensible thing about the world is that it is comprehensible. Indeed, we appear to live in a cosmos that can be described or approximated by compact mathematical expressions and physical laws. However, beyond discovering these laws of nature, physicists often delve into some of the most profound and mind-boggling concepts that humans have ever contemplated—topics ranging from relativity and quantum mechanics to string theory and the nature of the Big Bang. Quantum mechanics gives us a glimpse of a world that is so strangely counterintuitive that it raises questions about space, time, information, and cause and effect. However, despite the mysterious implications of quantum mechanics, this field of study is applied in numerous fields and in technologies that include the laser, the transistor, the microchip, and magnetic resonance imaging.

This book is also about the *people* behind many of the great ideas of physics. Physics is the foundation of modern science, and it has fascinated men and women for centuries. Some of the world's greatest and most intriguing minds, such as Isaac Newton, James Clerk Maxwell, Marie Curie, Albert Einstein, Richard Feynman, and Stephen Hawking, have dedicated themselves to the advancement of physics. These individuals have helped change the way we look at the cosmos.

Physics is among the most difficult of sciences. Our physics-centered description of the universe grows forever, but our brains and language skills remain entrenched. New kinds of physics are uncovered through time, but we need fresh ways to think and to understand. When German theoretical physicist Werner Heisenberg (1901–1976) worried that human beings might never truly understand atoms, Danish physicist Niels Bohr (1885–1962) provided optimism. In the early 1920s, he replied, "I think we may yet be able to do so, but in the process we may have to learn what the word understanding really means." Today, we use computers to help us reason beyond the limitations of our own intuition. In fact, experiments with computers are leading physicists to theories and insights never dreamed of before the ubiquity of computers.

A number of prominent physicists now suggest that universes exist that are parallel to ours, like layers in an onion or bubbles in a milkshake. In some theories of parallel universes, we might actually detect these universes by gravity leaks from one universe to an adjacent universe. For example, light from distant stars may be distorted by the gravity of invisible objects residing in parallel universes only millimeters away. The entire idea of multiple universes is not as far-fetched as it may sound. According to a poll of 72 leading physicists conducted by the American researcher David Raub and published in 1998, 58% of physicists (including Stephen Hawking) believe in some form of multiple universes theory.

The Physics Book ranges from theoretical and eminently practical topics to the odd and perplexing. In what other physics book will you find the 1964 hypothesis of the subatomic God Particle next to the 1965 ultra-bouncy Super Ball, which led to a craze that swept across America? We'll encounter mysterious dark energy, which may one day tear apart galaxies and end the universe in a terrible cosmic rip, and the blackbody radiation law, which started the science of quantum mechanics. We'll muse about the Fermi Paradox, which involves communication with alien life, and we'll ponder a prehistoric nuclear reactor discovered in Africa that has been operating for two billion years. We'll discuss the race to create the blackest black color ever created—more than 100 times darker than the paint on a black car! This "ultimate black" color may one day be used to more efficiently capture energy from the Sun or to design extremely sensitive optical instruments.

Each book entry is short—at most only a few paragraphs in length. This format allows readers to jump in to ponder a subject, without having to sort through a lot of verbiage. When was the first time humans glimpsed the far side of the moon? Turn to the entry on "Dark Side of the Moon" for a brief introduction. What is the riddle of the ancient Baghdad batteries, and what are black diamonds? We'll tackle these and other thought-provoking topics in the pages that follow. We'll wonder if reality could be nothing more than an artificial construct. As we learn more about the universe and are able to simulate complex worlds on computers, even serious scientists begin to question the nature of reality. Could we be living in a computer simulation?

In our own small pocket of the universe, we've already developed computers with the ability to simulate lifelike behaviors using software and mathematical rules. One day, we may create thinking beings that live in rich simulated spaces—in ecosystems as complex and vibrant as the Madagascar rain forest. Perhaps we'll be able to simulate reality itself, and it is possible that more advanced beings are already doing this elsewhere in the universe.

Purpose and Chronology

Examples of physical principles are all around us. My goal in writing *The Physics Book* is to provide a wide audience with a brief guide to important physics ideas and thinkers, with entries short enough to digest in a few minutes. Most entries are ones that interested me personally. Alas, not all of the great physics milestones are included in this book in order to prevent the book from growing too large. Thus, in celebrating the wonders of physics in this short volume, I have been forced to omit many important physics marvels. Nevertheless, I believe that I have included a majority of those with historical significance and that have had a strong influence on physics, society, or human thought. Some entries are practical and even fun, involving topics that range from pulleys, dynamite, and lasers to integrated circuits, boomerangs, and Silly Putty. Occasionally, I include several strange or even crazy-sounding philosophical concepts or oddities that are nonetheless significant, such as quantum immortality, the anthropic principle, or tachyons. Sometimes, snippets of information are repeated so that each entry can be read on its own. Occasional text in bold type points the reader to related entries. Additionally, a small "See also" section near the bottom of each entry helps weave entries together in a web of interconnectedness and may help the reader traverse the book in a playful quest for discovery.

The Physics Book reflects my own intellectual shortcomings, and while I try to study as many areas of physics as I can, it is difficult to become fluent in all aspects. This book clearly reflects my own personal interests, strengths, and weaknesses, as I am responsible for the choice of pivotal entries included in this book and, of course, for any errors and infelicities. Rather than being a comprehensive or scholarly dissertation, it is intended as recreational reading for students of science and mathematics as well as interested lay people. I welcome feedback and suggestions for improvement from readers, as I consider this an ongoing project and a labor of love.

This book is organized chronologically, according to the year associated with an entry. For most entries, I used dates that are associated with the discovery of a concept or property. However, in the "Setting the Stage" and "Closing the Curtain" sections, I have used dates associated with an actual (or hypothetical) happening, such as a cosmological or astronomical event.

Of course, dating of entries can be a question of judgment when more than one individual made a contribution. Often, I have used the earliest date where appropriate, but sometimes, after surveying colleagues and other scientists, I have decided to use the date when a concept gained particular prominence. For example, many dates could have been assigned to the entry "Black Holes," given that certain kinds of black holes may have formed as far back as during the Big Bang, about 13.7 billion years ago. However, the term black hole wasn't coined until 1967—by theoretical physicist John Wheeler. In the final analysis, I used the date when human ingenuity first allowed scientists to rigorously formulate the idea of black holes. Thus, the entry is dated as 1783, when geologist John Michell (1724–1793) discussed the concept of an object so massive that light could not escape. Similarly, I assign the date of 1933 to "Dark Matter," because during this year, Swiss astrophysicist Fritz Zwicky (1898–1974) observed the first evidence that implied the possible existence of mysterious, non-luminous particles. The year 1998 is assigned to "Dark Energy" because this was not only the year that the phrase was coined but also the time when observations of certain supernovae suggested that the expansion of the universe is accelerating.

Many of the older dates in this book, including the "B.C. dates," are only approximate (e.g. dates for the Baghdad Battery, the Archimedean Screw, and others). Rather than place the term "circa" in front of all of these older dates, I inform the reader here that both the ancient dates and the dates in the very far future are only rough estimates.

Readers may notice that a significant number of discoveries in basic physics also led to a range of medical tools and helped to reduce human suffering and save lives. Science writer John G. Simmons notes, "Medicine owes most of its tools for imaging the human body to twentieth-century physics. Within weeks of their discovery in 1895, the mysterious X-rays of Wilhelm Conrad Röntgen were used in diagnoses. Decades later, laser technology was a practical result of quantum mechanics. Ultrasonography emerged from problem solving in submarine detection, and CT scans capitalized on computer technology. Medicine's most significant recent technology, used for visualizing the interior of the human body in three-dimensional detail, is magnetic resonance imaging (MRI)."

Readers will also notice that a significant number of milestones were achieved in the twentieth century. To place the dates in perspective, consider the scientific revolution that occurred roughly during the period between 1543 and 1687. In 1543, Nicolaus Copernicus published his heliocentric theory of planetary motion. Between 1609 and 1619, Johannes Kepler established his three laws that described the paths of the planets about the Sun, and in 1687, Isaac Newton published his fundamental laws of motion and gravity. A second scientific revolution occurred between 1850 and 1865, when scientists introduced and refined various concepts concerning energy and entropy. Fields of study such as thermodynamics, statistical mechanics, and the kinetic theory of gases began to blossom. In the twentieth century, quantum theory and special and general relativity were among the most important insights in science to change our views of reality.

In the entries for this book, sometimes science reporters or famous researchers are quoted in the main entries, but purely for brevity I don't immediately list the source of the quote or the author's credentials. I apologize in advance for this occasional compact approach; however, references in the back of the book will help to make the author's identity clearer.

Because this book has entries ordered chronologically, be sure to use the index when hunting for a favorite concept, which may be discussed in entries that you might not have expected. For example, the concept of quantum mechanics is so rich and diverse that no single entry is titled "Quantum Mechanics." Rather, the reader can find intriguing and key aspects in such entries as Blackbody Radiation Law, Schrödinger's Wave Equation, Schrödinger's Cat, Parallel Universes, Bose-Einstein Condensate, Pauli Exclusion Principle, Quantum Teleportation, and more.

Who knows what the future of physics will offer? Toward the end of the nineteenth century, the prominent physicist William Thomson, also known as Lord Kelvin, proclaimed the end of physics. He could never have foreseen the rise of quantum mechanics and relativity—and the dramatic changes these areas would have on the field of physics. Physicist Ernest Rutherford, in the early 1930s, said of atomic energy: "Anyone who expects a source of power from the transformation of these atoms is talking moonshine." In short, predicting the future of the ideas and applications of physics is difficult, if not impossible.

In closing, let us note that discoveries in physics provide a framework in which to explore the subatomic and supergalactic realms, and the concepts of physics allow scientists to make predictions about the universe. It is a field in which philosophical speculation can provide a stimulus for scientific breakthroughs. Thus, the discoveries in this book are among humanity's greatest achievements. For me, physics cultivates a perpetual state of wonder about the limits of thoughts, the workings of the universe, and our place in the vast space-time landscape that we call home.

Acknowledgments

I thank J. Clint Sprott, Leon Cohen, Dennis Gordon, Nick Hobson, Teja Krašek, Pete Barnes, and Paul Moskowitz for their comments and suggestions. I would also like to especially acknowledge Melanie Madden, my editor for this book.

While researching the milestones and pivotal moments in this book, I studied a wide array of wonderful reference works and Web sites, many of which are listed in the "Notes and Further Reading" section at the end of this book. These references include Joanne Baker's *50 Physics Ideas You Really Need to Know*, James Trefil's *The Nature of Science*, and *Peter Tallack's The Science Book*. On-line sources such as Wikipedia (en.wikipedia.org) are a valuable starting point for readers and can be used as a launch-pad to more information.

I should also point out that some of my own previous books, such as *Archimedes to Hawking: Laws of Science and the Great Minds Behind Them*, provided background information for some of the entries dealing with physical laws, and the reader is urged to consult this book for additional information.

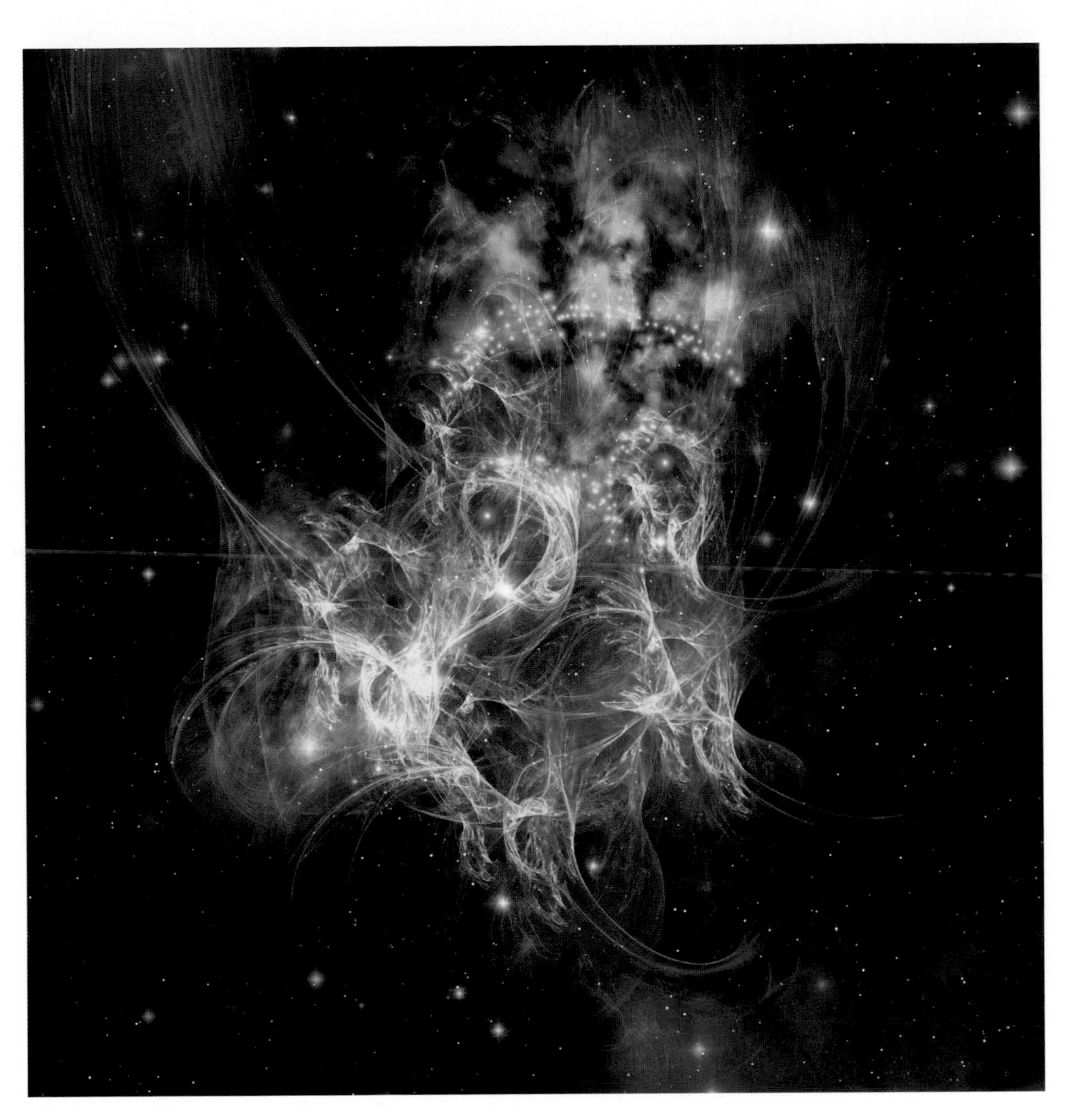

"I'll tell you what the Big Bang was, Lestat. It was when the cells of God began to divide."

—Anne Rice, *Tale of the Body Thief*

Big Bang

Georges Lemaître (1894–1966), **Edwin Hubble** (1889–1953), **Fred Hoyle** (1915–2001)

In the early 1930s, Belgian priest and physicist Georges Lemaître proposed what became known as the Big Bang theory, according to which our universe evolved from an extremely dense and hot state, and space has been expanding ever since. The Big Bang is believed to have occurred 13.7 billion years ago, and today most galaxies are still flying apart from one another. It is important to understand that galaxies are not like the flying debris from a bomb that has just exploded. Space itself is expanding. The distances between galaxies are increasing in a way that is reminiscent of black dots painted on the surface of a balloon that move away from one another when the balloon is inflated. It doesn't matter on which dot you reside in order to observe this expansion. Looking out from any dot, the other dots appear to be receding.

Astronomers who examine distant galaxies can directly observe this expansion, originally detected by U.S. astronomer Edwin Hubble in the 1920s. Fred Hoyle coined the term *Big Bang* during a 1949 radio broadcast. Not until about 400,000 years after the Big Bang did the universe cool sufficiently to permit protons and electrons to combine to form neutral hydrogen. The Big Bang created helium nuclei and the light elements in the first few minutes of the universe's existence, providing some raw material for the first generation of stars.

Marcus Chown, author of *The Magic Furnace*, suggests that soon after the Big Bang, clumps of gas began to congeal, and then the universe began to light up like a Christmas tree. These stars lived and died before our galaxy came into existence.

Astrophysicist Stephen Hawking has estimated that if the rate of the universe's expansion one second after the Big Bang had been smaller by even one part in a hundred thousand million million, the universe would have re-collapsed and no intelligent life could have evolved.

SEE ALSO Olbers' Paradox (1823), Hubble's Law of Cosmic Expansion (1929), CP Violation (1964), Cosmic Microwave Background (1965), Cosmic Inflation (1980), Hubble Telescope (1990), Cosmological Big Rip (36 Billion).

LEFT: *According to ancient Finnish creation mythology, the heavens and the Earth were formed during the breaking of a bird's egg.* RIGHT: *Artist representation of the Big Bang (topmost point). Time proceeds down the page. The universe undergoes a rapid period of initial expansion (up to the red spherical marker). The first stars appear after about 400 million years (denoted by the yellow sphere).*

Black Diamonds

"Glittering stars in the night sky aside," writes journalist Peter Tyson, "scientists have long known that there are diamonds in the heavens. . . . Outer space may also be the birthplace of the mysterious black diamonds known as carbonados."

Various theories continue to be debated with respect to the formation of carbonado diamonds, such as meteorite impacts that can cause extreme pressures needed to trigger the creation of the diamonds in a process known as *shock metamorphism*. In 2006, researchers Stephen Haggerty, Jozsef Garai, and colleagues reported on studies of carbonado porosity, the presence of various minerals and elements in carbonados, the melt-like surface patina, and other factors that suggested to them that these diamonds formed in carbon-rich exploding stars called supernovae. These stars may produce a high-temperature environment that is analogous to the environments used in chemical vapor deposition methods for producing synthetic diamonds in the laboratory.

The black diamonds are about 2.6 to 3.8 billion years old and may have originally crashed down from space in the form of a large asteroid at a time when South America and Africa were joined. Today, many of these diamonds are found in the Central African Republic and Brazil.

Carbonados are as hard as traditional diamonds, but they are opaque, porous, and made of numerous diamond crystals that are stuck together. Carbonados are sometimes used for cutting other diamonds. Brazilians first discovered the rare black diamonds around 1840 and named them carbonados for their carbonized or burned appearance. In the 1860s, carbonados became useful for tipping drills used to penetrate rocks. The largest carbonado ever found has a mass corresponding to about one and a half pounds (3,167 carats, or 60 carats more massive than the largest clear diamond).

Other forms of naturally occurring (non-carbonado) "black diamonds" include more traditional-looking diamonds that have a smoky, dark coloration resulting from mineral inclusions such as iron oxides or sulfide compounds that cloud the stone. The gorgeous Spirit of de Grisogono at 0.137 pound (312.24 carats) is the world's largest cut black diamond of this variety.

SEE ALSO Stellar Nucleosynthesis (1946).

According to one theory, exploding stars called supernovae produced the high-temperature environment and carbon required for carbonado formation. Shown here is the Crab Nebula, a remnant of a star's supernova explosion.

2 Billion B.C.

Prehistoric Nuclear Reactor

Francis Perrin (1901–1992)

"Creating a nuclear reaction is not simple," write technologists at the U.S. Department of Energy. "In power plants, it involves splitting uranium atoms, and that process releases energy as heat and neutrons that go on to cause other atoms to split. This splitting process is called nuclear fission. In a power plant, sustaining the process of splitting atoms requires the involvement of many scientists and technicians."

In fact, it was not until the late 1930s that physicists Enrico Fermi and Leó Szilárd fully appreciated that uranium would be the element capable of sustaining a chain reaction. Szilárd and Fermi conducted experiments at Columbia University and discovered significant **Neutron** (subatomic particle) production with uranium, proving that the chain reaction was possible and enabling nuclear weapons. Szilárd wrote on the night of the discovery, "there was very little doubt in my mind that the world was headed for grief."

Because of the complexity of the process, the world was stunned in 1972 when French physicist Francis Perrin discovered that nature had created the world's first nuclear reactor two billion years before humankind, beneath Oklo in Gabon, Africa. This natural reactor formed when a uranium-rich mineral deposit came in contact with groundwater, which slowed the neutrons (subatomic particles) ejected from the uranium so that they could interact with and split other atoms. Heat was produced, turning the water to steam, thus temporarily slowing the chain reaction. The environment cooled, water returned, and the process repeated.

Scientists estimate that this prehistoric reactor ran for hundreds of thousands of years, producing the various isotopes (atomic variants) expected from such reactions that scientists detected at Oklo. The nuclear reactions in the uranium in underground veins consumed about five tons of radioactive uranium-235. Aside from the Oklo reactors, no other natural nuclear reactors have been identified. Roger Zelazny creatively speculates in his novel *Bridges of Ashes* that an alien race created the Gabon mine in order to cause mutations that eventually led to the human race.

SEE ALSO Radioactivity (1896), Neutron (1932), Energy from the Nucleus (1942), Little Boy Atomic Bomb (1945).

Nature created the world's first nuclear reactor in Africa. Billions of years later, Leó Szilárd and Enrico Fermi held U.S. Patent 2,708,656 on the nuclear reactor. Tank 355 is filled with water to act as a shield for radiation.

May 17, 1955 E. FERMI ET AL 2,708,656

NEUTRONIC REACTOR

Filed Dec. 19, 1944 27 Sheets-Sheet 25

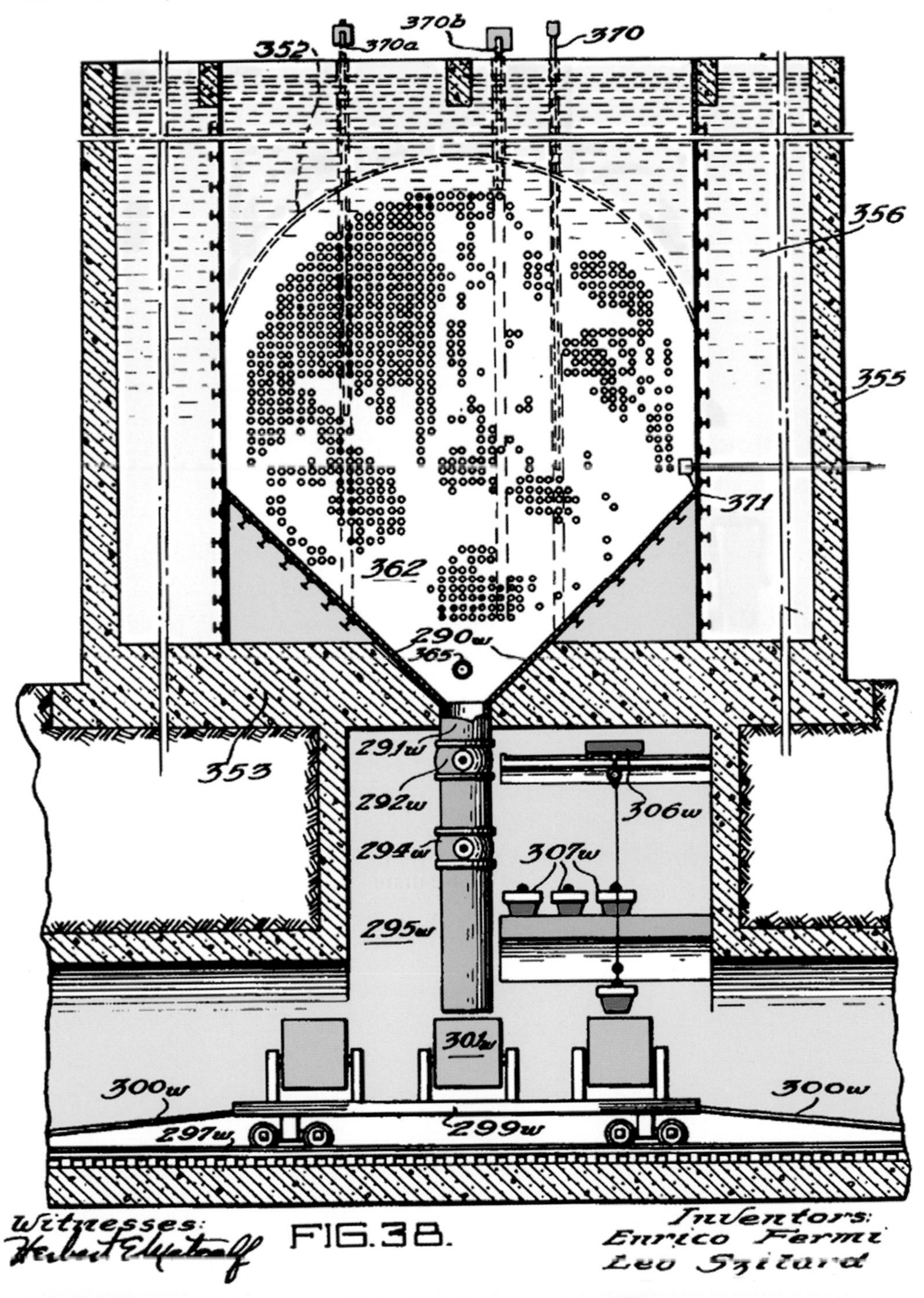

30,000 B.C.

Atlatl

At various locations around the world, ancient cultures discovered the physics of killing through the use of an ingenious device called the atlatl. The tool resembles a wooden rod or board with a cup or spur at one end, and it makes use of leverage and other simple physical principles to enable users to hurl a large arrow at a target with a tremendous range (over 328 feet [100 meters]) and speeds exceeding 92 miles per hour (150 kilometers/hour). In some sense, the atlatl functions as an extra arm segment.

A 27,000-year-old atlatl made of reindeer antler was discovered in France. Native Americans used the device 12,000 years ago. Australian Aborigines called the device a woomera. East Africans and Native Alaskans also used atlatl-like devices. The Mayans and Aztecs (who actually called it an *atlatl*) were fans of the instruments, and the Aztecs surprised the Spanish conquistadors by using atlatls to completely pierce plate armor. Prehistoric hunters could use the atlatl for killing animals as large as the mammoth.

Today, the World Atlatl Association supports both national and international competitions that attract engineers, hunters, and anyone interested in understanding the secrets of prehistoric technology.

One version of the atlatl resembles a two-foot long stick, although it underwent technological evolution over thousands of years. A dart (four to six feet long, or about 1.5 meters) fits into a spur at the back of the atlatl and is initially positioned parallel to the board. The atlatl user launches the dart with a sweeping arm and wrist motion, similar to a tennis serve.

As the atlatl evolved, users discovered that flexible atlatl boards could effectively store and release energy (like a diver on a diving board), and small stone weights were added to the device. The purpose of these weights has been debated through the years. Many feel that the weights add stability and distance to the throw by adjusting the timing and flexibility. It's also possible that the weights reduce the sound so that the launch is less obvious.

SEE ALSO Crossbow (341 B.C.), Trebuchet (1200), Hypersonic Whipcracking (1927).

Figure from the Fejérváry-Mayer Aztec Codex of central Mexico, depicting a god holding three arrows and an atlatl. This codex dates to a time prior to Hernán Cortés' destruction of the Aztec capital Tenochtitlan in 1521.

20,000 B.C.

Boomerang

I recall a silly song from childhood by the English singer Charlie Drake (1925–2006), about a sad Australian Aborigine who laments that "My boomerang won't come back." In practice, this may have been no problem, because the boomerangs used for hunting kangaroos or in war were heavy curved throwing sticks meant to break the bones of the quarry and not to return. A hunting boomerang, dated to around 20,000 B.C., has been found in a cave in Poland.

Today, when most of us think of boomerangs, we think of boomerangs shaped like the letter V. These shapes probably evolved from the non-returning boomerangs, perhaps when hunters noticed that particular branch shapes were more stable in flight or exhibited interesting flight patterns. The return boomerang is actually used for hunting to scare game birds into flight, though we don't know when such boomerangs were first invented. Each wing of this kind of boomerang is shaped like an airfoil, similar to the shape of an airplane wing, which is rounded on one side and flatter on the other. Air travels faster over one side of the wing than the other, which assists in providing lift. Unlike an airplane wing, the boomerang has "leading edges" on opposite sides of the V, given that the boomerang spins as it flies. This means that the airfoils face in different directions for the lead wing and the trailing wing.

The boomerang is initially thrown at a slightly off-vertical direction with the open part of the V facing forward. As the boomerang spins in the direction of the throw, the top wing of the boomerang advances faster than the bottom—which also contributes to the lift. Gyroscopic precession, which is a change in the orientation of the rotation axis of a rotating body, allows the boomerang to return to the thrower when thrown correctly. The combination of these factors creates the complex, circular flight path of the boomerang.

SEE ALSO Crossbow (341 B.C.), Trebuchet (1200), Gyroscope (1852).

Boomerangs have been used as weapons and for sport. Their shapes vary and depend on their geographic origins and function.

3000 B.C.

Sundial

"Hide not your talents. They for use were made. What's a sundial in the shade?"

—Ben Franklin

For centuries, people have wondered about the nature of time. Much of ancient Greek philosophy was concerned with understanding the concept of eternity, and the subject of time is central to all the world's religions and cultures. Angelus Silesius, a seventeenth-century mystic poet, actually suggested that the flow of time could be suspended by mental powers: "Time is of your own making; its clock ticks in your head. The moment you stop thought, time too stops dead."

One of the oldest of time-keeping devices is the sundial. Perhaps ancient humans noticed that the shadows they cast were long in the early morning, grew progressively shorter, and then grew longer again as the evening approached. The earliest known sundial dates to about 3300 B.C. and is found engraved in a stone in the Knowth Great Mound in Ireland.

A primitive sundial can be made from a vertical stick in the ground. In the northern hemisphere, the shadow rotates around the stick in a clockwise direction, and the shadow's position can be used to mark the passage of time. The accuracy of such a crude instrument is improved if the stick is slanted so that it points to the Celestial North Pole, or roughly toward the position of the Pole Star. With this modification, the pointer's shadow will not change with the seasons. One common form of sundial has a *horizontal* dial, sometimes used as an ornament in a garden. Because the shadow does not rotate uniformly around the face of this sundial, the marks for each hour are not spaced equally. Sundials may not be accurate for various reasons, including the variable speed of the Earth orbiting the Sun, the use of daylight savings time, and the fact that clock times today are generally kept uniform within time zones. Before the days of wristwatches, people sometimes carried a folding sundial in their pockets, attached to a small magnetic compass to estimate true north.

SEE ALSO Antikythera Mechanism (125 B.C.), Hourglass (1338), Anniversary Clock (1841), Time Travel (1949), Atomic Clocks (1955).

People have always wondered about the nature of time. One of the most ancient of timekeeping devices is the sundial.

GROW
ALONG
OLD
WITH
IS YET

Truss

Trusses are structures usually composed of several triangular units that are made from straight pieces of metal or wood connected by joints, or nodes. If all members of the truss lie in a plane, the truss is called a planar truss. For centuries, truss structures allowed builders to construct sturdy structures in an economical way, in terms of cost and use of materials. The rigid framework of a truss allowed it to span great distances.

The triangular shape is particularly useful because the triangle is the only shape that cannot experience a change in shape without a change in the length of a side. This means that a triangular frame of strong beams fastened at static joints cannot be deformed. (For example, a square could, in principle, assume the shape of a rhombus, if the joints accidentally slid.) Another advantage of the truss is that its stability can often be predicted by assuming that the beams are stressed primarily in terms of tension and compression and that these forces act at the nodes. If a force tends to elongate a beam, it is a *tensile force*. If the force tends to shorten the beam, it is a *compressive force*. Since the nodes of a truss are static, the sum of all forces at each node equals zero.

Wooden trusses were used in ancient lake dwellings during the Early Bronze Age, circa 2500 B.C. Romans used wooden trusses for bridge constructions. In the 1800s, trusses were extensively used in covered bridges in the United States, and numerous patents were filed for various truss configurations. The first iron-truss bridge in the United States was the Frankfort Bridge on the Erie Canal, built in 1840, and the first steel-truss bridge spanned the Missouri River in 1879. After the Civil War, metal-truss railroad bridges were popular, as they offered greater stability than suspension bridges when subject to the moving load of heavy trains.

SEE ALSO Arch (1850 B.C.), I-Beams (1844), Tensegrity (1948), Leaning Tower of Lire (1955).

For centuries, triangular truss structures allowed builders to construct sturdy, cost-effective structures.

1850 B.C.

Arch

In architecture, an arch is a curved structure that spans a space while supporting weight. The arch has also become a metaphor for extreme durability created by the interaction of simple parts. The Roman philosopher Seneca wrote, "Human society is like an arch, kept from falling by the mutual pressure of its parts." According to an ancient Hindu proverb, "An arch never sleeps."

The oldest existing arched city gate is the Ashkelon gate in Israel, built c. 1850 B.C. of mud-brick with some calcareous limestone. Mesopotamian brick arches are even older, but the arch gained particular prominence in ancient Rome, where it was applied to a wide range of structures.

In buildings, the arch allows the heavy load from above to be channeled into horizontal and vertical forces on supporting columns. The construction of arches usually relies upon wedge-shaped blocks, called voussoirs, that precisely fit together. The surfaces of neighboring blocks conduct loads in a mostly uniform manner. The central voussoir, at the top of the arch, is called the keystone. To build an arch, a supporting wooden framework is often used until the keystone is finally inserted, locking the arch in place. Once inserted, the arch becomes self-supporting. One advantage of the arch over earlier kinds of supporting structures is its creation from easily transported voussoirs and its spanning of large openings. Another advantage is that gravitational forces are distributed throughout the arch and converted to forces that are roughly perpendicular to voussoirs' bottom faces. However, this means that the base of the arch is subject to some lateral forces, which must be counterbalanced by materials (e.g. a brick wall) located at the bottom sides of the arch. Much of the force of the arch is converted to compressional forces on the voussoirs—forces that stones, concrete, and other materials can easily withstand. Romans mostly constructed semicircular arches, although other shapes are possible. In Roman aqueducts, the lateral forces of neighboring arches served to counteract each other.

SEE ALSO Truss (2500 B.C.), I-Beams (1844), Tensegrity (1948), Leaning Tower of Lire (1955).

The arch allows the heavy load from above to be channeled into horizontal and vertical forces. Arches usually rely upon wedge-shaped blocks, called voussoirs, that fit closely together as in these ancient Turkish arches.

Olmec Compass

Michael D. Coe (b. 1929), **John B. Carlson** (b. 1945)

For many centuries, navigators have used compasses with magnetized pointers for determining the Earth's magnetic north pole. The *Olmec compass* in Mesomerica may represent the earliest known compass. The Olmecs were an ancient pre-Columbian civilization situated in south-central Mexico from around 1400 B.C. to 400 B.C. and famous for the colossal artwork in the form of heads carved from volcanic rock.

American astronomer John B. Carlson used **Radiocarbon Dating** methods of the relevant layers of an excavation to determine that a flattened, polished, oblong piece of hematite (iron oxide) had its origin about 1400–1000 B.C. Carlson has speculated that the Olmecs used such objects as direction indicators for astrology and geomancy, and for orienting burial sites. The Olmec compass is part of a polished lodestone (magnetized piece of the mineral) bar with a groove at one end that was possibly used for sighting. Note that the ancient Chinese invented the compass some time before the second century, and the compass was used for navigation by the eleventh century.

Carlson writes in "Lodestone Compass: Chinese or Olmec Primacy?"

> Considering the unique morphology (purposefully shaped polished bar with a groove) and composition (magnetic mineral with magnetic moment vector in the floating plane) of M-160, and acknowledging that the Olmec were a sophisticated people who possessed advanced knowledge and skill in working iron ore minerals, I would suggest for consideration that the Early Formative artifact M-160 was probably manufactured and used as what I have called a zeroth-order compass, if not a first-order compass. Whether such a pointer would have been used to point to something astronomical (zeroth-order compass) or to geomagnetic north-south (first-order compass) is entirely open to speculation.

In the late 1960s, Yale University archeologist Michael Coe found the Olmec bar at San Lorenzo in the Mexican state of Veracruz, and it was tested by Carlson in 1973. Carlson floated it on mercury or on water with a cork mat.

SEE ALSO *De Magnete* (1600), Ampère's Law of Electromagnetism (1825), Galvanometer (1882), Radiocarbon Dating (1949).

In the most general definition, a lodestone *refers to a naturally magnetized mineral, such as those used in fragments that ancient people used to create magnetic compasses. Shown here is a lodestone in the Hall of Gems at the National Museum of Natural History, administered by The Smithsonian Institution.*

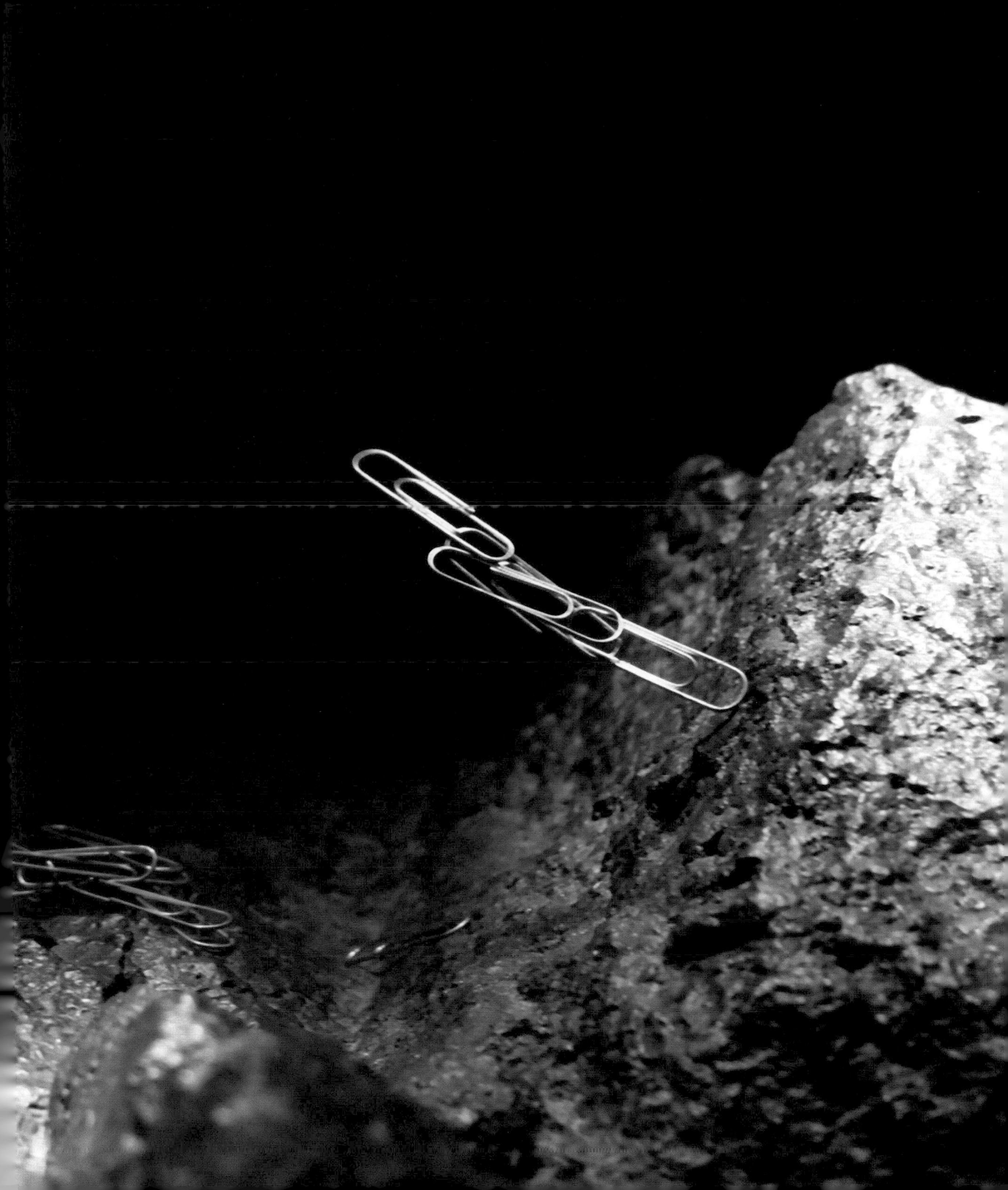

341 B.C.

Crossbow

Through the centuries, the crossbow was a weapon that employed the laws of physics to wreak military havoc and pierce armor. The crossbow changed the odds of victory in war during the Middle Ages. One of the first reliable records of crossbow use in war dates to the battle of Ma-Ling (341 B.C.) in China, but even older crossbows have been found in Chinese tombs.

Early crossbows generally were bows mounted to wooden tillers, or stocks. The short, heavy, arrow-like projectile called a bolt traveled along a groove through the tiller. As the crossbow evolved, various mechanisms were used to pull back the string and then hold the string in place until it was ready to be fired. Early crossbows had stirrups for holding an archer's foot as he pulled back the string with both hands or with a hook.

Physics improved these killing machines in several ways. A traditional bow and arrow required that the archer be very strong to draw the bow and hold it while aiming. However, with a crossbow, a weaker person could use his leg muscles to assist drawing the string. Later, various levers, gears, pulleys, and cranks were used to amplify the user's strength when pulling back the string. In the fourteenth century, European crossbows were made of steel and employed *crannequins*—a toothed wheel attached to a crank. An archer would turn the crank to pull the bowstring.

The penetrating power of a crossbow and ordinary bow comes from energy stored when bending the bow. Like a spring that is pulled and held, energy is stored in the elastic potential energy of the bow. When released, the potential energy is converted to the kinetic energy of movement. The amount of firing power the bow delivers depends on the bow's *draw weight* (amount of force needed to draw the bow) and *draw length* (the distance between the bowstring's resting position and its drawn position).

SEE ALSO Atlatl (30,000 B.C.), Boomerang (20,000 B.C.), Trebuchet (1200), Conservation of Energy (1843).

Around 1486, Leonardo da Vinci drew several designs for a colossal crossbow. This weapon was cranked using gears. One of its firing mechanisms employed a holding pin that was released by striking it with a mallet.

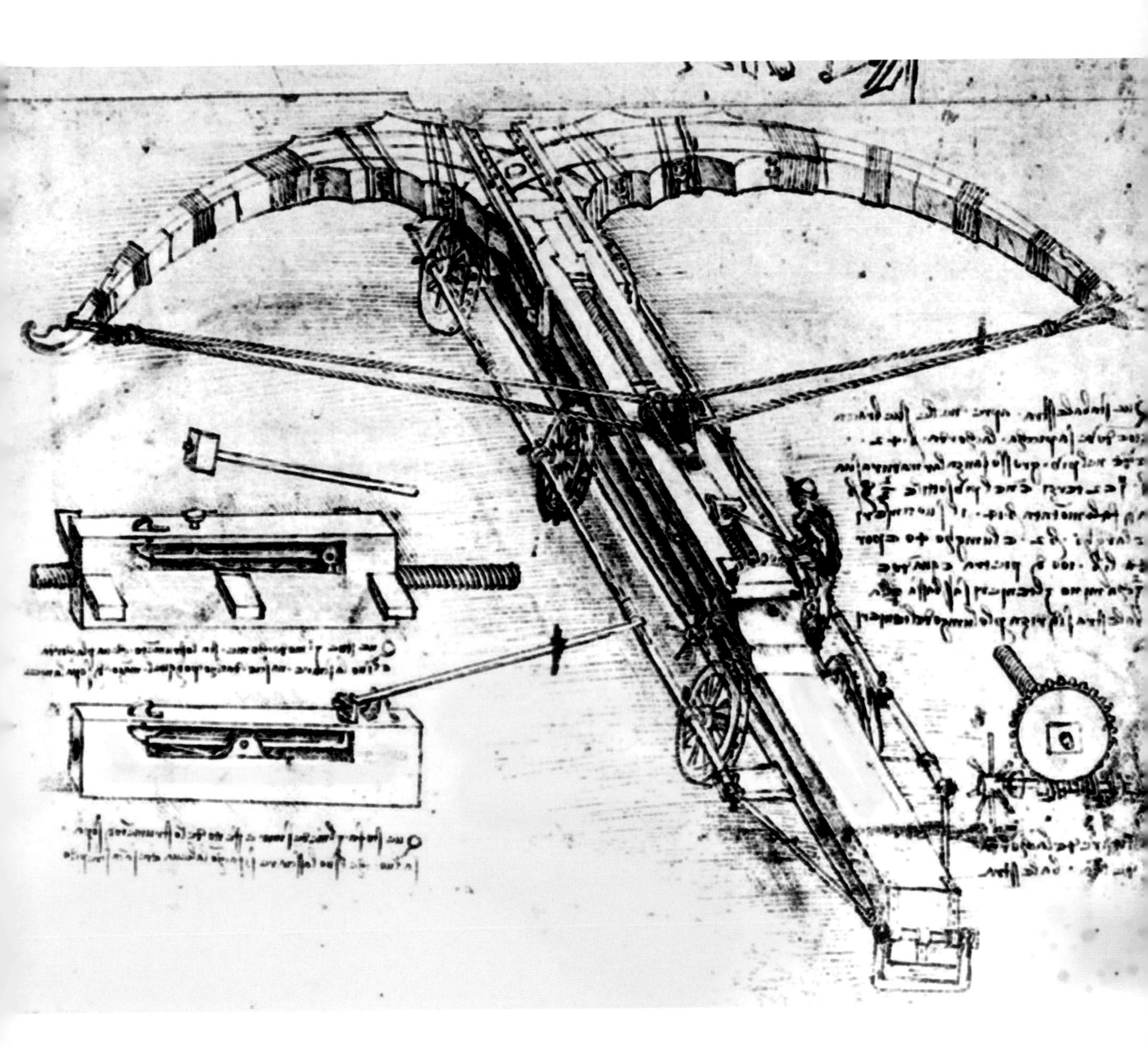

250 B.C.

Baghdad Battery

Alessandro Giuseppe Antonio Anastasio Volta (1745–1827)

In 1800, Italian physicist Alessandro Volta invented what has been traditionally considered to be the first electric battery when he stacked several pairs of alternating copper and zinc discs separated by cloth soaked in salt water. When the top and bottom of the pile were connected by a wire, an electric current began to flow. However, the discovery of certain archeological artifacts may suggest batteries predate this discovery by more than a millennium.

"Iraq has a rich national heritage," writes the BBC News. "The Garden of Eden and the Tower of Babel are said to have been sited in this ancient land." In 1938, while in Baghdad, German archeologist Wilhelm König discovered a five-inch-long (13 cm) clay jar containing a copper cylinder that surrounded an iron rod. The jar showed signs of corrosion and seemed to have once contained a mild acid, such as vinegar or wine. König believed these vessels to be galvanic cells, or parts of batteries, possibly used for electroplating gold onto silver objects. The acidic solution would serve as an electrolyte, or conducting medium. The dates of the artifacts are obscure. König dated them to around 250 B.C. to A.D. 224, while others have suggested a range of A.D. 225–640. Subsequent researchers have demonstrated that replicas of the Baghdad Battery do indeed produce electrical current when filled with grape juice or vinegar.

Referring to the batteries in 2003, metallurgist Dr. Paul Craddock notes, "They are a one-off. As far as we know, nobody else has found anything like these. They are odd things; they are one of life's enigmas." Various other uses for the Baghdad batteries have been suggested, ranging from their use to produce electrical currents for the purpose of acupuncture or to impress worshipers of idols. If wires or conductors are ever discovered along with other ancient batteries, this would support the idea that these objects functioned as batteries. Of course, even if the vessels did produce electric currents, this does not imply that the ancient people understood how the objects actually worked.

SEE ALSO Von Guericke's Electrostatic Generator (1660), Battery (1800), Fuel Cell (1839), Leyden Jar (1744), Solar Cells (1954).

The ancient Baghdad battery is composed of a clay jar with a stopper made of asphalt. Protruding through the asphalt is an iron rod surrounded by a copper cylinder. When filled with vinegar, the jar produces about 1.1 volts. (Photo courtesy of Stan Sherer.)

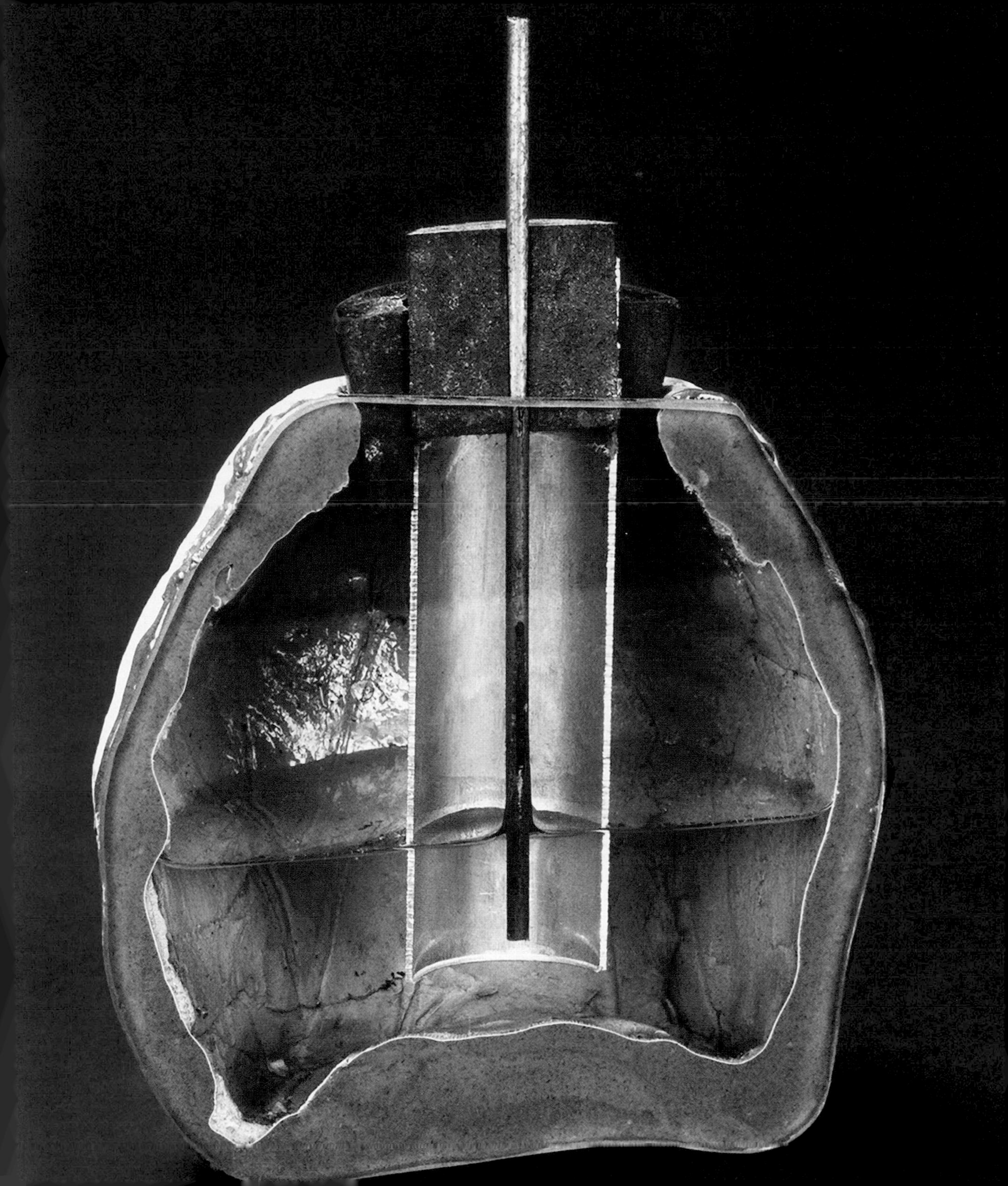

250 B.C.

Siphon

Ctesibius (285–222 B.C.)

A siphon is a tube that allows liquid to drain from a reservoir to another location. An intermediate location in the tube may actually be higher than the reservoir, yet the siphon still operates. No pump is required to maintain the flow of liquid since the flow is driven by differential hydrostatic pressures. The discovery of the siphon principle is often attributed to the Greek inventor and mathematician Ctesibius.

The liquid in a siphon can rise in the tube before flowing downward, partly because the weight of the liquid in the longer "outlet" tube is pulled downward by gravity. In fascinating laboratory experiments, some siphons have been demonstrated to work in a vacuum. The maximum height of the top "crest" of a traditional siphon is limited by atmospheric pressure because if the crest is too high, the pressure in the liquid may drop below the vapor pressure of a liquid, causing bubbles to form in the crest of the tube.

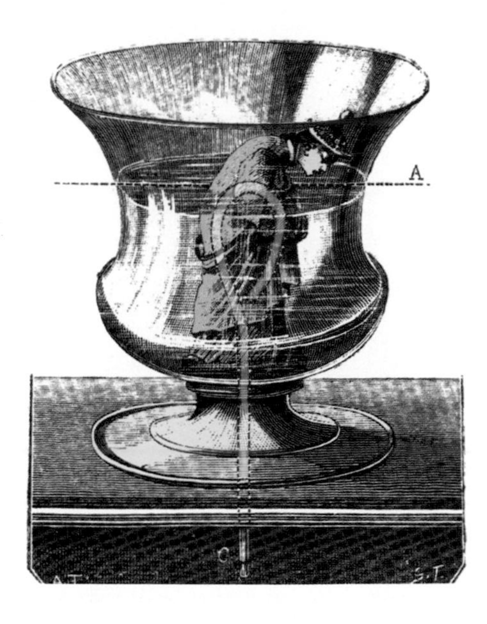

Interestingly, the end of the siphon does not need to be below the mouth of the tube, but it does need to be below the surface of the water in the reservoir. Although siphons are used in countless practical applications for draining liquids, my favorite application is the fanciful Tantalus Cup. In one form, the statue of a little man sits in the cup. A siphon is hidden in the statue with the crest near the level of the man's chin. As liquid is poured into the cup, it rises to the chin, and then the cup promptly continues to drain most of the cup's contents through the siphon's end that is hidden at the bottom! Tantalus is, therefore, thirsty forever. . . .

SEE ALSO Archimedean Screw (250 B.C.), Barometer (1643), Bernoulli's Law of Fluid Dynamics (1738), Drinking Bird (1945).

LEFT: *A Tantalus Cup, with hidden siphon depicted in blue.* RIGHT: *A simple siphon with liquid flowing between containers.*

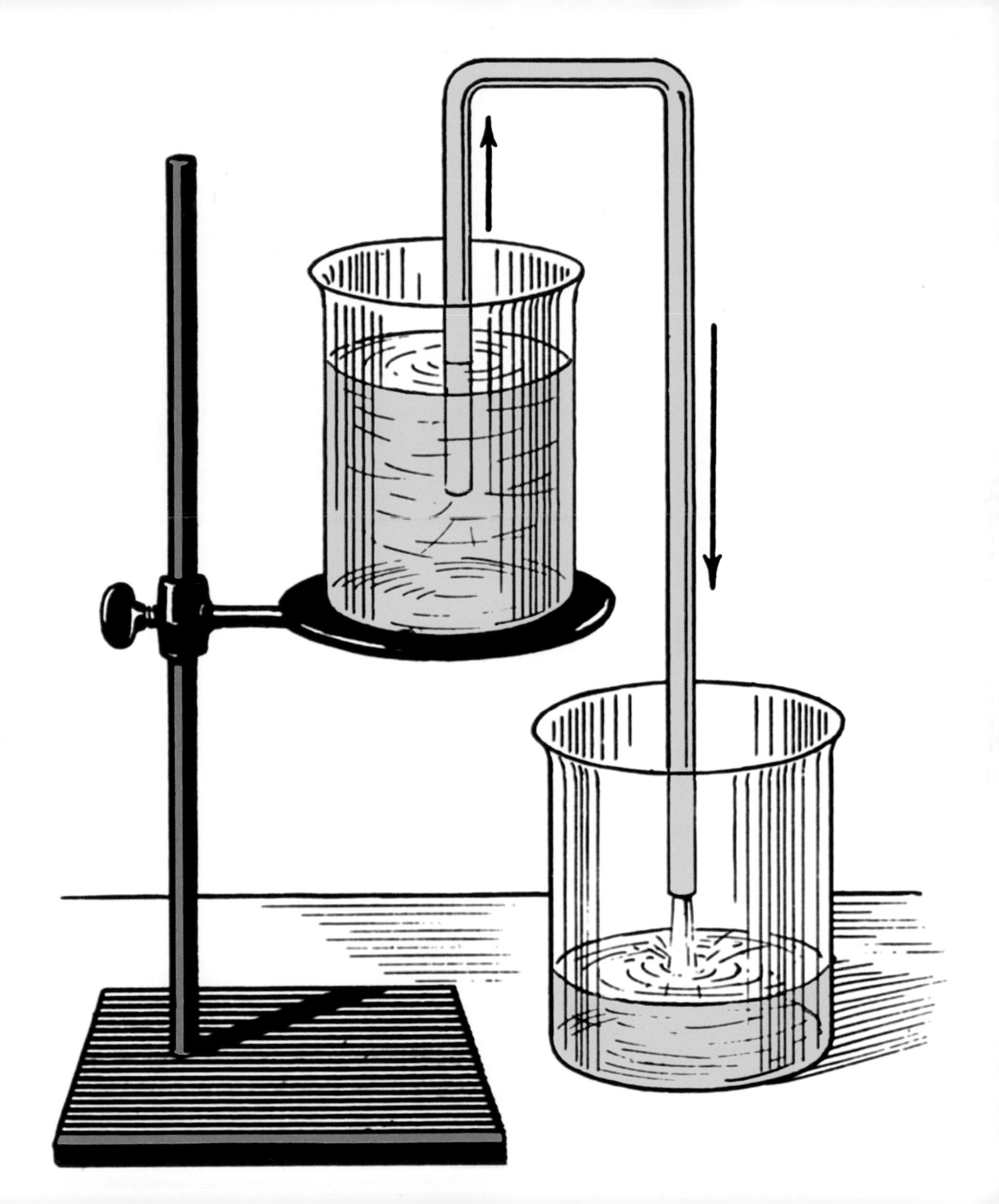

Archimedes' Principle of Buoyancy

Archimedes (c. 287 B.C.–c. 212 B.C.)

Imagine that you are weighing an object—like a fresh, uncooked egg—that is submerged in a kitchen sink. If you weigh the egg by hanging it from a scale, the egg would weigh less while in the water, according to the scale, than when the egg is lifted out of the sink and weighed. The water exerts an upward force that partially supports the weight of the egg. This force is more obvious if we perform the same experiment with an object of lower density, such as a cube made out of cork, which floats while being partially submerged in the water.

The force exerted by the water on the cork is called a buoyant force, and for a cork held under water, the upward force is greater than its weight. This buoyant force depends on the density of the liquid and the volume of the object, but not on the shape of the object or the material of which the object is composed. Thus, in our experiment, it doesn't matter if the egg is shaped like a sphere or a cube. An equal volume of egg or wood would experience the same buoyant force in water.

According to Archimedes' Principle of Buoyancy, named after the Greek mathematician and inventor famous for his geometric and hydrostatic studies, a body wholly or partially submerged in liquid is buoyed up by a force equal to the weight of displaced liquid.

As another example, consider a small pellet of lead placed in a bathtub. The pellet weighs more than the tiny weight of water it displaces, so the pellet sinks. A wooden rowboat is buoyed up by the large weight of water that it displaces, and hence the rowboat floats. A submarine floating underwater displaces a volume of water that has a weight that is precisely equal to the submarine's weight. In other words, the total weight of the submarine—which includes the people, the metal hull, and the enclosed air—equals the weight of displaced seawater.

SEE ALSO Archimedean Screw (250 B.C.), Stokes' Law of Viscosity (1851), Lava Lamp (1963).

LEFT: *An egg in water experiences an upward force equal to the weight of water it displaces.* RIGHT: *When plesiosaurs (extinct reptiles) floated within the sea, their total weights equaled the weights of the water they displaced. Gastrolith stones discovered in the stomach region of plesiosaur skeletons may have helped in controlling buoyancy and flotation.*

250 B.C.

Archimedean Screw

Archimedes (c. 287 B.C.–c. 212 B.C.), **Marcus Vitruvius Pollio** (c. 87 B.C.–c. 15 B.C.)

Archimedes, the ancient Greek geometer, is often regarded as the greatest mathematician and scientist of antiquity and one of the four greatest mathematicians to have walked the earth—together with Isaac Newton, Leonhard Euler, and Carl Friedrich Gauss.

The invention of the water snail, or Archimedean screw, to raise water and help irrigate crops was attributed to Archimedes by the Greek historian Diodorus Siculus in the first century B.C. The Roman engineer Vitruvius gives a detailed description of its operation for lifting water, which required intertwined helical blades. In order to lift water, the bottom end of the screw is immersed in a pond, after which the act of rotating the screw raises water from the pond to a higher elevation. Archimedes may also have designed a related helical pump, a corkscrew-like device used to remove water from the bottom of a large ship. The Archimedean screw is still used today in societies without access to advanced technology and works well even if the water is filled with debris. It also tends to minimize damage to aquatic life. Modern Archimedean screw-like devices are used to pump sewage in water-treatment plants.

Author Heather Hassan writes, "Some Egyptian farmers are still using the Archimedean screw to irrigate their fields. It ranges in size from a quarter of an inch to 12 feet (0.6 centimeters to 3.7 meters) in diameter. The screw is also used in the Netherlands, as well as in other countries, where unwanted water needs to be drained from the surface of the land."

Additional intriguing modern examples exist. Seven Archimedes screws are used to pump wastewater in a treatment plant in Memphis, Tennessee. Each of these screws is 96 inches (2.44 meters) in diameter and can lift 19,900 gallons (about 75,000 liters) per minute. According to mathematician Chris Rorres, an Archimedean screw the diameter of a pencil eraser is used in a Hemopump cardiac assist system that maintains blood circulation during heart failure, coronary bypass surgery, and other surgical procedures.

SEE ALSO Siphon (250 B.C.), Archimedes' Principle of Buoyancy (250 B.C.).

Archimedean screw, from Chambers's Encyclopedia, *1875.*

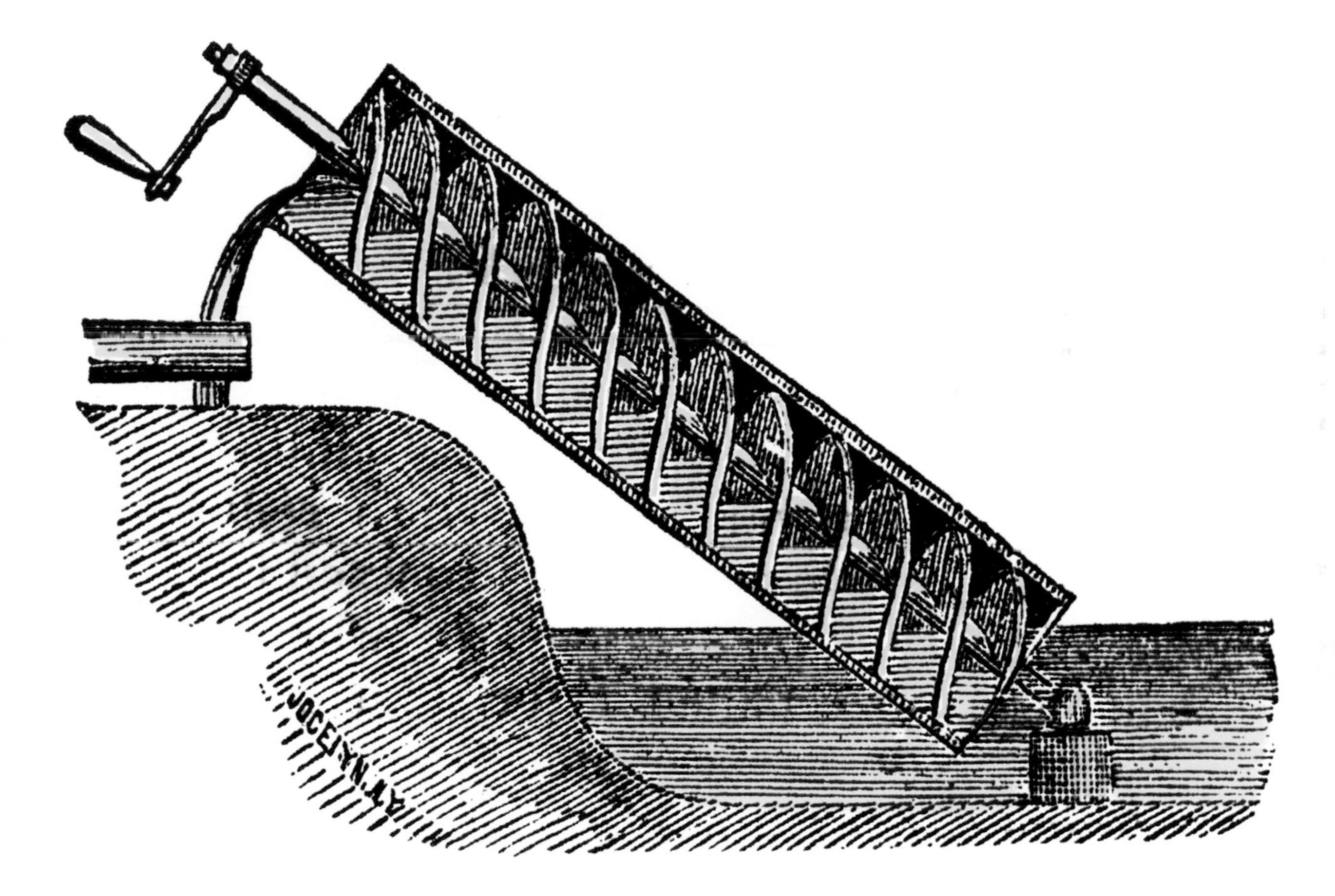
JOCELYN N.Y.

240 B.C.

Eratosthenes Measures the Earth

Eratosthenes of Cyrene (c. 276 B.C.–c. 194 B.C.)

According to author Douglas Hubbard, "Our first mentor of measurement did something that was probably thought by many in his day to be impossible. An ancient Greek named Eratosthenes made the first recorded measurement of the circumference of Earth. . . . [He] didn't use accurate survey equipment, and he certainly didn't have lasers and satellites. . . ." However, Eratosthenes knew of a particular deep well in Syene, a city in southern Egypt. The bottom of this well was entirely lit by the noon Sun one day out of the year, and thus the sun was directly overhead. He also was aware that, at the same time in the city of Alexandria, objects cast a shadow, which suggested to Eratosthenes that the Earth was spherical, not flat. He assumed that the Sun's rays were essentially parallel, and he knew that the shadow made an angle that was 1/50th of a circle. Thus, he determined that the circumference of the Earth must be approximately 50 times the known distance between Alexandria and Syene. Assessments of Eratosthenes' accuracy vary, due to the conversion of his ancient units of measure to modern units, along with other factors, but his measurements are usually deemed to be within a few percent of the actual circumference. Certainly, his estimation was more accurate than other estimates of his day. Today, we know that the circumference of the Earth at the equator is about 24,900 miles (40,075 kilometers). Curiously, if Columbus had not ignored the results of Eratosthenes, thereby underestimating the circumference of the Earth, the goal of reaching Asia by sailing west might have been considered to be an impossible task.

Eratosthenes was born in Cyrene (now in Libya) and later was a director of the great Library of Alexandria. He is also famous for founding scientific chronology (a system that endeavors to fix dates of events at correctly proportioned intervals), along with developing a simple algorithm for identifying prime numbers (numbers such as 13, divisible only by themselves and 1). In old age, Eratosthenes became blind and starved himself to death.

SEE ALSO Pulley (230 B.C.), Measuring the Solar System (1672), Black Drop Effect (1761), Stellar Parallax (1838), Birth of the Meter (1889).

Eratosthenes' map of the world (1895 reconstruction). Eratosthenes measured the circumference of the Earth without leaving Egypt. Ancient and medieval European scholars often believed that the world was spherical, although they were not aware of the Americas.

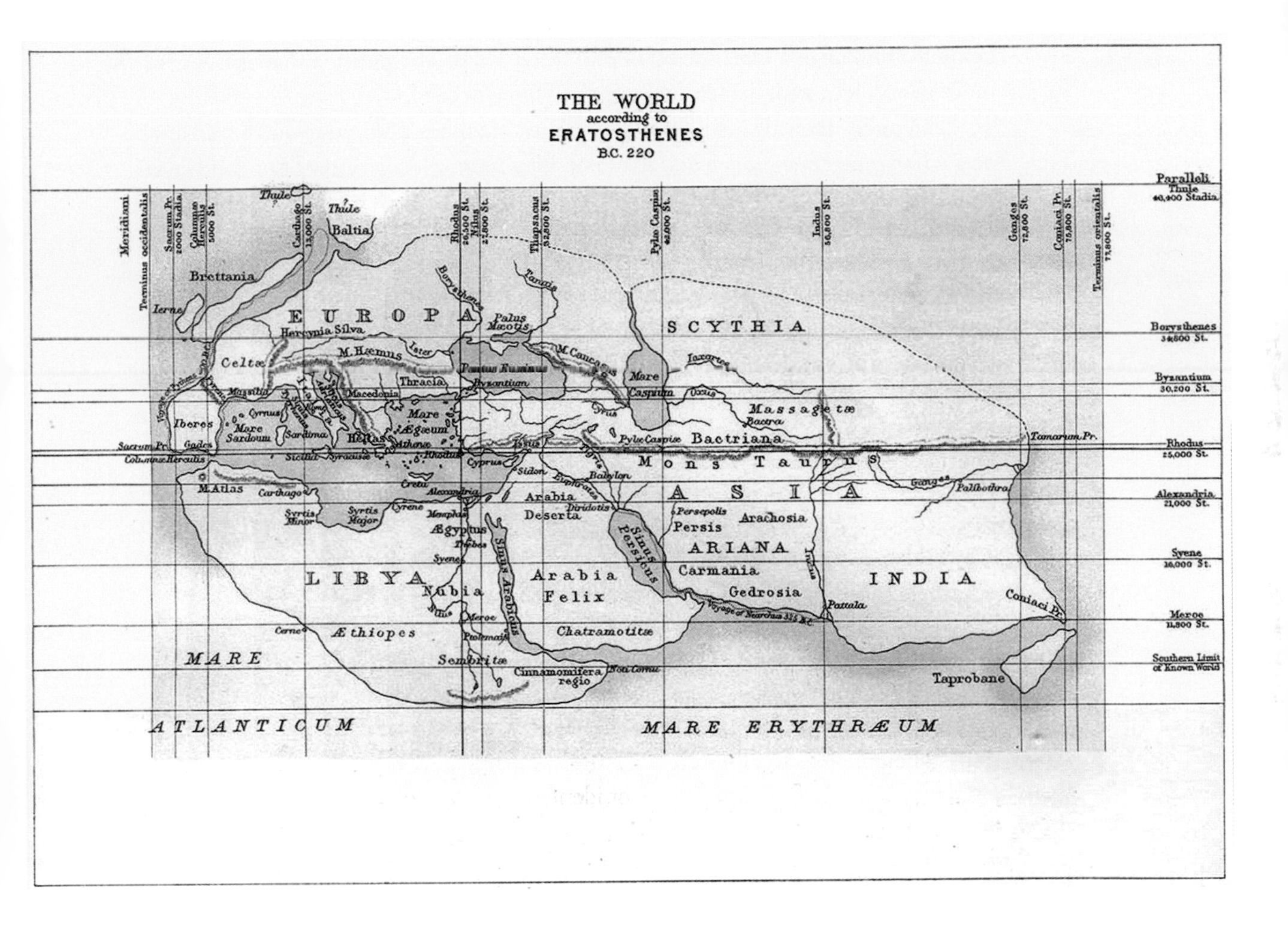
THE WORLD
according to
ERATOSTHENES
B.C. 220
Paralleli
Thule
40,400 Stadia
Borysthenes
34,800 St.
Byzantium
30,200 St.
Rhodus
25,000 St.
Alexandria
21,000 St.
Syene
16,000 St.
Meroe
11,800 St.
Southern Limit of Known World
Meridiani
Terminus occidentalis
Sacrum Pr. 2000 Stadia
Columnae Herculis 5000 St.
Carthago 13,000 St.
Rhodus 26,500 St.
Nilus 27,800 St.
Thapsacus 32,800 St.
Pylae Caspiae 42,000 St.
Indus 56,800 St.
Ganges 72,800 St.
Coniaci Pr. 75,800 St.
Terminus orientalis 77,800 St.
Thule
Baltia
Brettania
Ierne
EUROPA
Hercynia Silva
Celtae
M. Haemus
Ister
Palus Maeotis
Tanais
Borysthenes
SCYTHIA
Pontus Euxinus
Byzantium
Thracia
Macedonia
Massilia
Iberes
Mare Sardoum
Sardinia
Gades
Sacrum Pr.
Columnae Herculis
Sicilia
Syracusae
Hellas
Athenae
Mare Aegaeum
Rhodus
Creta
Cyprus
Sidon
M. Caucasus
Mare Caspium
Iaxartes
Oxus
Massagetae
Bactra
Pylae Caspiae
Bactriana
Mons Taurus
Tamarum Pr.
ASIA
Ganges
Palibothra
Babylon
Tigris
Euphrates
Arabia Deserta
Diridotis
Persepolis
Persis
Arachosia
ARIANA
Carmania
Gedrosia
Sinus Persicus
INDIA
Pattala
Voyage of Nearchus 325 B.C.
Coniaci Pr.
Taprobane
M. Atlas
Carthago
Syrtis Minor
Syrtis Major
Cyrene
Alexandria
Memphis
Aegyptus
Thebes
Syene
Sinus Arabicus
LIBYA
Nubia
Arabia Felix
Meroe
Ptolemais
Chatramotitae
Cerne
Aethiopes
Sembritae
Cinnamomifera regio
Noti Cornu
MARE
ATLANTICUM
MARE ERYTHRAEUM

230 B.C.

Pulley

Archimedes (c. 287 B.C.–c. 212 B.C.)

A pulley is a mechanism that usually consists of a wheel on an axle. A rope runs over the wheel so that the pulley can change the direction of an applied force, for example, when helping a human or a machine lift or pull heavy loads. The pulley also makes it easier to move a load because it decreases the applied force needed.

The pulley probably had its birth in prehistoric times when someone tossed a rope over a horizontal tree branch and used it to lift a heavy object. Author Kendall Haven writes, "By 3000 B.C., such pulleys with grooved wheels (so that the rope wouldn't slip off) existed in Egypt and Syria. The Greek mathematician and inventor Archimedes gets credit for inventing the *compound* pulley in about 230 B.C. . . . in which a number of wheels and ropes combine to lift a single object . . . to multiply the lifting power of a person. Modern block and tackle systems are examples of compound pulleys."

Pulleys almost seem magical in the way they can decrease the width and strength of the rope required, and of the force needed, to lift heavy objects. In fact, according to legends and the writings of the Greek historian Plutarch, Archimedes may have used a compound pulley to help move heavy ships with minimal effort. Of course, no laws of nature are violated. Work, which is defined as force times the distance moved, remains the same—pulleys allow one to pull with less force but over a longer distance. In practice, more pulleys increase the sliding friction, and, thus, a system of pulleys may become decreasingly efficient after a certain number are employed. When performing computations to estimate the effort needed to use a pulley system, engineers often assume that the pulley and rope weigh very little compared to the weight that is being moved. Through history, block-and-tackle systems were particularly common on sailing ships, where motorized aids were not always available.

SEE ALSO Atlatl (30,000 B.C.), Crossbow (341 B.C.), Foucault's Pendulum (1851).

Close-up of a pulley system on a vintage yacht. Ropes in pulleys travel over wheels so that the pulley can change the direction of applied forces and make it easier to move a load.

212 B.C.

Archimedes' Burning Mirrors

Archimedes (c. 287 B.C.–c. 212 B.C.)

The tale of Archimedes' burning mirrors has fascinated historians through the centuries. In 212 B.C., Archimedes is said to have made a "death ray" consisting of a set of mirrors that focused sunlight on Roman ships, setting them afire. Various individuals have tried to test the practical use of such mirrors and declared their use to have been unlikely. However, mechanical engineer David Wallace of MIT encouraged his students in 2005 to build an oak replica of a Roman warship and focus sunlight on it, using 127 flat mirrors, each with an edge one foot (0.3 meter) in length. The ship was about 100 feet (30 meters) away. After ten minutes of exposure to the focused light, the warship burst into flames!

In 1973, a Greek engineer employed 70 flat mirrors (each about five feet by three feet in length, or 1.5 × 0.9 meter) in order to focus sunlight onto a rowboat. In this experiment, the rowboat also quickly burst into flames. However, while it is possible to set a ship afire with mirrors, this task would have probably been very difficult for Archimedes if ships were moving.

As an interesting aside, Arthur C. Clarke's short story, "A Slight Case of Sunstroke" describes the fate of a disliked soccer referee. When the referee makes an unpopular decision, the spectators focus sunlight onto the referee using their shiny souvenir programs that they hold in their hands. The shiny surfaces act like Archimedes' mirror, and the poor man is burned to ashes.

Archimedes developed other weapons. According to Greek historian Plutarch, Archimedes' ballistic weaponry was used effectively against the Romans in the siege of 212 B.C. For example, Plutarch wrote, "When Archimedes began to ply his engines, he at once shot all sorts of missile weapons against the land forces, and immense masses of stone that came down with incredible noise and violence, against which no man could stand; for they knocked down those upon whom they fell in heaps. . . ."

SEE ALSO Fiber Optics (1841), Solar Cells (1954), Laser (1960), Unilluminable Rooms (1969).

LEFT: *Wood engraving of a burning mirror from F. Marion's* The Wonders of Optics *(1870).* RIGHT: *The largest solar furnace in the world is in Odeillo, France. An array of plane mirrors (not shown) reflects sunlight onto the large curved mirror, which focuses light onto a small area that reaches temperatures of 5,430 °F (3,000 °C).*

125 B.C.

Antikythera Mechanism

Valerios Stais (1857–1923)

The Antikythera mechanism is an ancient geared computing device that was used to calculate astronomical positions and that mystified scientists for over a century. Discovered around 1902 by archaeologist Valerios Stais in a shipwreck off the coast of the Greek island Antikythera, the device is thought to have been built about 150–100 B.C. Journalist Jo Marchant writes, "Among the salvaged hoard subsequently shipped to Athens was a piece of formless rock that no one noticed at first, until it cracked open, revealing bronze gearwheels, pointers, and tiny Greek inscriptions. . . . A sophisticated piece of machinery consisting of precisely cut dials, pointers and at least thirty interlocking gear wheels, nothing close to its complexity appears again in the historical record for more than a thousand years, until the development of astronomical clocks in medieval Europe."

A dial on the front of the device probably carried at least three hands, one indicating the date and the other two indicating the positions of the Sun and the Moon. The device was also probably used to track dates of ancient Olympic games, predict solar eclipses, and indicate other planetary motions.

Of special delight to physicists, the Moon mechanism uses a special train of bronze gears, two of them linked with a slightly offset axis, to indicate the position and phase of the moon. As is known today from **Kepler's Laws of Planetary Motion**, the moon travels at different speeds as it orbits the Earth (e.g. faster when it is closer to the Earth), and this speed differential is modeled by the Antikythera mechanism, even though the ancient Greeks were not aware of the actual elliptical shape of the orbit. Additionally, the Earth travels faster when it is closer to the Sun than when it is far away.

Marchant writes, "By turning the handle on the box you could make time pass forwards or backwards, to see the state of the cosmos today, tomorrow, last Tuesday or a hundred years in the future. Whoever owned this device must have felt like master of the heavens."

SEE ALSO Sundial (3000 B.C.), Gears (50), Kepler's Laws of Planetary Motion (1609).

The Antikythera mechanism is an ancient geared computing device that was used to calculate astronomical positions. X-ray radiographs of the mechanism have revealed information about the device's internal configuration. (Photo courtesy of Rien van de Weijgaert.)

50

Hero's Jet Engine

Hero (or Heron) of Alexandria (c. 10–c. 70 A.D.), **Marcus Vitruvius Pollio** (c. 85 B.C.–c. 15 B.C.), **Ctesibius** (c. 285–c. 222 B.C.)

The history of modern rockets can be traced back through countless experiments to the ancient Greek mathematician and engineer Hero of Alexandria, who invented a rocket-like device called an aeolipile that used steam for propulsion. Hero's engine consisted of a sphere mounted on top of a kettle of water. A fire below the kettle produced steam that traveled through pipes into the sphere. Steam escaped through two bent tubes on opposite sides of the sphere, providing sufficient thrust so that the sphere would rotate. Because of friction in the bearings, Hero's engine does not spin increasingly fast but achieves a steady-state speed.

Both Hero and the Roman engineer Vitruvius were fascinated by these kinds of steam-powered devices, as well as the earlier Greek inventor Ctesibius. Historians of science are not certain as to whether Hero's engine served any practical purpose at the time. According to the 1865 *Quarterly Journal of Science*, "From the time of Hero we hear nothing more of the application of steam till the beginning of the seventeenth century. In a work published about 1600, Hero's engine is recommended to be used for turning spits, its great advantages being that the partakers of roasted meat may feel confident 'that the haunch has not been pawed by the turnspit (in the absence of the housewife's eye), for the pleasure of licking his unclean fingers'."

Jet and rocket engines rely on Newton's Third Law of Motion, which states for every action (force in one direction) there is an equal and opposite reaction (force in the opposite direction). One can see this principle in operation when an inflated, untied balloon is released into the air. The first jet-propelled aircraft was the German *Heinkel He 178*, which first flew in 1939.

SEE ALSO Newton's Laws of Motion and Gravitation (1687), Charles' Gas Law (1787), Tsiolkovsky Rocket Equation (1903).

John R. Bentley created and photographed his replica of Hero's Engine that spins almost silently at 1,500 rpm with a steam pressure of only 1.8 pounds per square inch, producing surprisingly little visible exhaust.

50

Gears

Hero (or Heron) of Alexandria (c. 10–c. 70 A.D.)

Rotating gears, with their intermeshed teeth, have played a crucial role in the history of technology. Not only are gear mechanisms important for increasing the applied twisting force, or torque, but gears are also useful for changing the speed and direction of force. One of the oldest machines is a potter's wheel, and primitive gears associated with these kinds of wheels probably existed for thousands of years. In the fourth century B.C., Aristotle wrote about wheels using friction between smooth surfaces to convey motions. Built around 125 B.C., the **Antikythera Mechanism** employed toothed gears for calculating astronomical positions. One of the earliest written references to toothed gears was made by Hero of Alexandria, c. 50 A.D. Through time, gears have played a crucial role in mills, clocks, bicycles, cars, washing machines, and drills. Because they are so useful in amplifying forces, early engineers used them for lifting heavy construction loads. The speed-changing properties of gear assemblies were put to use when ancient textile machines were powered by the movement of horses or water. The rotational speed of these power supplies was often insufficient, so a set of wooden gears was used to increase the speed for textile production.

When two gears are intermeshed, the rotational speed ratio s_1/s_2 is simply the reciprocal ratio of the number n of teeth on the two gears: $s_1/s_2 = n_2/n_1$. Thus, a small gear turns faster than its larger partner. The torque ratio has an opposite relationship. The larger gear experiences greater torque, and the higher torque implies lower velocity. This is useful, for example, for electric screwdrivers, in which the motor can produce a small amount of torque at high speed, but we wish to have a slow *output* speed with increased torque.

Among the simplest gears are *spur gears*, with their straight-cut teeth. *Helical gears* in which the teeth are set at an angle have the advantage of running more smoothly and quietly and usually being able to handle greater torques.

SEE ALSO Pulley (230 B.C.), Antikythera Mechanism (125 B.C.), Hero's Jet Engine (50).

Gears have played important roles in history. Gear mechanisms can increase the applied force or torque and are also useful for changing the speed and direction of a force.

78

St. Elmo's Fire

Gaius Plinius Secundus (Pliny the Elder) (23–79)

"Everything is in flames," exclaimed Charles Darwin aboard his sailing ship, "the sky with lighting—the water with luminous particles, and even the masts are pointed with an blue flame." What Darwin was experiencing was St. Elmo's fire, a natural phenomenon that has ignited superstitions for millennia. The Roman philosopher Pliny the Elder makes reference to the "fire" in his *Naturalis Historia* around 78 A.D.

Often described as a ghostly blue-white dancing flame, this is actually an electrical weather phenomenon in which a glowing **Plasma**, or ionized gas, emits light. The plasma is created by atmospheric electricity, and the eerie glow often appears at the tips of pointed objects such as church towers or the masts of ships during stormy weather. St. Elmo was the patron saint of Mediterranean sailors, who regarded St. Elmo's fire as a good omen because the glow was often brightest near the end of a storm. Sharp pointed objects encourage the formation of the "fire" because electric fields are more concentrated in areas of high curvature. Pointed surfaces discharge at a lower voltage level than surfaces without points. The color of the fire arises from the nitrogen and oxygen composition of the air and their associated fluorescence. If the atmosphere were neon, the fire would be orange, just like a **Neon Sign**.

"On dark stormy nights," writes scientist Philip Callahan, "St. Elmo's fire is probably responsible for more ghost stories and tales of apparitions than any other natural phenomenon." In *Moby Dick*, Herman Melville describes the fire during a typhoon: "All the yard-arms were tipped with a pallid fire; and touched at each tri-pointed lightning-rod-end with three tapering white flames, each of the three tall masts was silently burning in that sulphurous air, like three gigantic wax tapers before an altar. . . . 'The corpusants have mercy on us all!' . . . In all my voyagings seldom have I heard a common oath when God's burning finger has been laid on the ship."

SEE ALSO Aurora Borealis (1621), Ben Franklin's Kite (1752), Plasma (1879), Stokes' Fluorescence (1852), Neon Signs (1923).

"St. Elmo's Fire on Mast of Ship at Sea" in The Aerial World, *by Dr. G. Hartwig, London, 1886.*

1132

Cannon

Niccolò Fontana Tartaglia (1500–1557), **Han Shizhong** (1089–1151)

The cannon, which generally fires heavy projectiles using gunpowder, focused Europe's most brilliant minds on questions regarding the forces and laws of motion. "Ultimately it was the effects of gunpowder on science rather than on warfare that were to have the greatest influence in bringing about the Machine Age," writes historian J. D. Bernal. "Gunpowder and cannon not only blew up the medieval world economically and politically; they were major forces in destroying its system of ideas." Author Jack Kelley remarks, "Both gunners and natural philosophers wanted to know: What happens to the cannonball after it leaves the barrel of the gun? The search for a definitive answer took four hundred years and required the creation of entirely new fields of science."

The first documented use of gunpowder artillery in war is the 1132 capturing of a city in Fujian, China, by Chinese general Han Shizhong. During the Middle Ages, cannons became standardized and more effective against soldiers and fortifications. Later, cannons transformed war at sea. In the American Civil War, the howitzer had an effective range of over 1.1 miles (1.8 kilometers), and in World War I, cannons caused the majority of all deaths in action.

In the 1500s, it was realized that gunpowder produced a large volume of hot gas that exerted pressure on the cannon ball. Italian engineer Niccolò Tartaglia assisted gunners in determining that an elevation of 45 degrees gave a cannon the longest shot (which we know today is only an approximation, due to the effects of air resistance). Galileo's theoretical studies showed that gravity constantly accelerated the fall of a cannonball, producing an idealized trajectory in the shape of a parabolic curve, which every cannonball traces regardless of its mass or initial firing angle. Although air resistance and other factors play a complicating role in cannon firing, the cannon "provided a focus for the scientific investigation of reality," writes Kelly, "one that overturned long-standing error and laid the foundation for a rational age."

SEE ALSO Atlatl (30,000 B.C.), Crossbow (341 B.C.), Trebuchet (1200), Tsiolkovsky Rocket Equation (1903), Golf Ball Dimples (1905).

Medieval cannon on the bastions of the Citadel in Gozo, Malta.

1150

Perpetual Motion Machines

Bhaskara II (1114–1185), **Richard Phillips Feynman** (1918–1988)

Although the construction of apparent perpetual motion machines may seem like an unlikely topic for a book on milestones in physics, important advances in physics often involve ideas at the fringes of physics, especially when scientists work to determine why a device violates the laws of physics.

Perpetual motion machines have been proposed for centuries, such as in 1150 when the Indian mathematician-astronomer Bhaskara II described a wheel with mercury-filled containers that he believed would turn forever as the mercury moved within the containers, keeping the wheel heavier on one side of the axle. More generally, *perpetual motion* often refers to any device or system that either: 1) forever produces more energy than it consumes (a violation of the law of **Conservation of Energy**) or 2) spontaneously extracts heat from its surroundings to produce mechanical work (a violation of the **Second Law of Thermodynamics**).

My favorite perpetual motion machine is Richard Feynman's Brownian Ratchet, discussed in 1962. Imagine a tiny ratchet attached to a paddlewheel that is submerged in water. Because of the one-way ratchet mechanism, when molecules randomly collide with the paddle wheel, the wheel can turn only in one direction and can presumably be used to perform work, such as lifting a weight. Thus, through using a simple ratchet, which might consist of a pawl that engages sloping teeth of a gear, the paddle spins forever. Amazing!

However, Feynman himself showed that his Brownian Ratchet must have a very tiny pawl to respond to the molecular collisions. If the temperature T of the ratchet and pawl were the same as that of the water, the tiny pawl would intermittently fail, and no net motion would occur. If T was less than the temperature of the water, the paddle wheel might be made to go only in one direction, but in doing so, it would make use of energy from the temperature gradient, which does not violate the Second Law of Thermodynamics.

SEE ALSO Brownian Motion (1827), Conservation of Energy (1843), Second Law of Thermodynamics (1850), Maxwell's Demon (1867), Superconductivity (1911), Drinking Bird (1945).

Cover of the October, 1920 issue of Popular Science *magazine, painted by American illustrator Norman Rockwell (1894–1978), showing an inventor researching a perpetual motion machine.*

OCTOBER
Reg. U.S. Pat. Off.
25 CENTS
Popular Science
Founded MON
Perpetual Motion?
See page 26
INVENTION
Norman Rockwell

1200

Trebuchet

The dreaded trebuchet employs simple laws of physics to cause mayhem. In the Middle Ages, this catapult-like device was used to smash through walls by hurling projectiles, using the principle of the lever and centrifugal force to hold a sling taut. Sometimes, the bodies of dead soldiers or rotting animals were shot over the walls of castles in order to spread disease.

Traction trebuchets, which required men to pull on a rope trigger, were used in the Greek world and in China in the fourth century B.C. The *counterweight trebuchet* (which we will simply call *trebuchet* hereafter) replaced the men with a heavy weight and did not appear in China, with certainty, until about 1268. The trebuchet has one part that resembles a seesaw. A heavy weight is attached to one end. At the other end is a ropelike sling that holds a projectile. As the counterweight descends, the sling swings to a vertical position, at which point a mechanism releases the projectile in the direction of the target. This approach is much more powerful than traditional catapults without the sling, in terms of projectile speed and distance. Some trebuchets placed the counterweight closer to the fulcrum (pivot point) of the swing in order to achieve a mechanical advantage. The force of the counterweight is large, and the load is small—like dropping an elephant on one side and transferring energy rapidly to a brick on the other.

The trebuchet was used by the crusaders and also by Islamic armies at various times in history. In 1421, the future Charles VII of France had his engineers build a trebuchet that could shoot a 1,760-pound (800-kilogram) stone. An average range was around 984 feet (300 meters).

Physicists have studied the mechanics of the trebuchet because although it may seem simple, the system of differential equations that governs its motion is very nonlinear.

SEE ALSO Atlatl (30,000 B.C.), Boomerang (20,000 B.C.), Crossbow (341 B.C.), Cannon (1132).

Trebuchet in the Castle of Castelnaud, a medieval fortress in Castelnaud-la-Chapelle, overlooking the Dordogne River in Périgord, southern France.

1304

Explaining the Rainbow

Abu Ali al-Hasan ibn al-Haytham (965–1039), **Kamal al-Din al-Farisi** (1267–c. 1320), **Theodoric of Freiberg** (c. 1250–c. 1310)

"Who among us has not admired the majestic beauty of a rainbow arching silently across a storm's wake?" write authors Raymond Lee, Jr. and Alistair Fraser. "Vivid and compelling, this image recalls childhood memories, venerable folklore, and perhaps some half-remembered science lessons. . . . Some societies see the rainbow as an ominous serpent arching across the sky, while others imagine it to be a tangible bridge between the gods and humanity." The bow spans some modern divides between the arts and sciences.

Today we know that the attractive colors of the rainbow are a result of sunlight that is first refracted (experiences a change in direction) as it enters the surface of the raindrop, and then is reflected off the back of the drop toward the observer, and refracted a second time as it leaves the drop. The separation of white light into various colors occurs because different wavelengths—which correspond to various colors—refract at different angles.

The first correct explanations of rainbows, which involved two refractions and a reflection, were independently provided at about the same time by Kamal al-Din al-Farisi and Theodoric of Freiberg. Al-Farisi was a Persian Muslim scientist born in Iran who conducted experiments using a transparent sphere filled with water. Theodoric of Freiberg, a German theologian and physicist, used a similar experimental setup.

It is fascinating the degree to which simultaneous discoveries appear in great works of science and mathematics. For example, various gas laws, the Möbius strip, calculus, the theory of evolution, and hyperbolic geometry were all developed simultaneously by different individuals. Most likely, such simultaneous discoveries have occurred because the time was "ripe" for such discoveries, given humanity's accumulated knowledge at the time the discoveries were made. Sometimes, two scientists, working separately, are stimulated by reading the same preliminary research. In the case of the rainbow, both Theodoric and al-Farisi relied on the *Book of Optics* by the Islamic polymath Ibn al-Haytham (Alhazen).

SEE ALSO Snell's Law of Refraction (1621), Newton's Prism (1672), Rayleigh Scattering (1871), Green Flash (1882).

LEFT: *In the Bible, God shows Noah a rainbow as a sign of God's covenant (painting by Joseph Anton Koch [1768–1839]).* RIGHT: *Rainbow colors result from sunlight that is refracted and reflected within water droplets.*

1338

Hourglass

Ambrogio Lorenzetti (1290–1348)

The French author Jules Renard (1864–1910) once wrote, "Love is like an hourglass, with the heart filling up as the brain empties." Hourglasses, also known as sandglasses, measure time using fine sand that flows through a narrow neck from an upper reservoir. The measured length of time depends on many factors, including the volume of sand, the shape of the bulbs, the width of the neck, and the type of sand used. Although hourglasses were probably in use in the third century B.C., the first recorded evidence of hourglasses did not appear until the 1338 fresco *Allegory of Good Government* by Italian painter Ambrogio Lorenzetti. Interestingly, the sailing ships of Ferdinand Magellan retained 18 hourglasses per ship as he attempted to circumnavigate the globe. One of the largest hourglasses—39 feet (11.9 meters) in height—was built in 2008 in Moscow. Through history, hourglasses were used in factories and to control the duration of sermons in church.

In 1996, British researchers at the University of Leicester determined that the rate of flow depends only on the few centimeters above the neck and not on the bulk of sand above that. They also found that small glass beads known as ballotini gave the most reproducible results. "For a given volume of ballotini," the researchers write, "the period is controlled by their size, the size of the orifice, and the shape of the reservoir. Provided the aperture is at least 5 times the particle diameter, the period P is given by the expression $P = KV(D - d)^{-2.5}$, where P is measured in seconds, V denotes the bulk volume of ballotini in ml, d the maximum bead diameter in mm . . . , and D the diameter of a circular orifice in mm. The constant of proportionality K depends on the shape of the reservoir." For example, the researchers found different values of K for conical container shapes and hourglass shapes. Any disturbance to the hourglass lengthened the period of time, but changes in temperature produced no observable effect.

SEE ALSO Sundial (3000 B.C.), Anniversary Clock (1841), Time Travel (1949), Atomic Clocks (1955).

Hourglasses were probably in use in the third century. Ferdinand Magellan's sailing ships had 18 hourglasses per ship when he attempted to circumnavigate the globe.

1543

Sun-Centered Universe

Nicolaus Copernicus (1473–1543)

"Of all discoveries and opinions," wrote the German polymath Johann Wolfgang von Goethe in 1808, "none may have exerted a greater effect on the human spirit than the doctrine of Copernicus. The world had scarcely become known as round and complete in itself when it was asked to waive the tremendous privilege of being the center of the universe. Never, perhaps, was a greater demand made on mankind—for by this admission so many things vanished in mist and smoke! What became of our Eden, our world of innocence, piety and poetry; the testimony of the senses; the conviction of a poetic-religious faith?"

Nicolaus Copernicus was the first individual to present a comprehensive heliocentric theory that suggested the Earth was not the center of the universe. His book, *De revolutionibus orbium coelestium* (*On the Revolutions of the Celestial Spheres*) was published in 1543, the year he died, and put forward the theory that the Earth revolved around the Sun. Copernicus was a Polish mathematician, physician, and classical scholar—astronomy was something he studied in his spare time—but it was in the field of astronomy that he changed the world. His theory relied on a number of assumptions: that the Earth's center is not the center of the universe, that the distance from the Earth to the Sun is miniscule when compared with the distance to the stars, that the rotation of the Earth accounts for the apparent daily rotation of the stars, and that the apparent retrograde motion of the planets (in which they appear to briefly stop and reverse directions at certain times when viewed from the Earth) is caused by the motion of the Earth. Although Copernicus' proposed circular orbits and epicycles of planets were incorrect, his work motivated other astronomers, such as Johannes Kepler, to investigate planetary orbits and later discover their elliptical nature.

Interestingly, it was not until many years later, in 1616, that the Roman Catholic Church proclaimed that Copernicus' heliocentric theory was false and "altogether opposed to Holy Scripture."

SEE ALSO *Mysterium Cosmographicum* (1596), Telescope (1608), Kepler's Laws of Planetary Motion (1609), Measuring the Solar System (1672), Hubble Telescope (1990).

Orreries are mechanical devices that show positions and motions of the planets and moons in a heliocentric model of the solar system. Shown here is a device constructed in 1766 by instrument-maker Benjamin Martin (1704–1782) and used by astronomer John Winthrop (1714–1779) to teach astronomy at Harvard University. On display at the Putnam Gallery in the Harvard Science Center.

1596

Mysterium Cosmographicum

Johannes Kepler (1571–1630)

Throughout his life, German astronomer Johannes Kepler attributed his scientific ideas and motivation to his quest for understanding the mind of God. For example, in his work *Mysterium Cosmographicum* (*The Sacred Mystery of the Cosmos*, 1596), he wrote, "I believe that Divine Providence intervened so that by chance I found what I could never obtain by my own efforts. I believe this all the more because I have constantly prayed to God that I might succeed."

Kepler's initial vision of the universe rested upon his studies of symmetrical, three-dimensional objects known as Platonic solids. Centuries before Kepler, the Greek mathematician Euclid (325 B.C.–265 B.C.) showed that only five such solids exist with identical faces: the cube, dodecahedron, icosahedron, octahedron, and tetrahedron. Although Kepler's sixteenth-century theory seems strange to us today, he attempted to show that the distances from the planets to the Sun could be found by studying spheres inside these regular polyhedra, which he drew nested in one another like layers of an onion. For example, the small orbit of Mercury is represented by the innermost sphere in his models. The other planets known at his time were Venus, Earth, Mars, Jupiter, and Saturn.

In particular, an outer sphere surrounds a cube. Inside the cube is a sphere, followed by a tetrahedron, followed by another sphere, followed by a dodecahedron, followed by a sphere, an icosahedron, sphere, and finally a small inner octahedron. A planet may be imagined as being embedded in each sphere that defines an orbit of a planet. With a few subtle compromises, Kepler's scheme worked fairly well as a rough approximation to what was known about planetary orbits at the time. Owen Gingerich writes, "Although the principal idea of the *Mysterium Cosmographicum* was erroneous, Kepler established himself as the first . . . scientist to demand physical explanations for celestial phenomena. Seldom in history has so wrong a book been so seminal in directing the future course of science."

SEE ALSO Sun-Centered Universe (1543), Kepler's Laws of Planetary Motion (1609), Measuring the Solar System (1672), Bode's Law of Planetary Distances (1766).

Kepler's initial vision of the universe relied on his studies of symmetrical, three-dimensional objects known as Platonic solids. This diagram is from his Mysterium Cosmographicum*, published in 1596.*

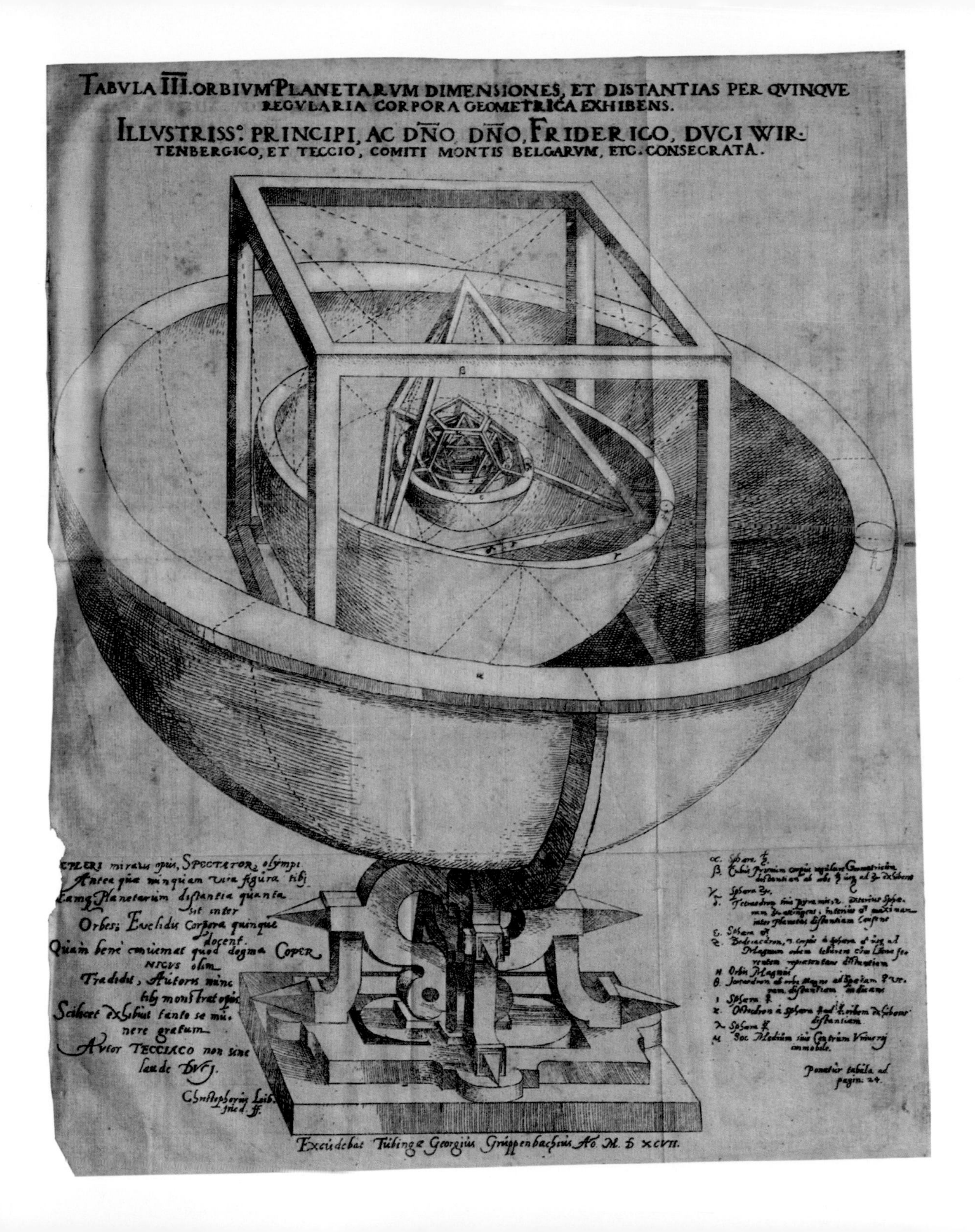
TABVLA III. ORBIVM PLANETARVM DIMENSIONES, ET DISTANTIAS PER QVINQVE REGVLARIA CORPORA GEOMETRICA EXHIBENS.
ILLVSTRISS°. PRINCIPI, AC DÑO DÑO, FRIDERICO, DVCI WIRTENBERGICO, ET TECCIO, COMITI MONTIS BELGARVM, ETC. CONSECRATA.
Orbes; Euclidis Corpora quinque docent.
Quam bene conveniat quod dogma COPERNICVS olim
Tradidit, Autoris nunc tibi monstrat opus.
Avtor TECCIACO non sine laude DVCI.
Christophorus Leibfried. ff.
Ponatur tabula ad pagin: 24.
Excudebat Tubingæ Georgius Gruppenbachius Ao. M. D XCVII.

1600

De Magnete

William Gilbert (1544–1603)

William Gilbert's book *De Magnete*, published in 1600, is considered to be the first great work on physical science to be produced in England, and much of European science had its roots in Gilbert's initial theories and fondness for experiments. A personal physician to Queen Elizabeth I, Gilbert is one of the important fathers of the science of electricity and magnetism.

"In the 1500s," writes author and engineer Joseph F. Keithley, "strong feelings abounded that knowledge was God's domain, and therefore humans should not pry into it. Experiments were considered to be dangerous to intellectual and moral life. . . . Gilbert, however, broke with traditional ways of thinking and was impatient" with those who did not use experiments to probe the workings of reality.

In his research in terrestrial magnetism, Gilbert created a spherical lodestone, about a foot (0.3 meter) in diameter, which he called a *terrella* (little Earth). By moving a small magnetic needle on a pivot around the terrella's surface, he showed that the terrella had a North and South Pole, and that the needle dipped as it neared a pole, which mimics the dipping of compass needles on the Earth as they approach the poles. He postulated that the Earth was like a giant lodestone. British ships had depended on the magnetic compass, but its working was a mystery. Some had thought that the Pole Star was the actual cause of the magnetic attraction of a compass needle. Others thought there was a magnetic mountain or island at the North Pole, which ships had better avoid because the iron nails of sailing ships would be torn out. Scientists Jacqueline Reynolds and Charles Tanford write, "Gilbert's demonstration that the Earth and not the heavens held the power went far beyond the force of magnetism and influenced all thinking about the physical world."

Gilbert correctly argued that the center of the Earth was iron. He believed, incorrectly, that quartz crystals were a solid form of water—something akin to compressed ice. Gilbert died in 1603, most likely from the bubonic plague.

SEE ALSO Olmec Compass (1000 B.C.), Von Guericke's Electrostatic Generator (1660), Ampère's Law of Electromagnetism (1825), Gauss and the Magnetic Monopole (1835), Galvanometer (1882), Curie's Magnetism Law (1895), Stern-Gerlach Experiment (1922).

William Gilbert suggested that the Earth generated its own magnetic fields. Today, we know that a magnetosphere, represented here as a violet bubble around the Earth, is formed when charged particles from the sun interact with and are deflected by the Earth's magnetic field.

1608

Telescope

Hans Lippershey (1570–1619), **Galileo Galilei** (1564–1642)

Physicist Brian Greene writes, "The invention of the telescope and its subsequent refinement and use by Galileo marked the birth of the modern scientific method and set the stage for a dramatic reassessment of our place in the cosmos. A technological device revealed conclusively that there is so much more to the universe than is available to our unaided senses." Computer scientist Chris Langton agrees, noting, "Nothing rivals the telescope. No other device has initiated such a thoroughgoing reconstruction of our world view. It has forced us to accept the earth (and ourselves) as merely a part of the larger cosmos."

In 1608, the German-Dutch lensmaker Hans Lippershey may have been the first to invent the telescope, and a year later, the Italian astronomer Galileo Galilei constructed a telescope with about a three-fold magnification. He later made others with up to a 30-fold magnification. Although the early telescopes were designed to observe remote objects using visible light, modern telescopes are a range of devices capable of utilizing other regions of the electromagnetic spectrum. *Refracting telescopes* employ lenses to form an image, while *reflecting telescopes* use an arrangement of mirrors for this purpose. *Catadioptric telescopes* use mirrors and lenses.

Interestingly, many important astronomical discoveries with telescopes have been largely unanticipated. Astrophysicist Kenneth Lang writes in *Science*, "Galileo Galilei turned his newly constructed spyglass to the skies, and thus began astronomers' use of novel telescopes to explore a universe that is invisible to the unaided eye. The search for the unseen has resulted in many important unexpected discoveries, including Jupiter's four large moons, the planet Uranus, the first asteroid Ceres, the large recession velocities of spiral nebulae, radio emission from the Milky Way, cosmic X-ray sources, **Gamma-Ray Bursts**, radio pulsars, the binary pulsar with its signature of gravitational radiation, and the **Cosmic Microwave Background** radiation. The observable universe is a modest part of a much vaster, undiscovered one that remains to be found, often in the least expected ways."

SEE ALSO Sun-Centered Universe (1543), Discovery of Saturn's Rings (1610), *Micrographia* (1665), Stellar Parallax (1838), Hubble Telescope (1990).

LEFT: *One antenna in the Very Large Array (VLA) used for studying signals from radio galaxies, quasars, pulsars, and more.* RIGHT: *Observatory staff astride the University of Pittsburgh's Thaw 30-inch refractor just before its completion in 1913. A man sits atop counterweights needed to keep the massive telescope in balance.*

1609

Kepler's Laws of Planetary Motion

Johannes Kepler (1571–1630)

"Although Kepler is remembered today chiefly for his three laws of planetary motion," writes astronomer Owen Gingerich, "these were but three elements in his much broader search for cosmic harmonies. . . . He left [astronomy] with a unified and physically motivated heliocentric [Sun-centered] system nearly 100 times more accurate."

Johannes Kepler was the German astronomer and theologian-cosmologist, famous for his laws that described the elliptical orbits of the Earth and other planets around the Sun. In order for Kepler to formulate his laws, he had to first abandon the prevailing notion that circles were the "perfect" curves for describing the cosmos and its planetary orbits. When Kepler first expressed his laws, he had no theoretical justification for them. They simply provided an elegant means by which to describe orbital paths obtained from experimental data. Roughly 70 years later, Newton showed that Kepler's Laws were a consequence of Newton's *Law of Universal Gravitation*.

Kepler's First Law (The Law of Orbits, 1609) indicated that all of the planets in our Solar System move in elliptical orbits, with the Sun at one focus. His Second Law (The Law of Equal Areas, 1618) showed that when a planet is far from the Sun, the planet moves more slowly than when it is close to the Sun. An imaginary line that connects a planet to the Sun sweeps out equal areas in equal intervals of time. Given Kepler's first two laws, planetary orbits and positions could now be easily calculated and with an accuracy that matched observations.

Kepler's Third Law (The Law of Periods, 1618) showed that for any planet, the square of the period of its revolution about the Sun is proportional to the cube of the semi-major axis of its elliptical orbit. Thus, planets far from the Sun have very long years. Kepler's Laws are among the earliest scientific laws to be established by humans, and, while unifying astronomy and physics, the laws provided a stimulus to subsequent scientists who attempted to express the behavior of reality in terms of simple formulas.

SEE ALSO Sun-Centered Universe (1543), *Mysterium Cosmographicum* (1596), Telescope (1608), Newton's Laws of Motion and Gravitation (1687).

Artistic representation of the Solar System. Johannes Kepler was the German astronomer and theologian-cosmologist, famous for his laws that described the elliptical orbits of the Earth and the other planets around the sun.

1610

Discovery of Saturn's Rings

Galileo Galilei (1564–1642), **Giovanni Domenico Cassini** (1625–1712), **Christiaan Huygens** (1629–1695)

"Saturn's rings seem almost immutable," writes science-reporter Rachel Courtland. "These planetary jewels, carved by moonlets and shaped by gravity, could well have looked much the same now as they did billions of years ago—but only from afar." In the 1980s, a mysterious event occurred that suddenly warped the planet's innermost rings into a ridged spiral pattern, "like the grooves on a vinyl record." Scientists hypothesize that the spiral winding could have been caused by a very large asteroid-like object or a dramatic shift in weather.

In 1610, Galileo Galilei became the first person to observe Saturn's rings; however, he described them as "ears." It was not until 1655 that Christiaan Huygens was able to use his higher-quality telescope and become the first person to describe the feature as an actual ring surrounding Saturn. Finally, in 1675, Giovanni Cassini determined that Saturn's "ring" was actually composed of subrings with gaps between them. Two such gaps have been carved out by orbiting moonlets, but other gaps remain unexplained. *Orbital resonances*, which result from the periodic gravitational influences of Saturn's moons, also affect the stability of the rings. Each subring orbits at a different speed around the planet.

Today, we know that the rings consist of small particles, made almost entirely of water ice, rocks, and dust. Astronomer Carl Sagan referred to the rings being a "vast horde of tiny ice worlds, each on its separate orbit, each bound to Saturn by the giant planet's gravity." The particles range in size from a sand grain to the size of a house. The ring structures also have a thin atmosphere composed of oxygen gas. The rings may have been created by the debris from the breakup of some ancient moon, comet, or asteroid.

In 2009, scientists at NASA discovered a nearly invisible ring around Saturn that is so large that it would take 1 billion Earths to fill it (or about 300 Saturns lined up side to side).

SEE ALSO Telescope (1608), Measuring the Solar System (1672), Discovery of Neptune (1846).

Image of Saturn and its rings made from a composite of 165 images taken by the wide-angle camera on the Cassini spacecraft. The colors in the image were created by using ultraviolet, infrared, and other photography.

1611

Kepler's "Six-Cornered Snowflake"

Johannes Kepler (1571–1630)

Philosopher Henry David Thoreau wrote of his awe of snowflakes, "How full of the creative genius is the air in which these are generated! I should hardly admire them more if real stars fell and lodged on my coat." Snow crystals with hexagonal symmetry have intrigued artists and scientists throughout history. In 1611, Johannes Kepler published the monograph "On the Six-Cornered Snowflake," which is among the earliest treatments of snowflake formations that sought a scientific understanding in contrast to a religious one. In fact, Kepler mused that it might be easier to understand the beautiful symmetry of snow crystals if they were considered as living entities with souls, each given a purpose by God. However, he considered it more likely that some kind of hexagonal packing of particles, smaller than his ability to discern, could provide an explanation of the marvelous snowflake geometries.

Snowflakes (or, more rigorously, *snow crystals*, given that actual flakes from the sky may consist of many crystals) often begin their lives as tiny dust particles on which water molecules condense at sufficiently low temperatures. As the growing crystal falls through different atmospheric humidities and temperatures, water vapor continues to condense into solid ice, and the crystal slowly obtains its shape. The sixfold symmetry commonly seen arises from the energetically favorable hexagonal crystal structure of ordinary ice. The six arms look so similar because they all form under similar conditions. Other crystal shapes can also form, including hexagonal columns.

Physicists study snow crystals and their formation partly because crystals are important in applications ranging from electronics to the science of self-assembly, molecular dynamics, and the spontaneous formation of patterns.

Since a typical snow crystal contains about 10^{18} water molecules, the odds are virtually zero that two crystals of typical size are identical. At the macroscopic level, it is unlikely that any two large, complex snowflakes have ever looked completely alike—since the first snowflake fell on the Earth.

SEE ALSO *Micrographia* (1665), Ice Slipperiness (1850), Quasicrystals (1982).

LEFT: *Rime frost on the two ends of a "capped column" snowflake.* RIGHT: *Hexagonal dendrite snowflake, magnified by a low-temperature scanning electron microscope. The central flake has been artificially colorized for emphasis.* (Agricultural Research Center)

1620

Triboluminescence

Francis Bacon (1561–1626)

Imagine you are traveling with the ancient Native American Ute shamans in the American Midwest, hunting for quartz crystals. After collecting the crystals and placing them into ceremonial rattles made of translucent buffalo skin, you wait for the nighttime rituals to begin to summon the spirits of the dead. When it is dark, you shake the rattles and they blaze with flashes of light as the crystals collide with one another. When participating in this ceremony, you are experiencing one of the oldest known applications of triboluminescence, a physical process through which light is generated when materials are crushed, rubbed, and ripped—as electrical charges are separated and reunited. The resultant electrical discharge ionizes the nearby air, triggering flashes of light.

In 1620, the English scholar Francis Bacon published the first known documentation of the phenomena, in which he mentions that sugar will sparkle when "broken or scraped" in the dark. Today, it is easy to experiment with triboluminescence in your own home by breaking sugar crystals or Wint-O-Green Life Savers candy in a dark room. The wintergreen oil (methyl salicylate) in the candy absorbs ultraviolet light produced by the crushing of sugar and reemits it as blue light.

The spectrum of light produced by sugar triboluminescence is the same as that for lightning. In both cases, electrical energy excites nitrogen molecules in the air. Most of the light emitted by nitrogen in the air is in the ultraviolet range that our eyes cannot see, and only a small fraction is emitted in the visible range. When the sugar crystals are stressed, positive and negative charges accumulate, finally causing electrons to jump across a crystal fracture and excite electrons in the nitrogen molecules.

If you peel Scotch tape in the dark, you may also be able to see emitted light from triboluminescence. Interestingly, the process of peeling such tape in a vacuum can produce **X-rays** that are sufficiently strong to create an x-ray image of the finger.

SEE ALSO Stokes' Fluorescence (1852), Piezoelectric Effect (1880), X-rays (1895), Sonoluminescence (1934).

The phenomenon of triboluminescence was first discovered in 1605 by Sir Francis Bacon from scratching sugar with a knife. Shown here is a photograph of the triboluminescence of N-acetylanthranilic acid crystals crushed between two transparent windows.

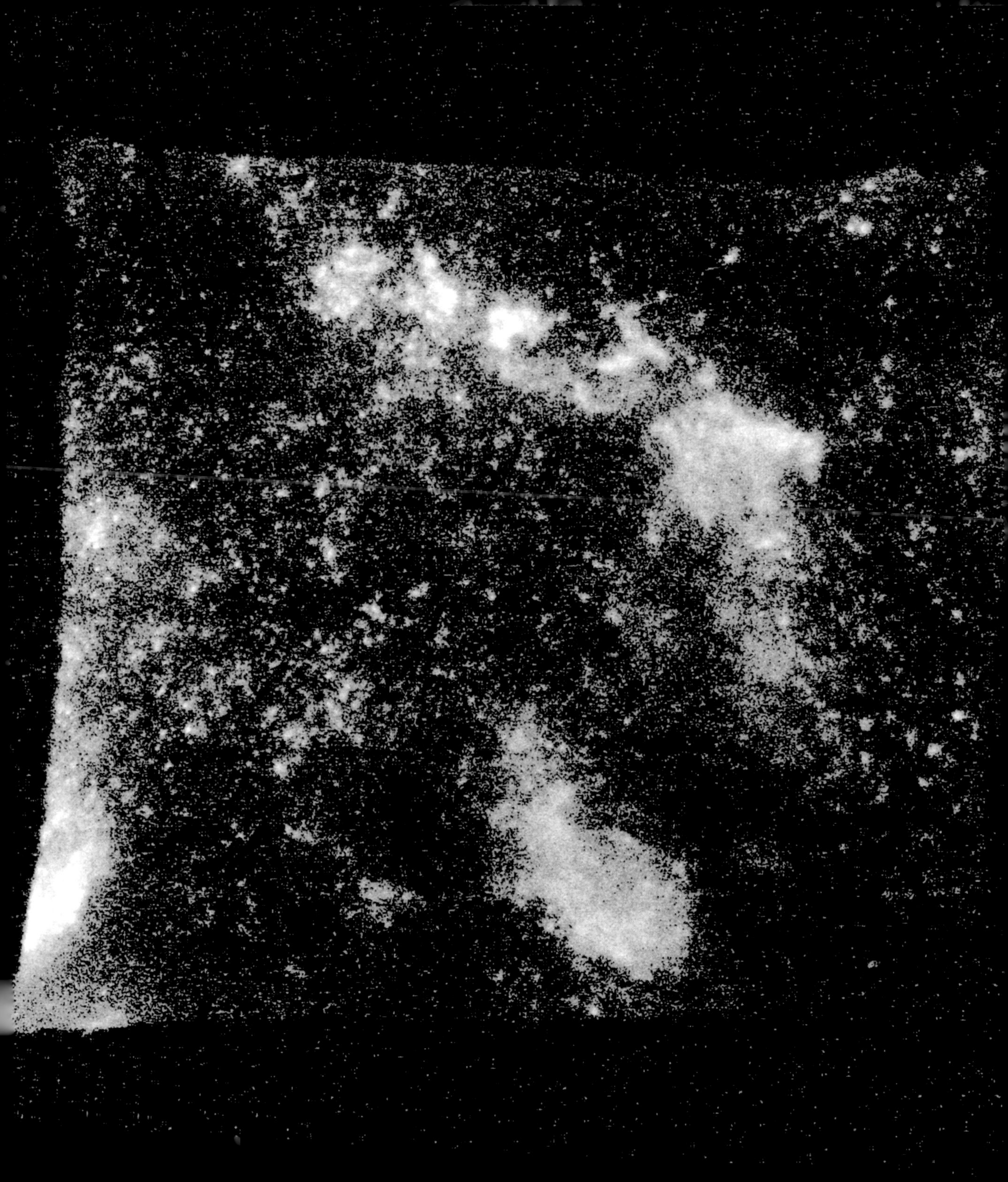

1621

Snell's Law of Refraction

Willebrord Snellius (1580–1626)

"Where art thou, beam of light?" wrote the poet James Macpherson, likely unaware of the physics of refraction. Snell's Law concerns the bending, or refracting, of light and other waves, for example, as they travel through the air and pass into another material, such as glass. When the waves are refracted, they experience a change in the direction of propagation due to a change in their velocities. You can see Snell's Law in action by placing a pencil into a glass of water and observing the apparent bend in the pencil. The law is expressed as $n_1\sin(\theta_1) = n_2\sin(\theta_2)$. Here, n_1 and n_2 are the refractive indices of media 1 and 2. The angle between the incident light and a line perpendicular to the interface between media is called the angle of incidence (θ_1). The light ray continues from medium 1 into medium 2, leaving the boundary between the media at an angle of θ_2 to a line that is perpendicular to the boundary. This second angle is known as the angle of refraction.

A convex lens makes use of refraction to cause parallel light rays to converge. Without the refraction of light by the lenses in our eyes, we couldn't see properly. Seismic waves—for example, the waves of energy caused by the sudden breaking of subterranean rock—change speed within the earth and are bent when they encounter interfaces between materials in accordance with Snell's Law.

When a beam of light is transmitted from a material with high index of refraction to one of low index, the beam can, under some conditions, be totally reflected. This optical phenomenon is often called *total internal reflection*, and it occurs when light is refracted at a medium boundary to such an extent that it is reflected back. This phenomenon is observed in certain optical fibers in which the light enters at one end and remains trapped inside until it emerges from the other end. Cut diamonds often exhibit total internal reflection when the diamond sparkles and emits light in the direction of the observer's eye.

Snell's Law was discovered independently by various investigators over the centuries but was named after the Dutch astronomer and mathematician, Willebrord Snellius.

SEE ALSO Explaining the Rainbow (1304), Newton's Prism (1672), Brewster's Optics (1815), Fiber Optics (1841), Green Flash (1882), Cherenkov Radiation (1934).

LEFT: *Diamonds exhibit total internal reflection.* RIGHT: *When an archerfish shoots streams of water at prey, the fish must compensate for light refraction when aiming. How an archerfish corrects for refraction is still not entirely explained.* (Photo courtesy of Shelby Temple.)

1621

Aurora Borealis

Pierre Gassendi (1592–1655), **Alfred Angot** (1848–1924), **Olof Petrus Hiorter** (1696–1750), **Anders Celsius** (1701–1744)

"The aurora borealis had become a source of terror," writes meteorologist Alfred Ango of people's reactions in the 1500s to the appearance of luminous curtains in the sky. "Bloody lances, heads separated from the trunk, armies in conflict, were clearly distinguished. At the sight of them people fainted . . . others went mad." Author George Bryson writes, "Ancient Scandinavians saw the northern lights as the recently departed souls of strong, beautiful women hovering in the air. . . . Electric green pierced by neon blue, shocking pink spinning into deep red, shimmering violet fading. . . ."

Energetic charged particles streaming from the solar wind of the Sun enter the Earth's atmosphere and are channeled toward the north and south magnetic poles of the Earth. As the particles spiral around the magnetic field lines, they collide with atmospheric oxygen and nitrogen atoms, forcing the atoms into excited states. When the electrons of the atoms return to their lower-energy ground states, they emit a light—for example, red and green for oxygen atoms—that is seen in amazing light displays near the Earth's polar regions and that take place in the ionosphere (the uppermost part of the atmosphere, charged by solar radiation). Nitrogen may contribute a blue tinge when a nitrogen atom regains an electron after it has been ionized. When near the North Pole, the light display is called the *aurora borealis*. The southern counterpart is the *aurora australis*.

Although Cro-Magnon cave paintings exist (c. 30,000 B.C.) that appear to depict ancient auroras, it wasn't until 1621 that the term *aurora borealis* was coined by Pierre Gassendi—French philosopher, priest, astronomer, and mathematician—after Aurora (the Roman goddess of dawn) and Boreas (the Greek name for "north wind").

In 1741, Swedish astronomers Olof Petrus Hiorter and Anders Celsius suggested that auroras were controlled by magnetic effects after observing fluctuations of compass needles when auroras were in the sky. Today, we know that other planets, such as Jupiter and Saturn, have magnetic fields stronger than those of the Earth; they also have auroras.

SEE ALSO St. Elmo's Fire (78), Rayleigh Scattering (1871), Plasma (1879), Green Flash (1882), HAARP (2007).

The Aurora Borealis, or Northern Lights, glows above Bear Lake, Eielson Air Force Base, Alaska.

1638

Acceleration of Falling Objects

Galileo Galilei (1564–1642)

"To appreciate the full nature of Galileo's discoveries," writes I. Bernard Cohen, "we must understand the importance of abstract thinking, of its use by Galileo as a tool that in its ultimate polish was a much more revolutionary instrument for science than even the telescope." According to legend, Galileo dropped two balls of different weights from the Leaning Tower of Pisa to demonstrate that they would both hit the ground at the same time. Although this precise experiment probably did not take place, Galileo certainly performed experiments that had a profound effect on contemporary understanding of the laws of motions. Aristotle had taught that heavy objects fall faster than light ones. Galileo showed that this was only an artifact of differing air resistances of the objects, and he supported his claims by performing numerous experiments with balls rolling down inclined planes. Extrapolating from these experiments, he demonstrated that if objects could fall without air resistance, all objects accelerate at the same rate. More precisely, he showed that the distance traveled by a constantly accelerating body starting at zero velocity is proportional to the square of the time falling.

Galileo also proposed the principle of inertia, in which an object's motion continues at the same speed and direction unless acted upon by another force. Aristotle had erroneously believed that a body could be kept in motion only by applying a force. Newton later incorporated Galileo's principle into his **Laws of Motion**. If it is not apparent to you that a moving object does not "naturally" stop moving without an applied force, you can imagine an experiment in which the face of a penny is sliding along an infinite smooth horizontal table that is so well oiled that there is no friction. Here, the penny would continue sliding along such an imaginary surface forever.

SEE ALSO Conservation of Momentum (1644), Tautochrone Ramp (1673), Newton's Laws of Motion and Gravitation (1687), Clothoid Loop (1901), Terminal Velocity (1960).

Imagine spheres, or any objects, of different masses released at the same height at the same time. Galileo showed that they must all fall together at the same speed, if we neglect any differences in air resistance.

1643

Barometer

Evangelista Torricelli (1608–1647), **Blaise Pascal** (1623–1662)

Although the barometer is extremely simple, its implications are profound, going far beyond its usefulness for predicting the weather. The device has helped scientists understand the nature of the atmosphere and to discover that the atmosphere is finite and does not reach to the stars.

Barometers are devices used to measure atmospheric pressure and are embodied in two main forms: *mercury* and *aneroid*. In the mercury barometer, a glass tube contains mercury and is sealed at the top, while the bottom of the tube opens into a mercury-filled reservoir. The level of mercury in the tube is controlled by the atmosphere that presses down on the mercury in the reservoir. For example, under conditions of high atmospheric pressure, the mercury rises higher in the tube than when atmospheric pressure is low. In the tube, the mercury adjusts until the weight of the mercury in the column balances the atmospheric force.

Italian physicist Evangelista Torricelli is generally credited with inventing the barometer in 1643, and he observed that the height of the mercury in a barometer changed slightly each day as a result of atmospheric pressure changes. He wrote, "We live submerged at the bottom of an ocean of elementary air, which is known by incontestable experiments to have weight." In 1648, Blaise Pascal used a barometer to show that there is less air pushing down at the top of a mountain then there is at the bottom; thus, the atmosphere did not extend forever.

In aneroid barometers, there is no moving liquid. Instead, a small flexible evacuated metal capsule is used. Inside the capsule is a spring. Small changes in atmospheric pressure cause the capsule to expand or contract. Lever mechanisms within the barometer amplify these small movements, allowing users to read pressure values.

If the atmospheric pressure is falling, this change often indicates that stormy weather is likely. Rising air pressure suggests that fair weather with no precipitation is likely.

SEE ALSO Siphon (250 B.C.), Buys-Ballot's Weather Law (1857), Fastest Tornado Speed (1999).

A barometer, indicating atmospheric pressure in units of millimeters of mercury and in hPa (hectopascals). A pressure of one atmosphere is exactly equal to 1013.25 hPa.

72
73
74
75
76
77
78
79
mm
960 hPa
970
980
990
1000
1010
1020
1030
1040
1050
hPa 1060
mm

Conservation of Momentum

René Descartes (1596–1650)

Since the time of the ancient Greek philosophers, humans have wondered about the first great question of physics: "How do things move?" The conservation of momentum, one of the great laws of physics, was discussed in an early form by philosopher and scientist René Descartes in his *Principia Philosophiae* (*Principles of Philosophy*), published in 1644.

In classical mechanics, *linear momentum* **P** is defined as the product of the mass *m* and velocity **v** of an object, $\mathbf{P} = m\mathbf{v}$, where **P** and **v** are vector quantities having a magnitude and direction. For a "closed" (i.e. isolated) system of interacting bodies, the total momentum $\mathbf{P}_T$ is conserved. In other words, $\mathbf{P}_T$ is constant even though the motion of the individual bodies may be changing.

For example, consider a motionless ice skater with a mass of 45 kilograms. A ball of mass 5 kilograms is tossed at her at 5 meters/second by a machine directly in front of her and a short distance away so that we can assume the ball's flight is nearly horizontal. She catches the ball, and the impact slides her backwards at 0.5 meter/second. Here, the momentum for the moving ball and motionless skater before collision is 5 kg × 5 m/s (ball) + 0 (skater), and the momentum after collision for the skater holding the ball is (45 + 5 kg) × 0.5 m/s, thus conserving momentum.

Angular momentum is a related concept that concerns rotating objects. Consider a point mass (e.g. roughly, a ball on a string) rotating with a momentum **P** along a circle of radius *r*. The angular momentum is essentially the product of **P** and *r*—and the greater the mass, velocity, or radius, the greater the angular momentum. Angular momentum of an isolated system is also conserved. For example, when a spinning ice skater pulls in her arms, *r* decreases, which then causes her to spin faster. Helicopters use two rotors (propellers) for stabilization because a single horizontal rotor would cause the helicopter's body to rotate in the opposite direction to conserve angular momentum.

SEE ALSO Acceleration of Falling Objects (1638), Newton's Laws of Motion and Gravitation (1687), and Newton's Cradle (1967).

Person being winched to an air-sea rescue helicopter. Without the tail rotor for stabilization, this helicopter's body would rotate in a direction opposite to the top rotor's direction of spin to conserve angular momentum.

1660

Hooke's Law of Elasticity

Robert Hooke (1635–1703), **Augustin-Louis Cauchy** (1789–1857)

I fell in love with Hooke's Law while playing with the Slinky toy consisting of a helical spring. In 1660, the English physicist Robert Hooke discovered what we now call Hooke's Law of Elasticity, which states that if an object, such as a metal rod or spring, is elongated by some distance, x, the restoring force F exerted by the object is proportional to x. This relationship is represented by the equation $F = -kx$. Here, k is a constant of proportionality that is often referred to as the spring constant when Hooke's Law is applied to springs. Hooke's Law is an approximation that applies for certain materials, like steel, which are called "Hookean" materials because they obey Hooke's Law under a significant range of conditions.

Students most often encounter Hooke's Law in their study of springs, where the law relates the force F, exerted by the spring, to the distance x that the spring is stretched. The spring constant k is measured in force per length. The negative sign in $F = -kx$ indicates that the force exerted by the spring opposes the direction of displacement. For example, if we were to pull the end of a spring to the right, the spring exerts a "restoring" force to the left. The displacement of the spring refers to its displacement from equilibrium position at $x = 0$.

We have been discussing movements and forces in one direction. French mathematician Augustin-Louis Cauchy generalized Hooke's Law to three-dimensional (3D) forces and elastic bodies, and this more complicated formulation relies on six components of stress and six components of strain. The stress-strain relationship forms a 36-component stress-strain tensor when written in matrix form.

If a metal is lightly stressed, a temporary deformation may be achieved by an elastic displacement of the atoms in the 3D lattice. Removal of the stress results in a return of the metal to its original shape and dimensions.

Many of Hooke's inventions have been kept in obscurity, partly due to Isaac Newton's dislike for him. In fact, Newton had Hooke's portrait removed from the Royal Society and attempted to have Hooke's Royal Society papers burned.

SEE ALSO Truss (2500 B.C.), *Micrographia* (1665), Super Ball (1965).

Chromed motorcycle suspension springs. Hooke's Law of Elasticity helps describe how springs and other elastic objects behave as their lengths change.

1660

Von Guericke's Electrostatic Generator

Otto von Guericke (1602–1686), **Robert Jemison Van de Graaff** (1901–1967)

Neurophysiologist Arnold Trehub writes, "The most important invention in the past two thousand years must be a seminal invention with the broadest and most significant consequences. In my opinion, it is the invention by Otto von Guericke of a machine that produced static electricity." Although electrical phenomena were known by 1660, von Guericke appears to have produced the forerunner of the first machine for generating electricity. His electrostatic generator employed a globe made of sulfur that could be rotated and rubbed by hand. (Historians are not clear if his device was continuously rotated, a feature that would make it more easy to label his generator a machine.)

More generally, an electrostatic generator produces static electricity by transforming mechanical work into electric energy. Toward the end of the 1800s, electrostatic generators played a key role in research into the structure of matter. In 1929, an electrostatic generator known as the *Van de Graaff generator* (VG) was designed and built by American physicist Robert Van de Graaff, and it has been used extensively in nuclear physics research. Author William Gurstelle writes, "The biggest, brightest, angriest, and most fulgent electrical discharges don't come from Wimshurst-style electrostatic machines [see **Leyden Jar**] . . . or **Tesla coils** either. They come from an auditorium-sized pair of tall cylindrical machines . . . called Van De Graaff generators, [which] produce cascades of sparks, electrical effluvia, and strong electric fields. . . ."

VGs employ an electronic power supply to charge a moving belt in order to accumulate high voltages, usually on a hollow metal sphere. To use VGs in a particle accelerator, a source of ions (charged particles) is accelerated by the voltage difference. The fact that the VG produces precisely controllable voltages allowed VGs to be used in studies of nuclear reactions during the designing of the atomic bomb.

Over the years, electrostatic accelerators have been used for cancer therapies, for semiconductor production (via ion implantation), for electron-microscope beams, for sterilizing food, and for accelerating protons in nuclear physics experiments.

SEE ALSO Baghdad Battery (250 B.C.), *De Magnete* (1600), Leyden Jar (1744), Ben Franklin's Kite (1752), Lichtenberg Figures (1777), Coulomb's Law of Electrostatics (1785), Battery (1800), Tesla Coil (1891), Electron (1897), Jacob's Ladder (1931), Little Boy Atomic Bomb (1945), Seeing the Single Atom (1955).

LEFT: *Von Guericke invented perhaps the first electrostatic generator, a version of which is illustrated in Hubert-François Gravelot's engraving (c. 1750).* RIGHT: *World's largest air-insulated Van de Graaff generator, originally designed by Van de Graaff for early atomic energy experiments and currently operating in the Boston Museum of Science.*

1662

Boyle's Gas Law

Robert Boyle (1627–1691)

"Marge, what's wrong?" asked Homer Simpson when noticing his wife's panic in an airplane. "Are you hungry? Gassy? Is it gas? It's gas, isn't it?" Perhaps Boyle's Law would have made Homer feel a bit more knowledgeable. In 1662, Irish chemist and physicist Robert Boyle studied the relationship between the pressure P and the volume V of a gas in a container held at a constant temperature. Boyle observed that the product of the pressure and volume are nearly constant: $P \times V = C$.

A hand-pumped bicycle pump provides a rough example of Boyle's Law. When you push down on the piston, you decrease the volume inside the pump, increasing the pressure and causing the air to be forced into the tire. A balloon inflated at sea level will expand as it rises in the atmosphere and is subject to decreased pressure. Similarly, when we inhale, the ribs are lifted and the diaphragm contracts, increasing the lung volume and reducing the pressure so that air flows into the lungs. In a sense, Boyle's Law keeps us alive with each breath we take.

Boyle's Law is most accurate for an *ideal gas*, which consists of identical particles of negligible volume, with no intermolecular forces and with atoms or molecules that collide elastically with the walls of the container. Real gases obey Boyle's Law at sufficiently low pressures, and the approximation is often accurate for practical purposes.

Scuba divers learn about Boyle's Law because it helps to explain what happens during ascent and descent with respect to the lungs, mask, and buoyancy control device (BCD). For example, as a person descends, pressure increases, causing any air volume to decrease. Divers notice that their BCDs appear to deflate, and the airspace behind the ears becomes decompressed. To equalize the ear space, air must flow through the diver's Eustachian tubes to compensate for the reduction in air volume.

Realizing that his results could be explained if all gases were made of tiny particles, Boyle attempted to formulate a universal *corpuscular theory* of chemistry. In his 1661 *The Sceptical Chymist*, Boyle denounced the Aristotelian theory of the four elements (earth, air, fire, and water) and developed the concept of primary particles that came together to produce corpuscles.

SEE ALSO Charles' Gas Law (1787), Henry's Gas Law (1803), Avogadro's Gas Law (1811), Kinetic Theory (1859).

Scuba divers should know about Boyle's Law. If divers hold their breaths during an ascent after breathing compressed air, the air in their lungs will expand as the ambient water pressure decreases, potentially causing lung damage.

1665

Micrographia

Robert Hooke (1635–1703)

Although microscopes had been available since about the late 1500s, English scientist Robert Hooke's use of the compound microscope (a microscope with more than one lens) represents a particularly notable milestone, and his instrument can be considered as an important optical and mechanical forerunner of the modern microscope. For an optical microscope with two lenses, the overall magnification is the product of the powers of the ocular (eyepiece lens), usually about 10×, and the objective lens, which is closer to the specimen.

Hooke's book *Micrographia* featured breathtaking microscopic observations and biological speculation on specimens that ranged from plants to fleas. The book also discussed planets, the wave theory of light, and the origin of fossils, while stimulating both public and scientific interest in the power of the microscope.

Hooke was first to discover biological cells and coined the word cell to describe the basic units of all living things. The word cell was motivated by his observations of plant cells that reminded him of "cellula," which were the quarters in which monks lived. About this magnificent work, the historian of science Richard Westfall writes, "Robert Hooke's *Micrographia* remains one of the masterpieces of seventeenth century science, [presenting] a bouquet of observations with courses from the mineral, animal and vegetable kingdoms."

Hooke was the first person to use a microscope to study fossils, and he observed that the structures of petrified wood and fossil seashells bore a striking similarity to actual wood and the shells of living mollusks. In *Micrographia*, he compared petrified wood to rotten wood, and concluded that wood could be turned to stone by a gradual process. He also believed that many fossils represented extinct creatures, writing, "There have been many other Species of Creatures in former Ages, of which we can find none at present; and that 'tis not unlikely also but that there may be divers new kinds now, which have not been from the beginning." More recent advances in microscopes are described in the entry "**Seeing the Single Atom**."

SEE ALSO Telescope (1608), Kepler's "Six-Cornered Snowflake" (1611), Brownian Motion (1827), Seeing the Single Atom (1955).

Flea, from Robert Hooke's Micrographia, *published in 1665.*

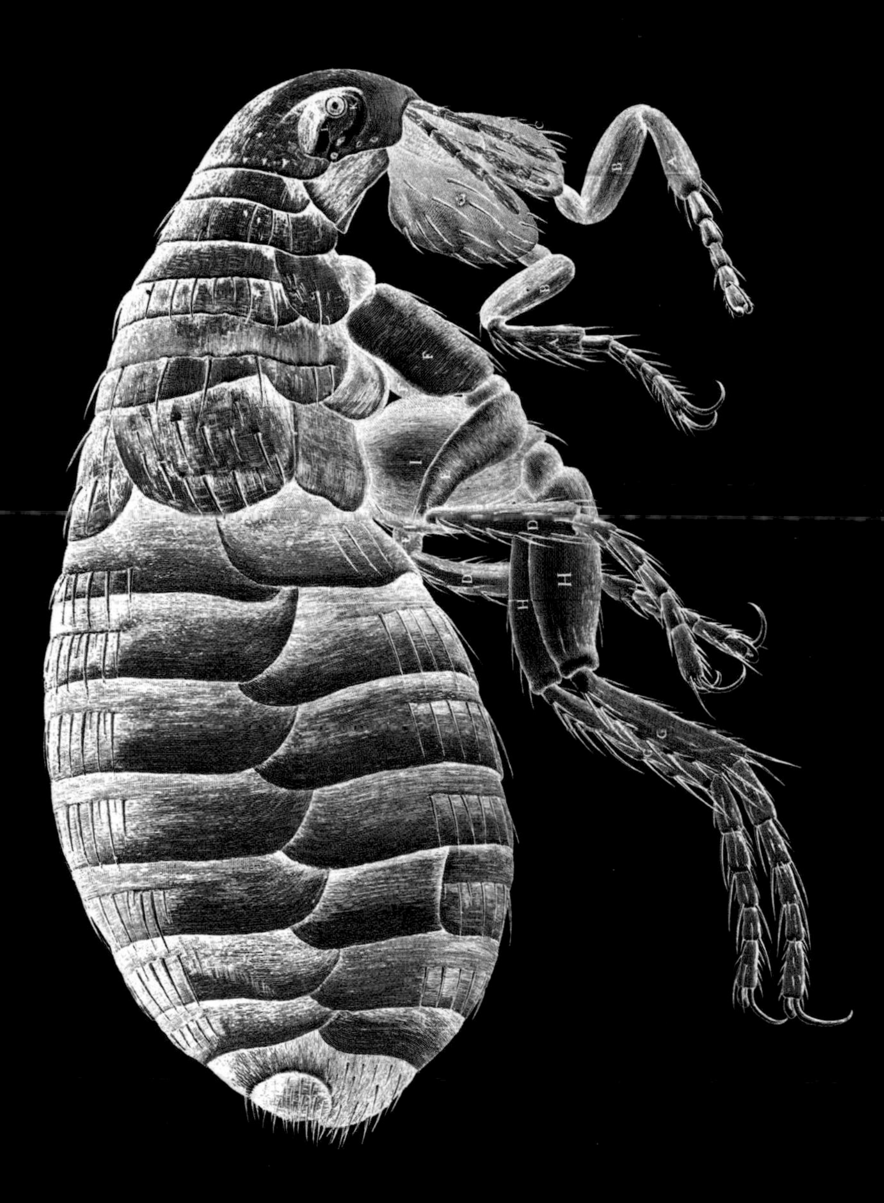

1669

Amontons' Friction

Guillaume Amontons (1663–1705), **Leonardo da Vinci** (1452–1519), **Charles-Augustin de Coulomb** (1736–1806)

Friction is a force that resists the sliding of objects with respect to each other. Although it is responsible for the wearing of parts and the wasting of energy in engines, friction is beneficial in our everyday lives. Imagine a world without friction. How would one walk, drive a car, attach objects with nails and screws, or drill cavities in teeth?

In 1669, French physicist Guillaume Amontons showed that the frictional force between two objects is directly proportional to the applied load (i.e., the force perpendicular to the surfaces in contact), with a constant of proportionality (a frictional coefficient) that is independent of the size of the contact area. These relationships were first suggested by Leonardo da Vinci and rediscovered by Amontons. It may seem counterintuitive that the amount of friction is nearly independent of the apparent area of contact. However, if a brick is pushed along the floor, the resisting frictional force is the same whether the brick is sliding on its larger or smaller face.

Several studies have been conducted in the early years of the twenty-first century to determine the extent to which Amontons' Law actually applies for materials at length scales from nanometers to millimeters—for example, in the area of MEMS (micro-electromechanical systems), which involves tiny devices such as those now used in inkjet printers and as accelerometers in car airbag systems. MEMS make use of microfabrication technology to integrate mechanical elements, sensors, and electronics on a silicon substrate. Amontons' Law, which is often useful when studying traditional machines and moving parts, may not be applicable to machines the size of a pinhead.

In 1779, French physicist Charles-Augustin de Coulomb began his research into friction and found that for two surfaces in relative motion, the *kinetic friction* is almost independent of the relative speed of the surfaces. For an object at rest, the *static frictional* force is usually greater than the resisting force for the same object in motion.

SEE ALSO Acceleration of Falling Objects (1638), Tautochrone Ramp (1673), Ice Slipperiness (1850), Stokes' Law of Viscosity (1851).

Devices such as wheels and ball bearings are used to convert sliding friction into a decreased form of rolling friction, thus creating less resistance to motion.

1672

Measuring the Solar System

Giovanni Domenico Cassini (1625–1712)

Before astronomer Giovanni Cassini's 1672 experiment to determine the size of the Solar System, there were some rather outlandish theories floating about. Aristarchus of Samos in 280 B.C. had said that the Sun was a mere 20 times farther from the Earth than the Moon. Some scientists around Cassini's time suggested that stars were only a few million miles away. While in Paris, Cassini sent astronomer Jean Richer to the city of Cayenne on the northeast coast of South America. Cassini and Richer made simultaneous measurements of the angular position of Mars against the distant stars. Using simple geometrical methods (see the entry "**Stellar Parallax**"), and knowing the distance between Paris and Cayenne, Cassini determined the distance between the Earth and Mars. Once this distance was obtained, he employed Kepler's Third Law to compute the distance between Mars and the Sun (see "**Kepler's Laws of Planetary Motion**"). Using both pieces of information, Cassini determined that the distance between the Earth and the Sun was about 87 million miles (140 million kilometers), which is only seven percent less than the actual average distance. Author Kendall Haven writes, "Cassini's discoveries of distance meant that the universe was millions of times bigger than anyone had dreamed." Note that it would be difficult to make direct measurements of the Sun without risking his eyesight.

Cassini became famous for many other discoveries. For example, he discovered four moons of Saturn and discovered the major gap in the rings of Saturn, which, today, is called the Cassini Gap in his honor. Interestingly, he was among the earliest scientists to correctly suspect that light traveled at a finite speed, but he did not publish his evidence for this theory because, according to Kendall Haven, "He was a deeply religious man and believed that light was of God. Light therefore had to be perfect and infinite, and not limited by a finite speed of travel."

Since the time of Cassini, our concept of the Solar System has grown, with the discovery, for example, of Uranus (1781), Neptune (1846), Pluto (1930), and Eris (2005).

SEE ALSO Eratosthenes Measures the Earth (240 B.C.), Sun-Centered Universe (1543), *Mysterium Cosmographicum* (1596), Kepler's Laws of Planetary Motion (1609), Discovery of Saturn's Rings (1610), Bode's Law of Planetary Distances (1766), Stellar Parallax (1838), Michelson-Morley Experiment (1887), Dyson Sphere (1960).

Cassini calculated the distance from Earth to Mars, and then the distance from Earth to the Sun. Shown here is a size comparison between Mars and Earth; Mars has approximately half the radius of Earth.

1672

Newton's Prism

Isaac Newton (1642–1727)

"Our modern understanding of light and color begins with Isaac Newton," writes educator Michael Douma, "and a series of experiments that he publishes in 1672. Newton is the first to understand the rainbow—he refracts white light with a prism, resolving it into its component colors: red, orange, yellow, green, blue and violet."

When Newton was experimenting with lights and colors in the late 1660s, many contemporaries thought that colors were a mixture of light and darkness, and that prisms colored light. Despite the prevailing view, he became convinced that white light was not the single entity that Aristotle believed it to be but rather a mixture of many different rays corresponding to different colors. The English physicist Robert Hooke criticized Newton's work on the characteristics of light, which filled Newton with a rage that seemed out of proportion to the comments Hooke had made. As a result, Newton withheld publication of his monumental book *Opticks* until after Hooke's death in 1703—so that Newton could have the last word on the subject of light and could avoid all arguments with Hooke. In 1704, Newton's *Opticks* was finally published. In this work, Newton further discusses his investigations of colors and the diffraction of light.

Newton used triangular glass prisms in his experiments. Light enters one side of the prism and is refracted by the glass into various colors (since their degree of separation changes as a function of the wavelength of the color). Prisms work because light changes speed when it moves from air into the glass of the prism. Once the colors were separated, Newton used a second prism to refract them back together to form white light again. This experiment demonstrated that the prism was not simply adding colors to the light, as many believed. Newton also passed only the red color from one prism through a second prism and found the redness unchanged. This was further evidence that the prism did not create colors, but merely separated colors present in the original light beam.

SEE ALSO Explaining the Rainbow (1304), Snell's Law of Refraction (1621), Brewster's Optics (1815), Electromagnetic Spectrum (1864), Metamaterials (1967).

Newton used prisms to show that white light was not the single entity that Aristotle believed it to be, but rather was a mixture of many different rays corresponding to different colors.

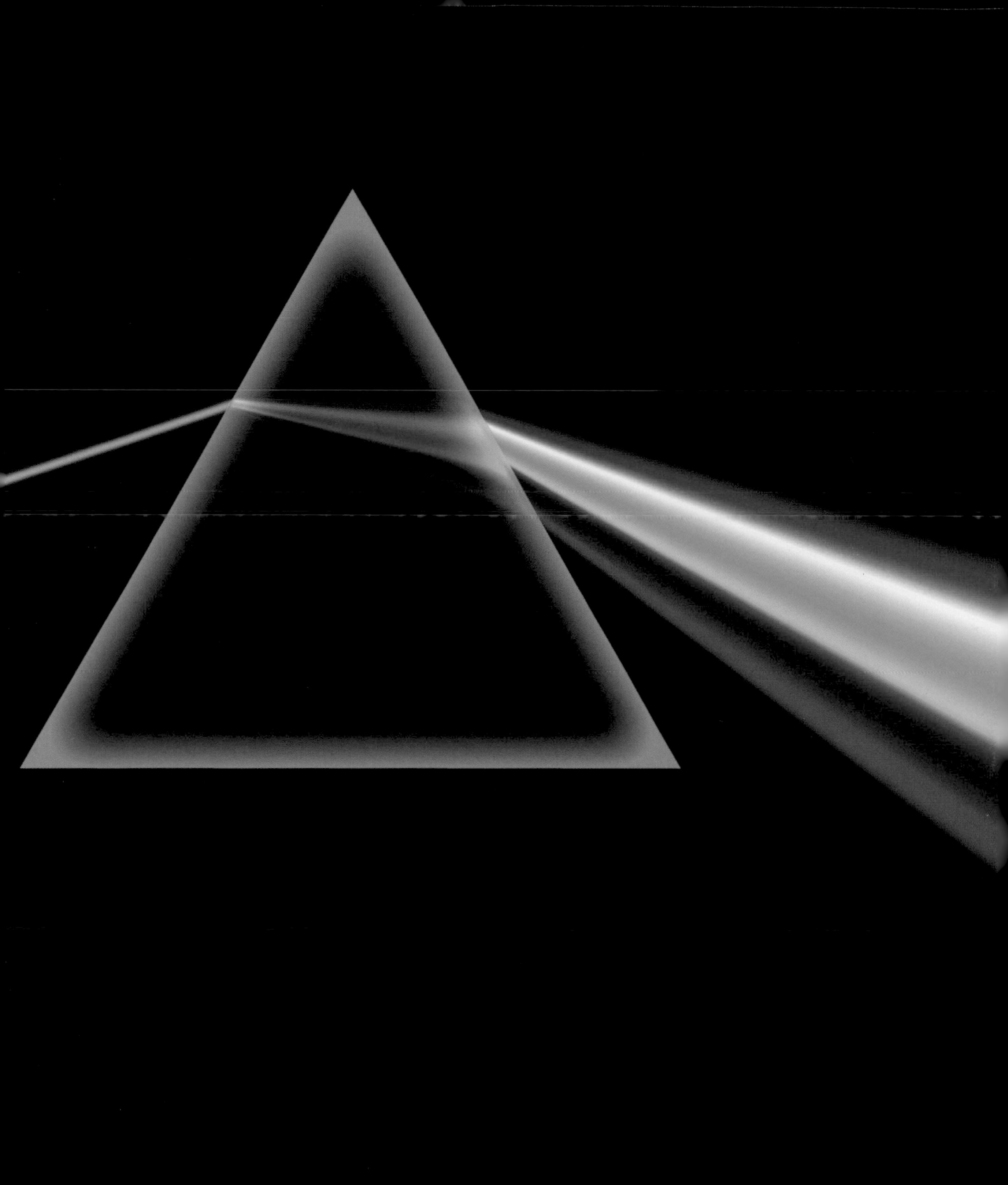

1673

Tautochrone Ramp

Christiaan Huygens (1629–1695)

Years ago, I wrote a tall tale of seven skateboarders who find a seemingly magical mountain road. Wherever on the road the skateboarders start their downhill coasting, they always reach the bottom in precisely the same amount of time. How could this be?

In the 1600s, mathematicians and physicists sought a curve that specified the shape of a special kind of ramp or road. On this special ramp, objects must slide down to the very bottom in the same amount of time, regardless of the starting position. The objects are accelerated by gravity, and the ramp is considered to have no friction.

Dutch mathematician, astronomer, and physicist Christiaan Huygens discovered a solution in 1673 and published it in his *Horologium Oscillatorium* (*The Pendulum Clock*). Technically speaking, the tautochrone is a cycloid—that is, a curve defined by the path of a point on the edge of circle as the circle rolls along a straight line. The tautochrone is also called the brachistochrone when referring to the curve that gives a frictionless object the fastest rate of descent when the object slides down from one point to another.

Huygens attempted to use his discovery to design a more accurate pendulum clock. The clock made use of inverted cycloid arcs near where the pendulum string pivoted to ensure that the string followed the optimum curve, no matter where the pendulum started swinging. (Alas, the friction caused by the bending of the string along the arcs introduced more error than it corrected.)

The special property of the tautochrone is mentioned in *Moby Dick* in a discussion on a try-pot, a bowl used for rendering blubber to produce oil: "[The try-pot] is also a place for profound mathematical meditation. It was in the left-hand try-pot of the *Pequod*, with the soapstone diligently circling round me, that I was first indirectly struck by the remarkable fact, that in geometry all bodies gliding along a cycloid, my soapstone, for example, will descend from any point in precisely the same time."

SEE ALSO Acceleration of Falling Objects (1638), Clothoid Loop (1901).

LEFT: *Christiaan Huygens, painted by Caspar Netscher (1639–1684).* RIGHT: *Under the influence of gravity, these billiard balls roll along the tautochrone ramp starting from different positions, yet the balls will arrive at the candle at the same time. The balls are placed on the ramp, one at a time.*

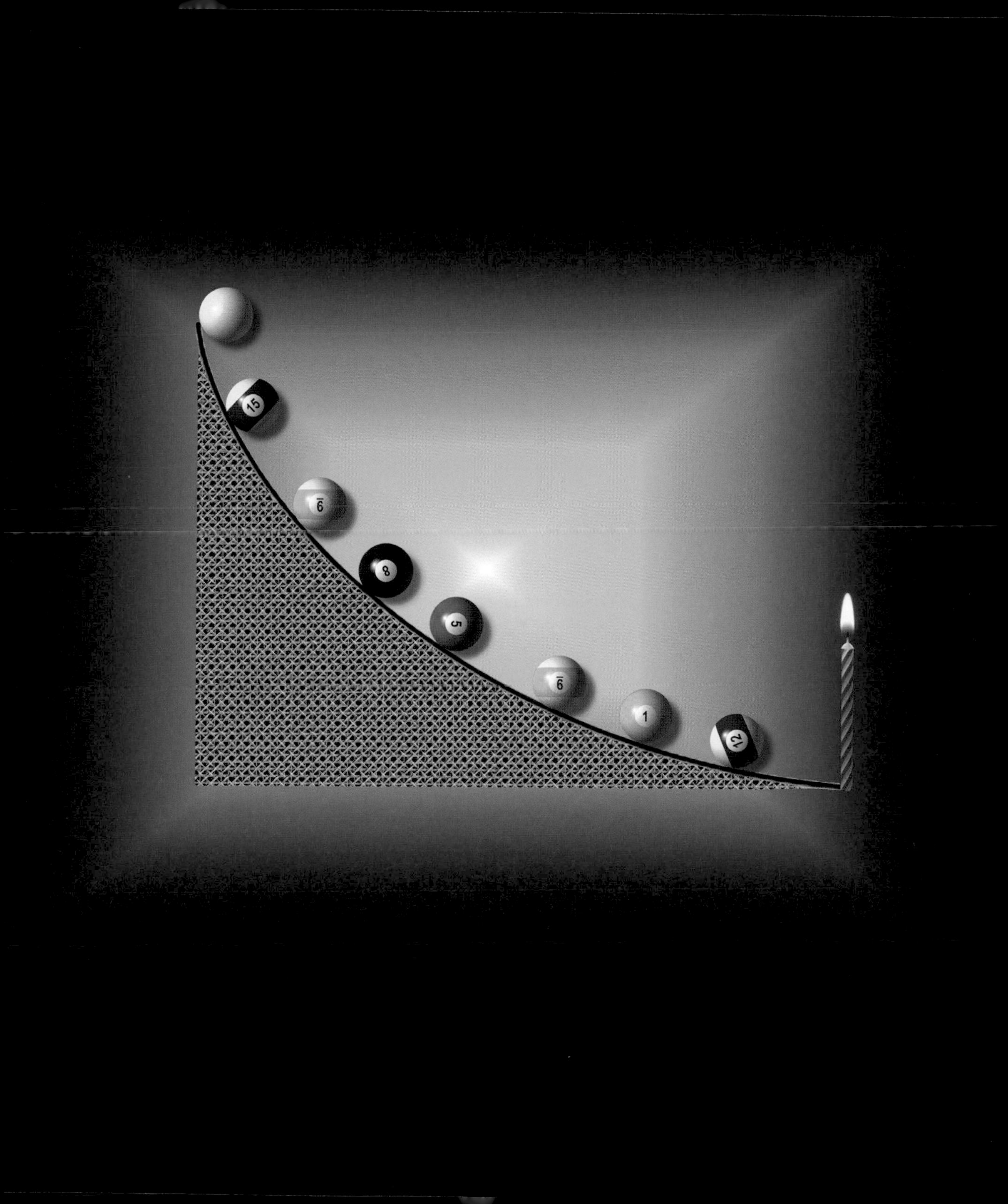

1687

Newton's Laws of Motion and Gravitation

Isaac Newton (1642–1727)

"God created everything by number, weight, and measure," wrote Isaac Newton, the English mathematician, physicist, and astronomer who invented calculus, proved that white light was a mixture of colors, explained the rainbow, built the first reflecting telescope, discovered the binomial theorem, introduced polar coordinates, and showed the force causing objects to fall is the same kind of force that drives planetary motions and produces tides.

Newton's Laws of Motion concern relations between forces acting on objects and the motion of these objects. His *Law of Universal Gravitation* states that objects attract one another with a force that varies as the product of the masses of the objects and inversely as the square of the distance between the objects. Newton's *First Law of Motion* (*Law of Inertia*) states that bodies do not alter their motions unless forces are applied to them. A body at rest stays at rest. A moving body continues to travel with the same speed and direction unless acted upon by a net force. According to Newton's *Second Law of Motion*, when a net force acts upon an object, the rate at which the momentum (mass × velocity) changes is proportional to the force applied. According to Newton's *Third Law of Motion*, whenever one body exerts a force on a second body, the second body exerts a force on the first body that is equal in magnitude and opposite in direction. For example, the downward force of a spoon on the table is equal to the upward force of the table on the spoon.

Throughout his life, Newton is believed to have had bouts of manic depression. He had always hated his mother and stepfather, and as a teenager threatened to burn them alive in their house. Newton was also author of treatises on biblical subjects, including biblical prophecies. Few are aware that he devoted more time to the study of the Bible, theology, and alchemy than to science—and wrote more on religion than he did on natural science. Regardless, the English mathematician and physicist may well be the most influential scientist of all time.

SEE ALSO Kepler's Laws of Planetary Motion (1609), Acceleration of Falling Objects (1638), Conservation of Momentum (1644), Newton's Prism (1672), Newton as Inspiration (1687), Clothoid Loop (1901), General Theory of Relativity (1915), Newton's Cradle (1967).

Gravity affects the motions of bodies in outer space. Shown here is an artistic depiction of a massive collision of objects, perhaps as large as Pluto, that created the dust ring around the nearby star Vega.

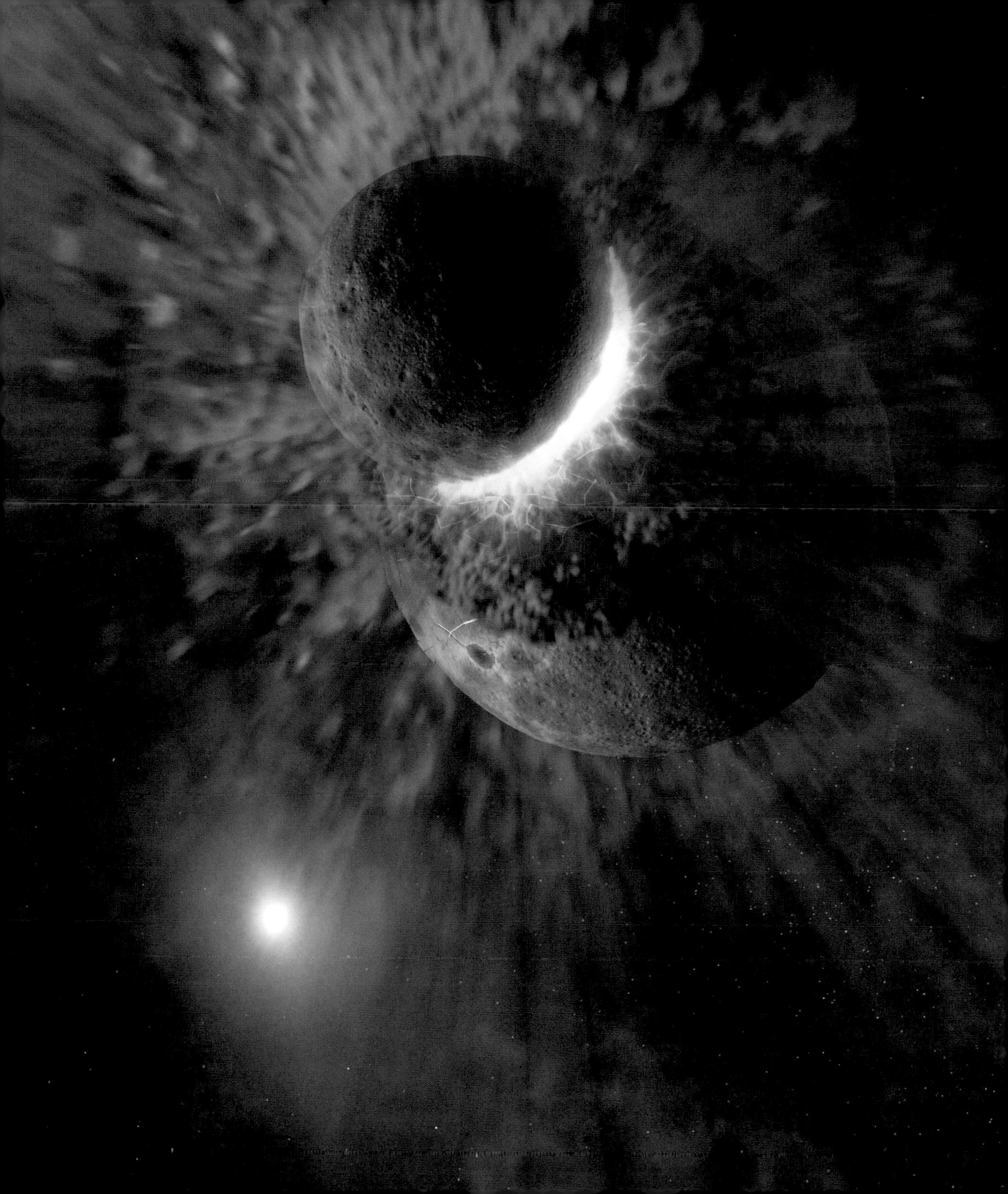

1687

Newton as Inspiration

Isaac Newton (1642–1727)

The chemist William H. Cropper writes, "Newton was the greatest creative genius that physics has ever seen. None of the other candidates for the superlative (Einstein, Maxwell, Boltzmann, Gibbs, and Feynman) has matched Newton's combined achievements as theoretician, experimentalist, and mathematician. . . . If you were to become a time traveler and meet Newton on a trip back to the seventeenth century, you might find him something like the performer who first exasperates everyone in sight and then goes on stage and sings like an angel. . . ."

Perhaps more than any other scientist, Newton inspired the scientists who followed him with the idea that the universe could be understood in terms of mathematics. Journalist James Gleick writes, "Isaac Newton was born into a world of darkness, obscurity, and magic . . . veered at least once to the brink of madness . . . and yet discovered more of the essential core of human knowledge than anyone before or after. He was chief architect of the modern world. . . . He made knowledge a thing of substance: quantitative and exact. He established principles, and they are called his laws."

Authors Richard Koch and Chris Smith note, "Some time between the 13th and 15th centuries, Europe pulled well ahead of the rest of the world in science and technology, a lead consolidated in the following 200 years. Then in 1687, Isaac Newton—foreshadowed by Copernicus, Kepler, and others—had his glorious insight that the universe is governed by a few physical, mechanical, and mathematical laws. This instilled tremendous confidence that everything made sense, everything fitted together, and everything could be improved by science."

Inspired by Newton, astrophysicist Stephen Hawking writes, "I do not agree with the view that the universe is a mystery. . . . This view does not do justice to the scientific revolution that was started almost four hundred years ago by Galileo and carried on by Newton. . . . We now have mathematical laws that govern everything we normally experience."

SEE ALSO Newton's Laws of Motion and Gravitation (1687), Einstein as Inspiration (1921), Stephen Hawking on *Star Trek* (1993).

Photograph of Newton's birthplace—Woolsthorpe Manor, England—along with an ancient apple tree. Newton performed many famous experiments on light and optics here. According to legend, Newton saw a falling apple here, which partly inspired his law of gravitation.

1711

Tuning Fork

John Shore (c. 1662–1752), **Hermann von Helmholtz** (1821–1894), **Jules Antoine Lissajous** (1822–1880), **Rudolph Koenig** (1832–1901)

Tuning forks—those Y-shaped metal devices that create a pure tone of constant frequency when struck—have played important roles in physics, medicine, art, and even literature. My favorite appearance in a novel occurs in *The Great Gatsby*, where Gatsby "knew that when he kissed this girl . . . his mind would never romp again like the mind of God. So he waited, listening for a moment longer to the tuning-fork that had been struck upon a star. Then he kissed her. At his lips' touch she blossomed for him like a flower. . . ."

The tuning fork was invented in 1711 by British musician John Shore. Its pure sinusoidal acoustic waveform makes it convenient for tuning musical instruments. The two prongs vibrate toward and away from one another, while the handle vibrates up and down. The handle motion is small, which means that the tuning fork can be held without significantly damping the sound. However, the handle can be used to amplify the sound by placing it in contact with a resonator, such as a hollow box. Simple formulas exist for computing the tuning-fork frequency based on parameters such as the density of the fork's material, the radius and length of the prongs, and the Young's modulus of the material, which is a measure of its stiffness.

In the 1850s, the mathematician Jules Lissajous studied waves produced by a tuning fork in contact with water by observing the ripples. He also obtained intricate Lissajous figures by successively reflecting light from one mirror attached to a vibrating tuning fork onto another mirror attached to a perpendicular vibrating tuning fork, then onto a wall. Around 1860, physicists Hermann von Helmholtz and Rudolph Koenig devised an electromagnetically driven tuning fork. In modern times, the tuning fork has been used by police departments to calibrate radar instruments for traffic speed control.

In medicine, these can be employed to assess a patient's hearing and sense of vibration on the skin, as well as for identifying bone fractures, which sometimes diminish the sound produced by a vibrating turning fork when it is applied to the body near the injury and monitored with a stethoscope.

SEE ALSO Stethoscope (1816), Doppler Effect (1842), War Tubas (1880).

Tuning forks have played important roles in physics, music, medicine, and art.

fp
Ped.
(mf)
Ped.
a)
b)
c)

1728

Escape Velocity

Isaac Newton (1642–1727)

Shoot an arrow straight up into the air, and it eventually comes down. Pull back the bow even farther, and the arrow takes longer to fall. The launch velocity at which the arrow would never return to the Earth is the escape velocity, v_e, and it can be computed with a simple formula: $v_e = [(2GM)/r]^{1/2}$, where G is the gravitational constant, and r is the distance of the bow and arrow from the center of the Earth, which has a mass of M. If we neglect air resistance and other forces and launch the arrow with some vertical component (along a radial line from the center of the Earth), then v_e = 6.96 miles per second (11.2 kilometers/second). This is surely one fast hypothetical arrow, which would have to be released at 34 times the speed of sound!

Notice that the mass of the projectile (e.g., whether it be an arrow or an elephant) does not affect its escape velocity, although it does affect the energy required to force the object to escape. The formula for v_e assumes a uniform spherical planet and a projectile mass that is much less than the planet's mass. Also, v_e relative to the Earth's surface is affected by the rotation of the Earth. For example, the arrow launched eastward while standing at the Earth's equator has a v_e equal to about 6.6 miles/second (10.7 kilometers/second) relative to the Earth.

Note that the v_e formula applies to a "one-time" vertical component of velocity for the projectile. An actual rocket ship does not have to achieve this speed because it may continue to fire its engines as it travels.

This entry is dated to 1728, the publication date for Isaac Newton's *A Treatise of the System of the World*, in which he contemplates firing a cannonball at different high speeds and considers ball trajectories with respect to the Earth. The escape velocity formula may be computed in many ways, including from Newton's Law of Universal Gravitation (1687), which states that objects attract one another with a force that varies as the product of the masses of the objects and inversely as the square of the distance between the objects.

SEE ALSO Tautochrone Ramp (1673), Newton's Laws of Motion and Gravitation (1687), Black Holes (1783), Terminal Velocity (1960).

Luna 1 was the first man-made object to reach the escape velocity of the Earth. Launched in 1959 by the Soviet Union, it was also the first spacecraft to reach the Moon.

1738

Bernoulli's Law of Fluid Dynamics

Daniel Bernoulli (1700–1782)

Imagine water flowing steadily through a pipe that carries the liquid from the roof of a building to the grass below. The pressure of the liquid will change along the pipe. Mathematician and physicist Daniel Bernoulli discovered the law that relates pressure, flow speed, and height for a fluid flowing in a pipe. Today, we write Bernoulli's Law as $v^2/2 + gz + p/\rho = C$. Here, v is the fluid velocity, g the acceleration due to gravity, z the elevation (height) of a point in the fluid, p the pressure, ρ the fluid density, and C is a constant. Scientists prior to Bernoulli had understood that a moving body exchanges its kinetic energy for potential energy when the body gains height. Bernoulli realized that, in a similar way, changes in the kinetic energy of a moving fluid result in a change in pressure.

The formula assumes a steady (non-turbulent) fluid flow in a closed pipe. The fluid must be incompressible. Because most liquid fluids are only slightly compressible, Bernoulli's Law is often a useful approximation. Additionally, the fluid should not be viscous, which means that the fluid should not have internal friction. Although no real fluid meets all these criteria, Bernoulli's relationship is generally very accurate for free flowing regions of fluids that are away from the walls of pipes or containers, and it is especially useful for gases and light liquids.

Bernoulli's Law often makes reference to a subset of the parameters in the above equation, namely that the decrease in pressure occurs simultaneously with an increase in velocity. The law is used when designing a *venturi throat*—a constricted region in the air passage of a carburetor that causes a reduction in pressure, which in turn causes fuel vapor to be drawn out of the carburetor bowl. The fluid increases speed in the smaller-diameter region, reducing its pressure and producing a partial vacuum via Bernoulli's Law.

Bernoulli's formula has numerous practical applications in the fields of aerodynamics, where it is considered when studying flow over airfoils, such as wings, propeller blades, and rudders.

SEE ALSO Siphon (250 B.C.), Poiseuille's Law of Fluid Flow (1840), Stokes' Law of Viscosity (1851), Kármán Vortex Street (1911).

Many engine carburetors have contained a venturi with a narrow throat region that speeds the air and reduces the pressure to draw fuel via Bernoulli's Law. The venture throat is labeled 10 in this 1935 carburetor patent.

April 9, 1935. C. N. POGUE 1,997,497

CARBURETOR

Filed Nov. 3, 1934 2 Sheets-Sheet 2

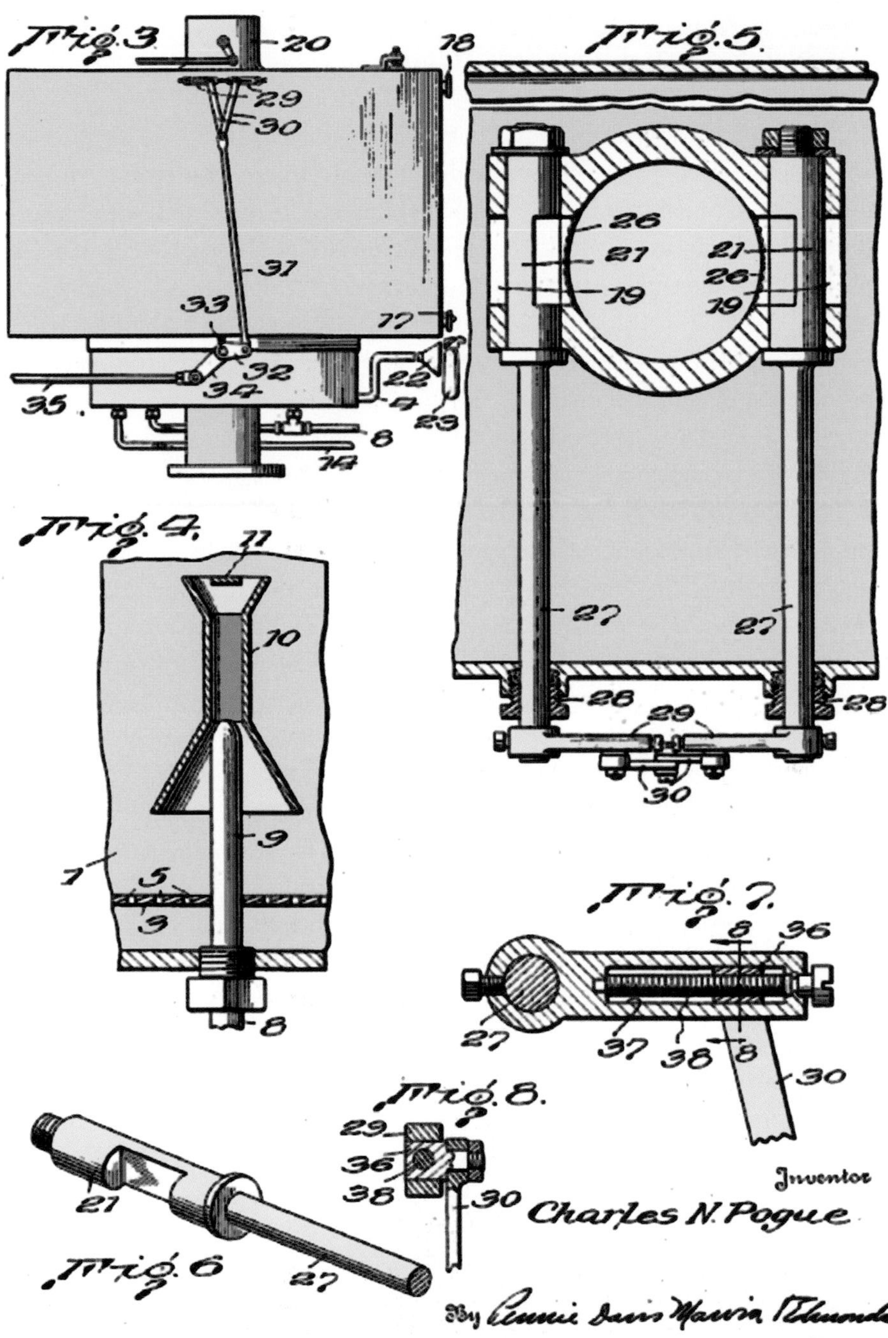

1744

Leyden Jar

Pieter van Musschenbroek (1692–1761), **Ewald Georg von Kleist** (1700–1748), **Jean-Antoine Nollet** (1700–1770), **Benjamin Franklin** (1706–1790)

"The Leyden jar was electricity in a bottle, an ingenious way to store a static electric charge and release it at will," writes author Tom McNichol. "Enterprising experimenters drew rapt crowds all over Europe... killing birds and small animals with a burst of stored electric charge.... In 1746, Jean-Antoine Nollet, a French clergyman and physicist, discharged a Leyden jar in the presence of King Louis XV, sending a current of static electricity rushing through a chain of 180 Royal Guards who were holding hands." Nollet also connected a row of a several hundred robed Carthusian monks, giving them the shock of their lives.

The Leyden jar is a device that stores static electricity between an electrode on the outside of a jar and another electrode on the inside. An early version was invented in 1744 by Prussian researcher Ewald Georg von Kleist. A year later, Dutch scientist Pieter van Musschenbroek independently invented a similar device while in Leiden (also spelled Leyden). The Leyden jar was important in many early experiments in electricity. Today, a Leyden jar is thought of as an early version of the *capacitor*, an electronic component that consists of two conductors separated by a dielectric (insulator). When a potential difference (voltage) exists across the conductors, an electric field is created in the dielectric, which stores energy. The narrower the separation between the conductors, the larger the charge that may be stored.

A typical design consists of a glass jar with conducting metal foils lining part of the outside and inside of the jar. A metal rod penetrates the cap of the jar and is connected to the inner metal lining by a chain. The rod is charged with static electricity by some convenient means—for example, by touching it with a silk-rubbed glass rod. If a person touches the metal rod, that person will receive a shock. Several jars may be connected in parallel to increase the amount of possible stored charge.

SEE ALSO Von Guericke's Electrostatic Generator (1660), Ben Franklin's Kite (1752), Lichtenberg Figures (1777), Battery (1800), Tesla Coil (1891), Jacob's Ladder (1931).

British inventor James Wimshurst (1832–1903) invented the Wimshurst Machine, *an electrostatic device for generating high voltages. A spark jumps across the gap formed by two metal spheres. Note the two Leyden jars for charge storage.*

1752

Ben Franklin's Kite

Benjamin Franklin (1706–1790)

Benjamin Franklin was an inventor, statesman, printer, philosopher, and scientist. Although he had many talents, historian Brooke Hindle writes, "The bulk of Franklin's scientific activities related to lightning and other electrical matters. His connection of lightning with electricity, through the famous experiment with a kite in a thunderstorm, was a significant advance of scientific knowledge. It found wide application in constructions of lightning rods to protect buildings in both the United States and Europe." Although perhaps not on par with many other physics milestones in this book, "Franklin's Kite" has often been a symbol of the quest for scientific truth and has inspired generations of school children.

In 1750, in order to verify that lightning is electricity, Franklin suggested an experiment that involved the flying of a kite in a storm that seemed likely to become a lightning storm. Although some historians have disputed the specifics of the story, according to Franklin, his experiments were conducted on June 15, 1752 in Philadelphia in order to successfully extract electrical energy from a cloud. In some versions of the story, he held a silk ribbon tied to a key at the end of the kite string to insulate himself from the electrical current that traveled down the string to the key and into the Leyden jar (a device that stores electricity between two electrodes). Other researchers did not take such precautions and were electrocuted when performing similar experiments. Franklin wrote, "When rain has wet the kite twine so that it can conduct the electric fire freely, you will find it streams out plentifully from the key at the approach of your knuckle, and with this key a . . . Leiden jar may be charged. . . ."

Historian Joyce Chaplin notes that the kite experiment was not the first to identify lighting with electricity, but the kite experiment verified this finding. Franklin was "trying to gauge whether the clouds were electrified and, if so, whether with a positive or a negative charge. He wanted to determine the presence of . . . electricity within nature, [and] it reduces his efforts considerably to describe them as resulting only in . . . the lightning rod."

SEE ALSO St. Elmo's Fire (78), Leyden Jar (1744), Lichtenberg Figures (1777), Tesla Coil (1891), Jacob's Ladder (1931).

"Benjamin Franklin Drawing Electricity from the Sky" (c. 1816), by Anglo-American painter Benjamin West (1738–1820). A bright electrical current appears to drop from the key to the jar in his hand.

1761

Black Drop Effect

Torbern Olof Bergman (1735–1784), **James Cook** (1728–1779)

Albert Einstein once suggested that the most incomprehensible thing about the world is that it is comprehensible. Indeed, we appear to live in a cosmos that can be described or approximated by compact mathematical expressions and physical laws. Even the strangest of astrophysical phenomena are often explained by scientists and scientific laws, although it can take many years to provide a coherent explanation.

The mysterious black drop effect (BDE) refers to the apparent shape assumed by Venus as it transits across the Sun when observed from the Earth. In particular, Venus appears to assume the shape of a black teardrop when visually "touching" the inside edge of the Sun. The tapered, stretched part of the teardrop resembles a fat umbilical cord or dark bridge, which made it impossible for early physicists to determine Venus' precise transit time across the Sun.

The first detailed description of the BDE came in 1761, when Swedish scientist Torbern Bergman described the BDE in terms of a "ligature" that joined the silhouette of Venus to the dark edge of the Sun. Many scientists provided similar reports in the years that followed. For example, British explorer James Cook made observations of the BDE during the 1769 transit of Venus.

Today, physicists continue to ponder the precise reason for the BDE. Astronomers Jay M. Pasachoff, Glenn Schneider, and Leon Golub suggest it is a "combination of instrumental effects and effects to some degree in the atmospheres of Earth, Venus, and Sun." During the 2004 transit of Venus, some observers saw the BDE while others did not. Journalist David Shiga writes, "So the 'black-drop effect' remains as enigmatic in the 21st century as in the 19th. Debate is likely to continue over what constitutes a 'true' black drop. . . . And it remains to be seen whether the conditions for the appearance of black drop will be nailed down as observers compare notes . . . in time for the next transit. . . ."

SEE ALSO Discovery of Saturn's Rings (1610), Measuring the Solar System (1672), Discovery of Neptune (1846), Green Flash (1882).

LEFT: *British explorer James Cook observed the Black Drop Effect during the 1769 transit of Venus, as depicted in this sketch by the Australian astronomer Henry Chamberlain Russell (1836–1907).* RIGHT: *Venus transits the Sun in 2004, exhibiting the black-drop effect.*

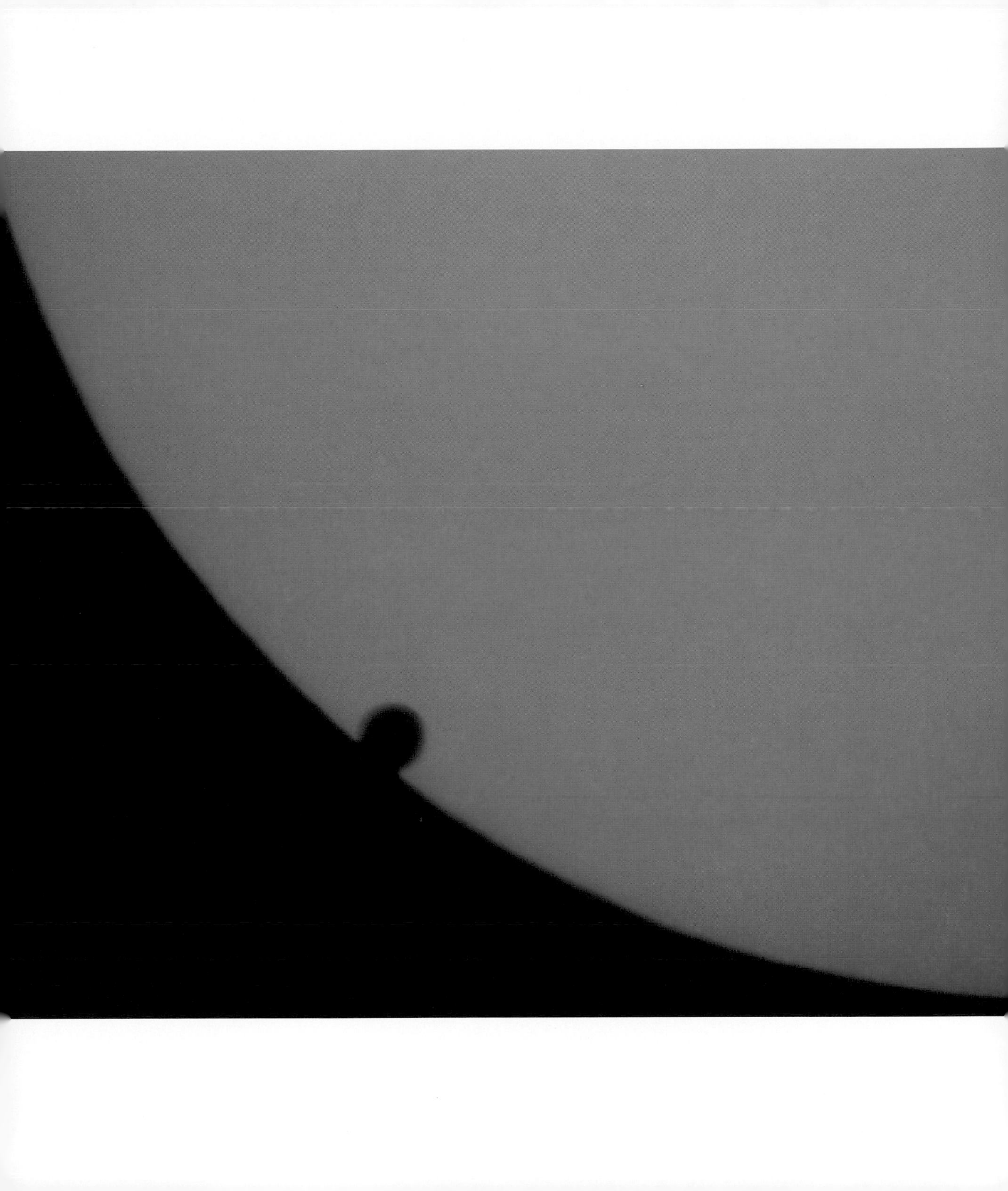

1766

Bode's Law of Planetary Distances

Johann Elert Bode (1747–1826), **Johann Daniel Titius** (1729–1796)

Bode's Law, also known as the Titius-Bode Law, is particularly fascinating because it seems like pseudo-scientific numerology and has intrigued both physicists and laypeople for centuries. The law expresses a relationship that describes the mean distances of the planets from the Sun. Consider the simple sequence 0, 3, 6, 12, 24, . . . in which each successive number is twice the previous number. Next, add 4 to each number and divide by 10 to form the sequence 0.4, 0.7, 1.0, 1.6, 2.8, 5.2, 10.0, 19.6, 38.8, 77.2, . . . Remarkably, Bode's Law provides a sequence that lists the mean distances D of many planets from the Sun, expressed in astronomical units (AU). An AU is the mean distance between the Earth and Sun, which is approximately 92,960,000 miles (149,604,970 kilometers). For example, Mercury is approximately 0.4 of an AU from the Sun, and Pluto is about 39 AU from the Sun.

This law was discovered by the German astronomer Johann Titius of Wittenberg in 1766 and published by Johann Bode six years later, though the relationship between the planetary orbits had been approximated by Scottish mathematician David Gregory in the early eighteenth century. At the time, the law gave a remarkably good estimate for the mean distances of the planets that were then known—Mercury (0.39), Venus (0.72), Earth (1.0), Mars (1.52), Jupiter (5.2), and Saturn (9.55). Uranus, discovered in 1781, has a mean orbital distance of 19.2, which also agrees with the law.

Today scientists have major reservations about Bode's Law, which is clearly not as universally applicable as other laws in this book. In fact, the relationship may be purely empirical and coincidental.

A phenomenon of "orbital resonances," caused by orbiting bodies that gravitationally interact with other orbiting bodies, can create regions around the Sun that are free of long-term stable orbits and thus, to some degree, can account for the spacing of planets. Orbital resonances can occur when two orbiting bodies have periods of revolution that are related in a simple integer ratio, so that the bodies exert a regular gravitational influence on each other.

SEE ALSO *Mysterium Cosmographicum* (1596), Measuring the Solar System (1672), Discovery of Neptune (1846).

According to Bode's Law, the mean distance of Jupiter to the Sun is 5.2 AU, and the actual measured value is 5.203 AU.

1777

Lichtenberg Figures

Georg Christoph Lichtenberg (1742–1799)

Among the most beautiful representations of natural phenomena are three-dimensional Lichtenberg figures, reminiscent of fossilized lightning trapped within a block of clear acrylic. These branching trails of electric discharges are named after German physicist Georg Lichtenberg, who originally studied similar electrical traces on surfaces. In the 1700s, Lichtenberg discharged electricity onto the surface of an insulator. Then, by sprinkling certain charged powders onto the surface, he was able to reveal curious tendrilous patterns.

Today, three-dimensional patterns can be created in acrylic, which is an insulator, or dielectric, meaning that it can hold a charge but that current cannot normally pass through it. First, the acrylic is exposed to a beam of high-speed electrons from an electron accelerator. The electrons penetrate the acrylic and are stored within. Since the acrylic is an insulator, the electrons are now trapped (think of a nest of wild hornets trying to break out of an acrylic prison). However, there comes a point where the electrical stress is greater than the dielectric strength of the acrylic, and some portions do suddenly become conductive. The escape of electrons can be triggered by piercing the acrylic with a metal point. As a result, some of the chemical bonds that hold the acrylic molecules together are torn apart. Within a fraction of a second, electrically conductive channels form within the acrylic as the electrical charge escapes from the acrylic, melting pathways along the way. Electrical engineer Bert Hickman speculates that these microcracks propagate faster than the speed of sound within the acrylic.

Lichtenberg figures are fractals, exhibiting branching self-similar structures at multiple magnifications. In fact, the fernlike discharge pattern may actually extend all the way down to the molecular level. Researchers have developed mathematical and physical models for the process that creates the dendritic patterns, which is of interest to physicists because such models may capture essential features of pattern formation in seemingly diverse physical phenomena. Such patterns may have medical applications as well. For example, researchers at Texas A&M University believe these feathery patterns may serve as templates for growing vascular tissue in artificial organs.

SEE ALSO Ben Franklin's Kite (1752), Tesla Coil (1891), Jacob's Ladder (1931), Sonic Booms (1947).

Bert Hickman's Lichtenberg figure in acrylic, created by electron-beam irradiation, followed by manual discharge. The specimen's internal potential prior to discharging was estimated to be around 2 million volts.

1779

Black Eye Galaxy

Edward Pigott (1753–1825), **Johann Elert Bode** (1747–1826), **Charles Messier** (1730–1817)

The Black Eye Galaxy resides in the constellation Coma Berenices and is about 24 million light years away from the Earth. Author and naturalist Stephen James O'Meara writes poetically of this famous galaxy with its "smooth silken arms [that] wrap gracefully around a porcelain core. . . . The galaxy resembles a closed human eye with a 'shiner.' The dark dust cloud looks as thick and dirty as tilled soil [but] a jar of its material would be difficult to distinguish from a perfect vacuum."

Discovered in 1779 by English astronomer Edward Pigott, it was independently discovered just twelve days later by German astronomer Johann Elert Bode and about a year later by French astronomer Charles Messier. As mentioned in the entry "**Explaining the Rainbow**," such nearly simultaneous discoveries are common in the history of science and mathematics. For example, British naturalists Charles Darwin and Alfred Wallace both developed the theory of evolution independently and simultaneously. Likewise, Isaac Newton and German mathematician Gottfried Wilhelm Leibniz developed calculus independently at about the same time. Simultaneity in science has led some philosophers to suggest that scientific discoveries are inevitable as they emerge from the common intellectual waters of a particular place and time.

Interestingly, recent discoveries indicate that the interstellar gas in the outer regions of the Black Eye Galaxy rotates in the opposite direction from the gas and stars in the inner regions. This differential rotation may arise from the Black Eye Galaxy having collided with another galaxy and having absorbed it over a billion years ago.

Author David Darling writes that the inner zone of the galaxy is about 3,000 light-years in radius and "rubs along the inner edge of an outer disk, which rotates in the opposite direction at about 300 km/s and extends out to at least 40,000 light-years. This rubbing may explain the vigorous burst of star formation that is currently taking place in the galaxy and is visible as blue knots embedded in the huge dust lane."

SEE ALSO Black Holes (1783), Nebular Hypothesis (1796), Fermi Paradox (1950), Quasars (1963), Dark Matter (1933).

The interstellar gas in the outer regions of the Black Eye Galaxy rotates in the opposite direction of the gas and stars in the inner regions. This differential rotation may arise from the galaxy having collided with another galaxy and having absorbed this galaxy over a billion years ago.

1783

Black Holes

John Michell (1724–1793), **Karl Schwarzschild** (1873–1916), **John Archibald Wheeler** (1911–2008), **Stephen William Hawking** (b. 1942)

Astronomers may not believe in Hell, but most believe in ravenous, black regions of space in front of which one would be advised to place a sign, "Abandon hope, all ye who enter here." This was Italian poet Dante Alighieri's warning when describing the entrance to the Inferno in his *Divine Comedy*, and, as astrophysicist Stephen Hawking has suggested, this would be the appropriate message for travelers approaching a black hole.

These cosmological hells truly exist in the centers of many galaxies. Such galactic black holes are collapsed objects having millions or even billions of times the mass of our Sun crammed into a space no larger than our Solar System. According to classical black hole theory, the gravitational field around such objects is so great that nothing—not even light—can escape from their tenacious grip. Anyone who falls into a black hole will plunge into a tiny central region of extremely high density and extremely small volume . . . and the end of time. When quantum theory is considered, black holes are thought to emit a form of radiation called Hawking radiation (see "Notes and Further Reading" and the entry "**Stephen Hawking on *Star Trek***").

Black holes can exist in many sizes. As some historical background, just a few weeks after Albert Einstein published his general relativity theory in 1915, German astronomer Karl Schwarzschild performed exact calculations of what is now called the Schwarzschild radius, or event horizon. This radius defines a sphere surrounding a body of a particular mass. In classical black-hole theory, within the sphere of a black hole, gravity is so strong that no light, matter, or signal can escape. For a mass equal to the mass of our Sun, the Schwarzschild radius is a few kilometers in length. A black hole with an event horizon the size of a walnut would have a mass equal to the mass of the Earth. The actual concept of an object so massive that light could not escape was first suggested in 1783 by the geologist John Michell. The term "black hole" was coined in 1967 by theoretical physicist John Wheeler.

SEE ALSO Escape Velocity (1728), General Theory of Relativity (1915), White Dwarfs and Chandrasekhar Limit (1931), Neutron Stars (1933), Quasars (1963), Stephen Hawking on *Star Trek* (1993), Universe Fades (100 Trillion).

LEFT: *Black holes and Hawking radiation are the stimulus for numerous impressionistic pieces by Slovenian artist Teja Krašek.* RIGHT: *Artistic depiction of the warpage of space in the vicinity of a black hole.*

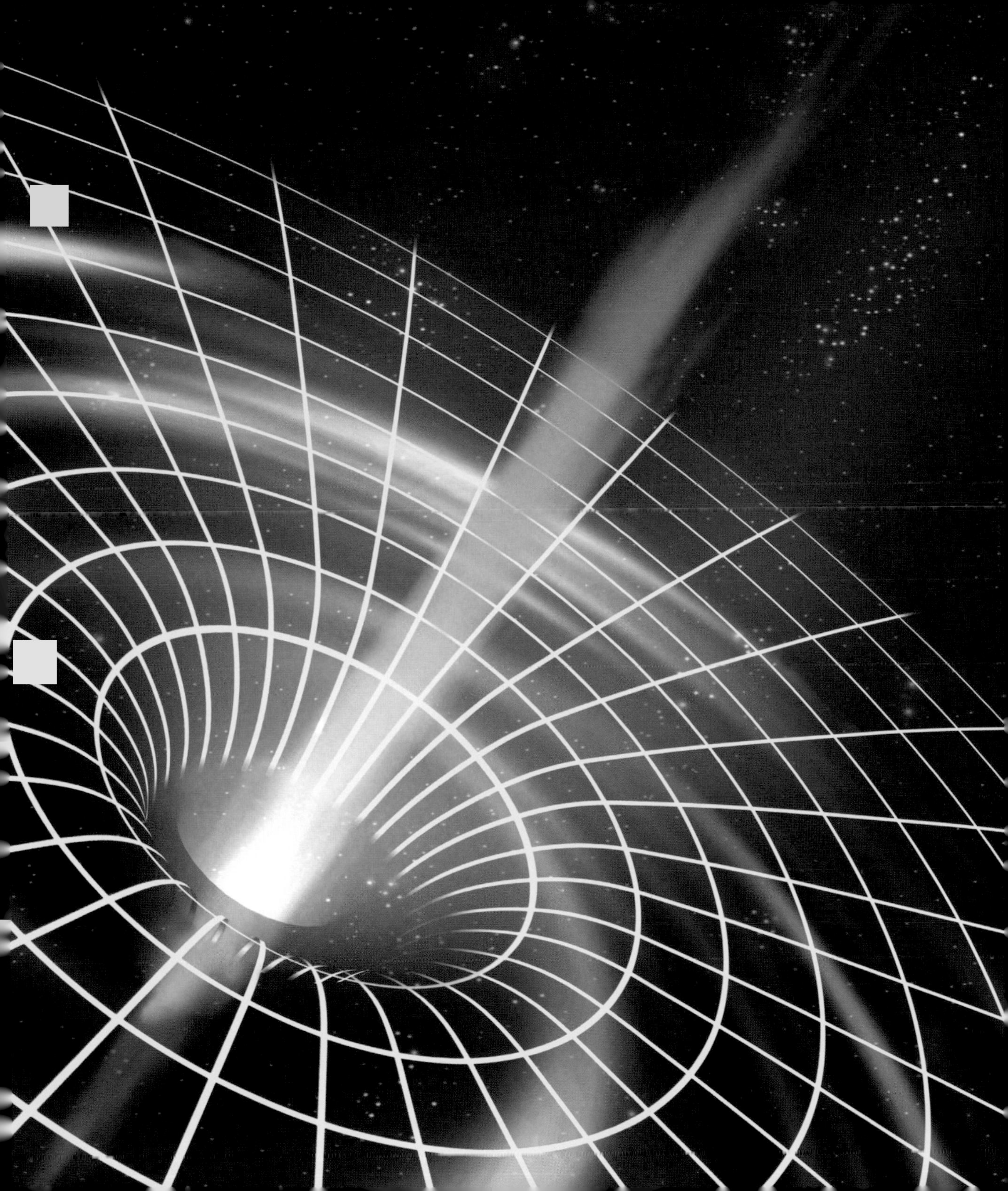

1785

Coulomb's Law of Electrostatics

Charles-Augustin Coulomb (1736–1806)

"We call that fire of the black thunder-cloud *electricity*," wrote essayist Thomas Carlyle in the 1800s, "but what is it? What made it?" Early steps to understand electric charge were taken by French physicist Charles-Augustin Coulomb, the preeminent physicist who contributed to the fields of electricity, magnetism, and mechanics. His Law of Electrostatics states that the force of attraction or repulsion between two electric charges is proportional to the product of the magnitude of the charges and inversely proportional to the square of their separation distance r. If the charges have the same sign, the force is repulsive. If the charges have opposite signs, the force is attractive.

Today, experiments have demonstrated that Coulomb's Law is valid over a remarkable range of separation distances, from as small as 10^{-16} meters (a tenth of the diameter of an atomic nucleus) to as large as 10^6 meters (where 1 meter is equal to 3.28 feet). Coulomb's Law is accurate only when the charged particles are stationary because movement produces magnetic fields that alter the forces on the charges.

Although other researchers before Coulomb had suggested the $1/r^2$ law, we refer to this relationship as Coulomb's Law in honor of Coulomb's independent results gained through the evidence provided by his torsional measuring. In other words, Coulomb provided convincing quantitative results for what was, up to 1785, just a good guess.

One version of Coulomb's torsion balance contains a metal and a non-metal ball attached to an insulating rod. The rod is suspended at its middle by a nonconducting filament or fiber. To measure the electrostatic force, the metal ball is charged. A third ball with similar charge is placed near the charged ball of the balance, causing the ball on the balance to be repelled. This repulsion causes the fiber to twist. If we measure how much force is required to twist the wire by the same angle of rotation, we can estimate the degree of force caused by the charged sphere. In other words, the fiber acts as a very sensitive spring that supplies a force proportional to the angle of twist.

SEE ALSO Maxwell's Equations (1861), Leyden Jar (1744), Eötvös' Gravitational Gradiometry (1890), Electron (1897), Millikan Oil Drop Experiment (1913).

Charles-Augustin de Coulomb's torsion balance, from his Mémoires sur l'électricité et le magnétisme *(1785–1789).*

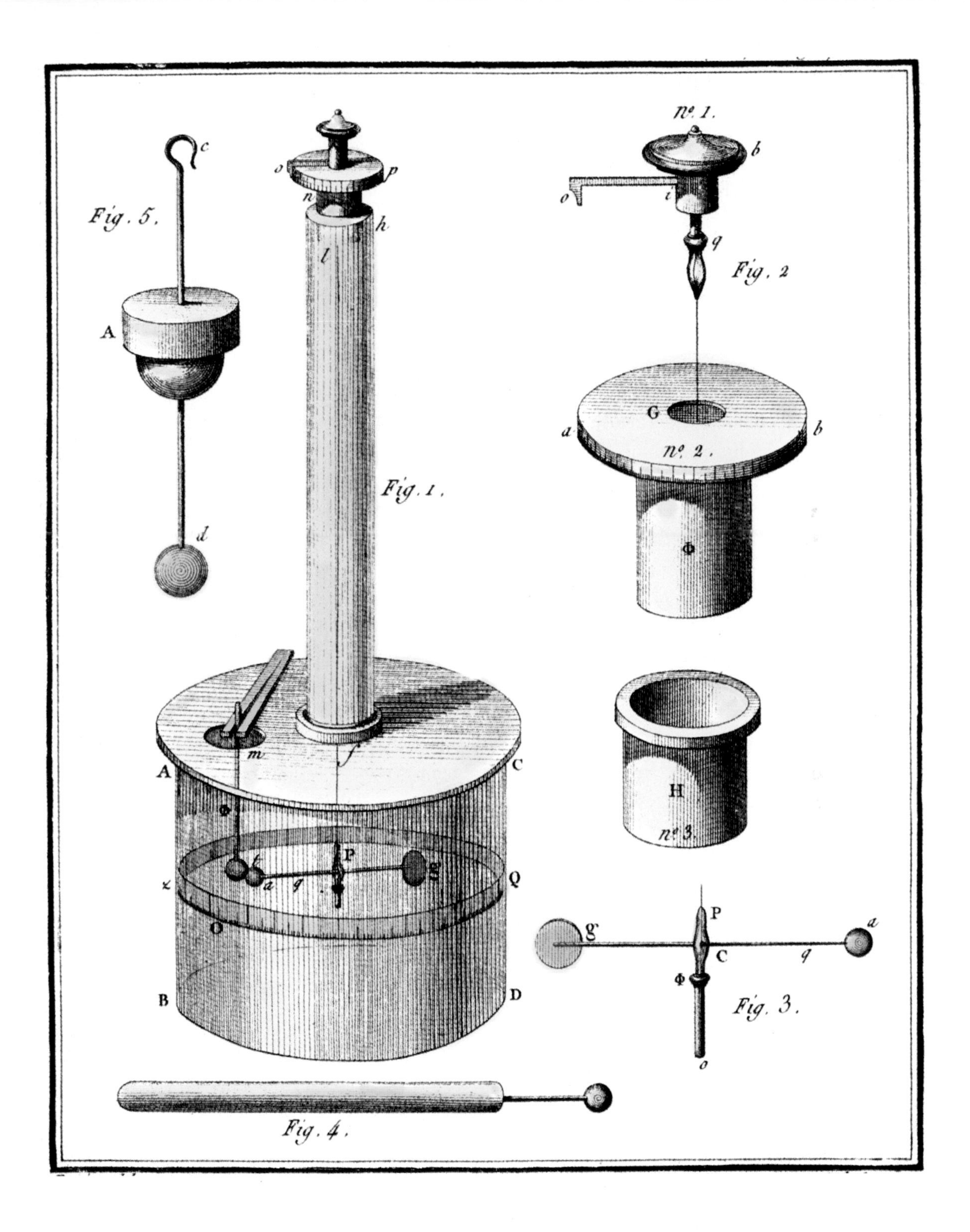
Fig. 5.
c
A
d
Fig. 1.
o
p
n
h
l
m
f
A
C
P
Q
a
g
O
B
D
Nº. 1.
b
o
i
q
Fig. 2
G
a
b
Nº. 2.
Φ
H
nº 3.
P
g
a
C
q
Φ
Fig. 3.
o
Fig. 4.

1787

Charles' Gas Law

Jacques Alexandre César Charles (1746–1823), **Joseph Louis Gay-Lussac** (1778–1850)

"It is our business to puncture gas bags and discover the seeds of truth," wrote essayist Virginia Woolf. On the other hand, the French balloonist Jacques Charles knew how to make "gas bags" soar to find truths. The gas law named in his honor states that the volume occupied by a fixed amount of gas varies directly with the absolute temperature (i.e., the temperature in kelvins). The law can be expressed as $V = kT$ where V is the volume at a constant pressure, T is the temperature, and k is a constant. Physicist Joseph Gay-Lussac first published the law in 1802, where he referenced unpublished work from around 1787 by Jacques Charles.

As the temperature of the gas increases, the gas molecules move more quickly and hit the walls of their container with more force—thus increasing the volume of gas, assuming that the container volume is able to expand. For a more specific example, consider warming the air within a balloon. As the temperature increases, the speed of the moving gas molecules increases inside the surface of the balloon. This in turn increases the rate at which the gas molecules bombard the interior surface. Because the balloon can stretch, the surface expands as a result of the increased internal bombardment. The volume of gas increases, and its density decreases. The act of cooling the gas inside a balloon will have the opposite effect, causing the pressure to be reduced and the balloon to shrink.

Charles was most famous to his contemporaries for his various exploits and inventions pertaining to the science of ballooning and other practical sciences. His first balloon journey took place in 1783, and an adoring audience of thousands watched as the balloon drifted by. The balloon ascended to a height of nearly 3,000 feet (914 meters) and seems to have finally landed in a field outside of Paris, where it was destroyed by terrified peasants. In fact, the locals believed that the balloon was some kind of evil spirit or beast from which they heard sighs and groans, accompanied by a noxious odor.

SEE ALSO Boyle's Gas Law (1662), Henry's Gas Law (1803), Avogadro's Gas Law (1811), Kinetic Theory (1859).

The first flight of Jacques Charles with co-pilot Nicolas-Louis Robert, 1783, who are seen waving flags to spectators. Versailles Palace is in the background. The engraving is likely created by Antoine François Sergent-Marceau, c. 1783.

1796

Nebular Hypothesis

Immanuel Kant (1724–1804), **Pierre-Simon Laplace** (1749–1827)

For centuries, scientists hypothesized that the Sun and planets were born from a rotating disk of cosmic gas and dust. The flat disk constrained the planets that formed from it to have orbits almost lying in the same plane. This *nebular theory* was developed in 1755 by the philosopher Immanuel Kant, and refined in 1796 by mathematician Pierre-Simon Laplace.

In short, stars and their disks form from the gravitational collapse of large volumes of sparse interstellar gas called solar nebulae. Sometimes a shock wave from a nearby supernova, or exploding star, may trigger the collapse. Gases in these protoplanetary disks (*proplyds*) of gas will be swirling more in one direction than the other, giving the gas cloud a net rotation.

Using the **Hubble Space Telescope**, astronomers have detected several proplyds in the Orion Nebula, a giant stellar nursery about 1,600 light-years away. The Orion proplyds are larger than the Sun's solar system and contain sufficient gas and dust to provide the raw material for future planetary systems.

The violence of the early Solar System was tremendous as huge chunks of matter bombarded one another. In the inner Solar System, the Sun's heat drove away the lighter-weight elements and materials, leaving Mercury, Venus, Earth, and Mars behind. In the colder outer part of the system, the solar nebula of gas and dust survived for some time and were accumulated by Jupiter, Saturn, Uranus, and Neptune.

Interestingly, Isaac Newton marveled at the fact that most of the objects that orbit the Sun are contained with an ecliptic plane offset by just a few degrees. He reasoned that natural processes could not create such behavior. This, he argued, was evidence of design by a benevolent and artistic creator. At one point, he thought of the Universe as "God's Sensorium," in which the objects in the Universe—their motions and their transformations—were the thoughts of God.

SEE ALSO Measuring the Solar System (1672), Black Eye Galaxy (1779), Hubble Telescope (1990).

Protoplanetary disk. This artistic depiction features a small young star encircled by a disk of gas and dust, the raw materials from which rocky planets such as Earth may form.

1798

Cavendish Weighs the Earth

Henry Cavendish (1731–1810)

Henry Cavendish was perhaps the greatest of all eighteenth-century scientists, and one of the greatest scientists who ever lived. Yet his extreme shyness—a trait that made the vast extent of his scientific writings secret until after his death—caused some of his important discoveries to be associated with the names of subsequent researchers. The huge number of manuscripts uncovered after Cavendish's death show that he conducted extensive research in literally all branches of physical sciences of his day.

The brilliant British chemist was so shy around women that he communicated with his housekeeper using only written notes. He ordered all his female housekeepers to keep out of sight. If they were unable to comply, he fired them. Once when he saw a female servant, he was so mortified that he built a second staircase for the servants' use so that he could avoid them.

In one of his most impressive experiments, Cavendish at the age of 70 "weighed" the world! In order to accomplish this feat, he didn't transform into the Greek god Atlas, but rather determined the density of the Earth using highly sensitive balances. In particular, he used a torsional balance consisting of two lead balls on either end of a suspended beam. These mobile balls were attracted by a pair of larger stationary lead balls. To reduce air currents, he enclosed the device in a glass case and observed the motion of the balls from far away, by means of a telescope. Cavendish calculated the force of attraction between the balls by observing the balance's oscillation period, and then computed the Earth's density from the force. He found that the Earth was 5.4 times as dense as water, a value that is only 1.3% lower than the accepted value today. Cavendish was the first scientist able to detect minute gravitational forces between small objects. (The attractions were 1/500,000,000 times as great as the weight of the bodies.) By helping to quantify Newton's Law of Universal Gravitation, he had made perhaps the most important addition to gravitational science since Newton.

SEE ALSO Newton's Laws of Motion and Gravitation (1687), Eötvös' Gravitational Gradiometry (1890), General Theory of Relativity (1915).

Close-up of a portion of the drawing of a torsion balance from Cavendish's 1798 paper "Experiments to Determine the Density of the Earth."

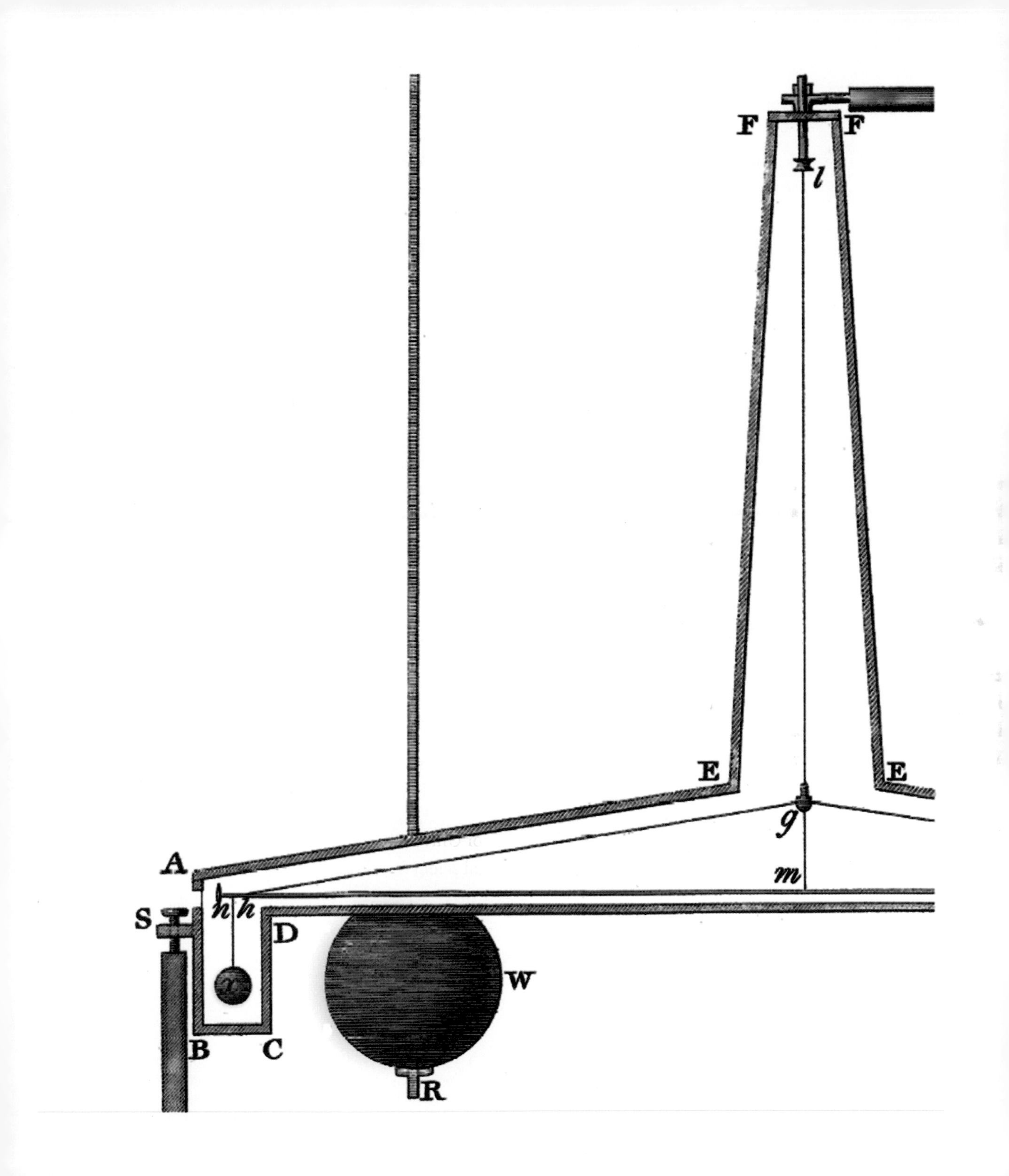
F
F
l
E
E
g
A
m
h h
S
D
W
B
C
R

1800

Battery

Luigi Galvani (1737–1798), **Alessandro Volta** (1745–1827), **Gaston Planté** (1834–1889)

Batteries have played an invaluable role in the history of physics, chemistry, and industry. As batteries evolved in power and sophistication, they facilitated important advances in electrical applications, from the emergence of telegraph communication systems to their use in vehicles, cameras, computers, and phones.

Around 1780, physiologist Luigi Galvani experimented with frogs' legs that he could cause to jerk when in contact with metal. Science-journalist Michael Guillen writes, "During his sensational public lectures, Galvani showed people how dozens of frogs' legs twitched uncontrollably when hung on copper hooks from an iron wire, like so much wet laundry strung out on a clothesline. Orthodox science cringed at his theories, but the spectacle of that chorus line of flexing frog legs guaranteed Galvani sell-out crowds in auditoriums the world over." Galvani ascribed the leg movement to "animal electricity." However, Italian physicist and friend Alessandro Volta believed that the phenomenon had more to do with the different metals Galvani employed, which were joined by a moist connecting substance. In 1800, Volta invented what has been traditionally considered to be the first electric battery when he stacked several pairs of alternating copper and zinc discs separated by cloth soaked in salt water. When the top and bottom of this *voltaic pile* were connected by a wire, an electric current began to flow. To determine that current was flowing, Volta could touch its two terminals to his tongue and experience a tingly sensation.

"A battery is essentially a can full of chemicals that produce electrons," write authors Marshall Brain and Charles Bryant. If a wire is connected between the negative and positive terminals, the electrons produced by chemical reactions flow from one terminal to the other.

In 1859, physicist Gaston Planté invented the rechargeable battery. By forcing a current through it "backwards," he could recharge his lead-acid battery. In the 1880s, scientists invented commercially successful dry cell batteries, which made use of pastes instead of liquid electrolytes (substances containing free ions that make the substances electrically conductive).

SEE ALSO Baghdad Battery (250 B.C.), Von Guericke's Electrostatic Generator (1660), Fuel Cell (1839), Leyden Jar (1744), Solar Cells (1954), Buckyballs (1985).

As batteries evolved, they facilitated important advances in electrical applications, ranging from the emergence of telegraph communication systems to their use in vehicles, cameras, computers, and phones.

1801

Wave Nature of Light

Christiaan Huygens (1629–1695), **Isaac Newton** (1642–1727), **Thomas Young** (1773–1829)

"What is light?" is a question that has intrigued scientists for centuries. In 1675, the famous English scientist Isaac Newton proposed that light was a stream of tiny particles. His rival, the Dutch physicist Christiaan Huygens, suggested that light consisted of waves, but Newton's theories often dominated, partly due to Newton's prestige.

Around 1800, the English researcher Thomas Young—also famous for his work on deciphering the Rosetta Stone—began a series of experiments that provided support for Huygens' wave theory. In a modern version of Young's experiment, a laser equally illuminates two parallel slits in an opaque surface. The pattern that the light makes as it passes through the two slits is observed on a distant screen. Young used geometrical arguments to show that the superposition of light waves from the two slits explains the observed series of equally spaced bands (fringes) of light and dark regions, representing constructive and destructive interference, respectively. You can think of these patterns of light as being similar to the tossing of two stones into a lake and watching the waves running into one another and sometimes canceling each other out or building up to form even larger waves.

If we carry out the same experiment with a beam of electrons instead of light, the resulting interference pattern is similar. This observation is intriguing, because if the electrons behaved only as particles, one might expect to simply see two bright spots corresponding to the two slits.

Today, we know that the behavior of light and subatomic particles can be even more mysterious. When single electrons are sent through the slits one at a time, an interference pattern is produced that is similar to that produced for waves passing through both holes at once. This behavior applies to all subatomic particles, not just photons (light particles) and electrons, and suggests that light and other subatomic particles have a mysterious combination of particle and wavelike behavior, which is just one aspect of the quantum mechanics revolution in physics.

SEE ALSO Maxwell's Equations (1861), Electromagnetic Spectrum (1864), Electron (1897), Photoelectric Effect (1905), Bragg's Law of Crystal Diffraction (1912), De Broglie Relation (1924), Schrödinger's Wave Equation (1926), Complementarity Principle (1927).

Simulation of the interference between two point sources. Young showed that the superposition of light waves from two slits explains the observed series of bands of light and dark regions, representing constructive and destructive interference, respectively.

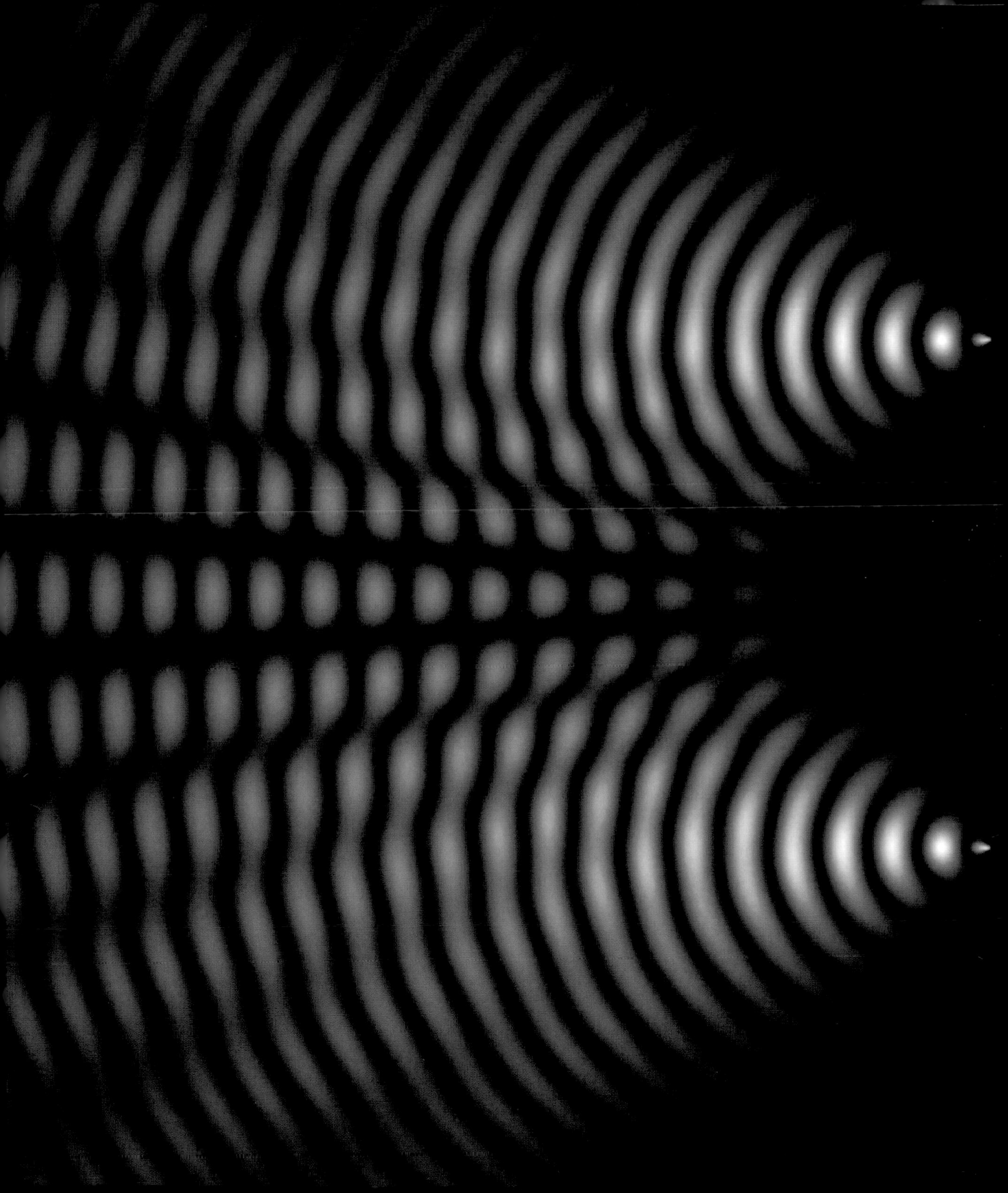

1803

Henry's Gas Law

William Henry (1775–1836)

Interesting physics is to be found even in the cracking of one's knuckles. Henry's Law, named after British chemist William Henry, states that the amount of a gas that is dissolved in a liquid is directly proportional to the pressure of the gas above the solution. It is assumed that the system under study has reached a state of equilibrium and that the gas does not chemically react with the liquid. A common formula used today for Henry's Law is $P = kC$, where P is the partial pressure of the particular gas above the solution, C is the concentration of the dissolved gas, and k is the Henry's Law constant.

We can visualize one aspect of Henry's Law by considering a scenario in which the partial pressure of a gas above a liquid increases by a factor of two. As a result, on the average, twice as many molecules will collide with the liquid surface in a given time interval, and, thus, twice as many gas molecules may enter the solution. Note that different gases have different solubilities, and these differences also affect the process along with the value of Henry's constant.

Henry's Law has been used by researchers to better understand the noise associated with "cracking" of finger knuckles. Gases that are dissolved in the synovial fluid in joints rapidly come out of solution as the joint is stretched and pressure is decreased. This cavitation, which refers to the sudden formation and collapse of low-pressure bubbles in liquids by means of mechanical forces, produces a characteristic noise.

In scuba diving, the pressure of the air breathed is roughly the same as the pressure of the surrounding water. The deeper one dives, the higher the air pressure, and the more air dissolves in the blood. When a diver ascends rapidly, the dissolved air may come out of solution too quickly in the blood, and bubbles in the blood may cause a painful and dangerous disorder known as decompression sickness ("the bends").

SEE ALSO Boyle's Gas Law (1662), Charles' Gas Law (1787), Avogadro's Gas Law (1811), Kinetic Theory (1859), Sonoluminescence (1934), Drinking Bird (1945).

Cola in a glass. When a soda can is opened, the reduced pressure causes dissolved gas to come out of solution according to Henry's Law. Carbon dioxide flows from the soda into the bubbles.

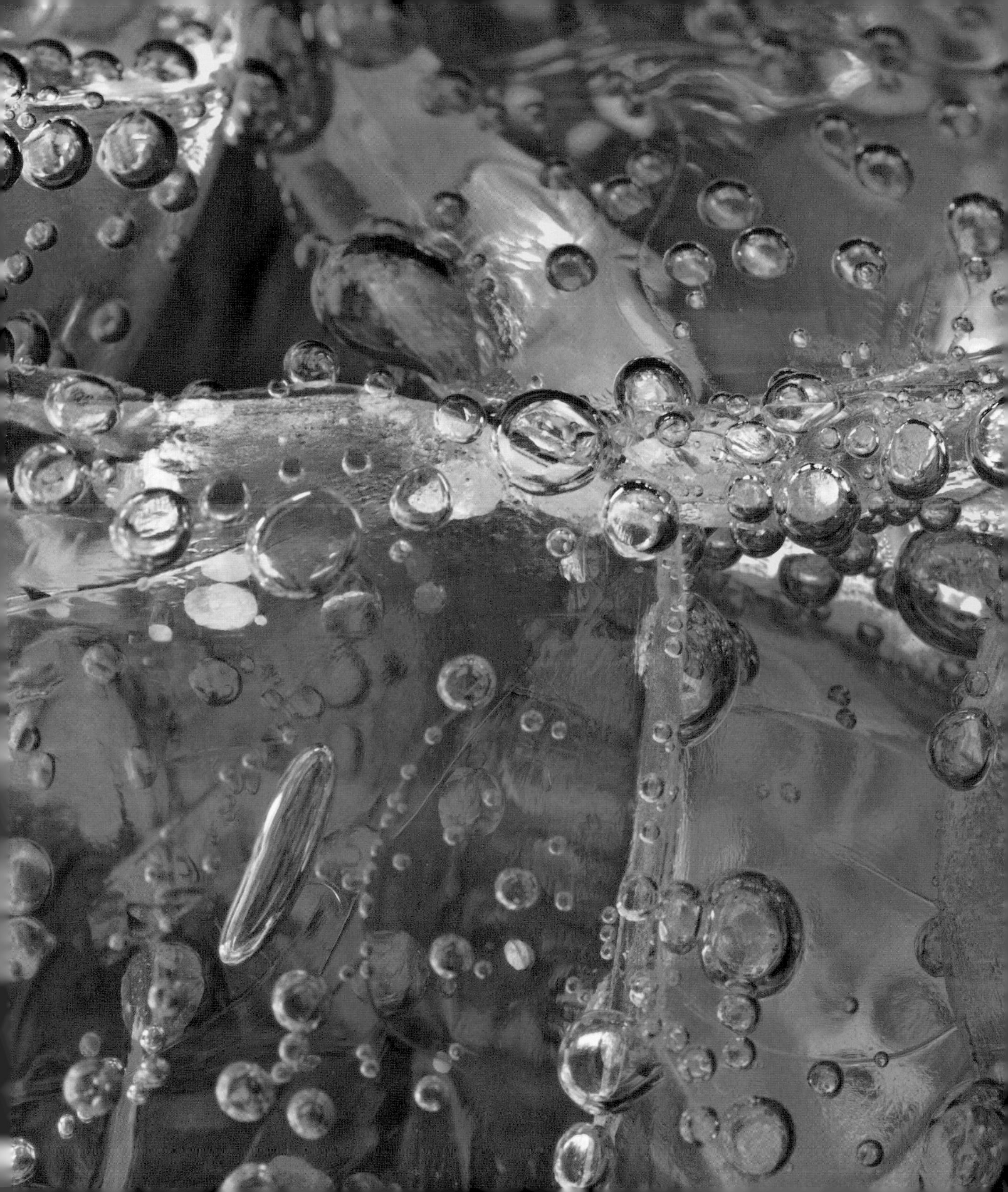

1807

Fourier Analysis

Jean Baptiste Joseph Fourier (1768–1830)

"The single most recurring theme of mathematical physics is Fourier analysis," writes physicist Sadri Hassani. "It shows up, for example, in classical mechanics . . . in electromagnetic theory and the frequency analysis of waves, in noise considerations and thermal physics, and in quantum theory"—virtually any field in which a frequency analysis is important. Fourier series can help scientists characterize and better understand the chemical composition of stars and quantify signal transmission in electronic circuits.

Before French mathematician Joseph Fourier discovered his famous mathematical series, he accompanied Napoleon on his 1798 expedition to Egypt, where Fourier spent several years studying Egyptian artifacts. Fourier's research on the mathematical theory of heat began around 1804 when he was back in France, and by 1807 he had completed his important memoir *On the Propagation of Heat in Solid Bodies*. One aspect of his fundamental work concerned heat diffusion in different shapes. For these problems, researchers are usually given the temperatures at points on the surface, as well as at its edges, at time $t = 0$. Fourier introduced a series with sine and cosine terms in order to find solutions to these kinds of problems. More generally, he found that any differentiable function can be represented to arbitrary accuracy by a sum of sine and cosine functions, no matter how bizarre the function may look when graphed.

Biographers Jerome Ravetz and I. Grattan-Guiness note, "Fourier's achievement can be understood by [considering] the powerful mathematical tools he invented for the solutions of the equations, which yielded a long series of descendants and raised problems in mathematical analysis that motivated much of the leading work in that field for the rest of the century and beyond." British physicist Sir James Jeans (1877–1946) remarked, "Fourier's theorem tells us that every curve, no matter what its nature may be, or in what way it was originally obtained, can be exactly reproduced by superposing a sufficient number of simple harmonic curves—in brief, every curve can be built up by piling up waves."

SEE ALSO Fourier's Law of Heat Conduction (1822), Greenhouse Effect (1824), Soliton (1834).

Portion of a jet engine. Fourier analysis methods are used to quantify and understand undesirable vibrations in numerous kinds of systems with moving parts.

1808

Atomic Theory

John Dalton (1766–1844)

John Dalton attained his professional success in spite of several hardships: He grew up in a family with little money; he was a poor speaker; he was severely color blind; and he was also considered to be a fairly crude or simple experimentalist. Perhaps some of these challenges would have presented an insurmountable barrier to any budding chemist of his time, but Dalton persevered and made exceptional contributions to the development of atomic theory, which states that all matter is composed of atoms of differing weights that combine in simple ratios in atomic compounds. During Dalton's time, atomic theory also suggested that these atoms were indestructible and that, for a particular element, all atoms were alike and had the same atomic weight.

He also formulated the *Law of Multiple Proportions*, which stated that whenever two elements can combine to form different compounds, the masses of one element that combine with a fixed mass of the other are in a ratio of small integers, such as 1:2. These simple ratios provided evidence that atoms were the building blocks of compounds.

Dalton encountered resistance to atomic theory. For example, the British chemist Sir Henry Enfield Roscoe (1833–1915) mocked Dalton in 1887, saying, "Atoms are round bits of wood invented by Mr. Dalton." Perhaps Roscoe was referring to the wood models that some scientists used in order to represent atoms of different sizes. Nonetheless, by 1850, the atomic theory of matter was accepted among a significant number of chemists, and most opposition disappeared.

The idea that matter was composed of tiny, indivisible particles was considered by the philosopher Democritus in Greece in the fifth century B.C., but this was not generally accepted until after Dalton's 1808 publication of *A New System of Chemical Philosophy*. Today, we understand that atoms are divisible into smaller particles, such as protons, neutrons, and electrons. Quarks are even smaller particles that combine to form other subatomic particles such as protons and neutrons.

SEE ALSO Kinetic Theory (1859), Electron (1897), Atomic Nucleus (1911), Seeing the Single Atom (1955), Neutrinos (1956), Quarks (1964).

LEFT: *Engraving of John Dalton, by William Henry Worthington (c. 1795–c. 1839).* RIGHT: *According to atomic theory, all matter is composed of atoms. Pictured here is a hemoglobin molecule with atoms represented as spheres. This protein is found in the red blood cell.*

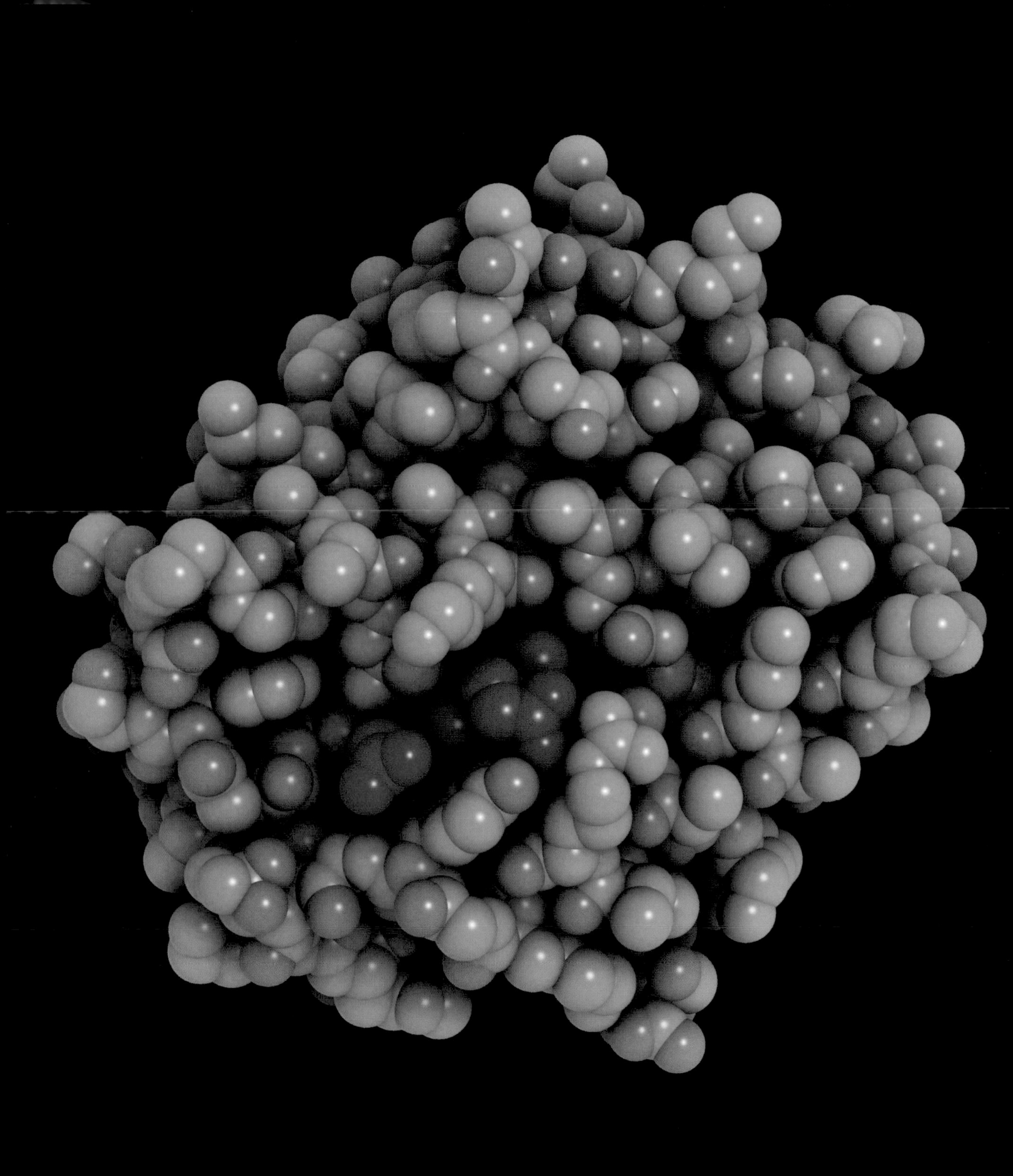

1811

Avogadro's Gas Law

Amedeo Avogadro (1776–1856)

Avogadro's Law, named after Italian physicist Amedeo Avogadro who proposed it in 1811, states that equal volumes of gases at the same temperature and pressure contain the same number of molecules, regardless of the molecular makeup of the gas. The law assumes that the gas particles are acting in an "ideal" manner, which is a valid assumption for most gases at pressures at or below a few atmospheres, near room temperature.

A variant of the law, also attributed to Avogadro, states that the volume of a gas is directly proportional to the number of molecules of the gas. This is represented by the formula $V = a \times N$, where a is a constant, V is the volume of the gas, and N is the number of gas molecules. Other contemporary scientists believed that such a proportionality should be true, but Avogadro's Law went further than competing theories because Avogadro essentially defined a molecule as the smallest characteristic particle of a substance—a particle that could be composed of several atoms. For example, he proposed that a water molecule consisted of two hydrogen atoms and one oxygen atom.

Avogadro's number, 6.0221367×10^{23}, is the number of atoms found in one mole of an element. Today we define Avogadro's number as the number of carbon-12 atoms in 12 grams of unbound carbon-12. A mole is the amount of an element that contains precisely the same number of grams as the value of the atomic weight of the substance. For example, nickel has an atomic weight of 58.6934, so there are 58.6934 grams in a mole of nickel.

Because atoms and molecules are so small, the magnitude of Avogadro's number is difficult to visualize. If an alien were to descend from the sky to deposit an Avogadro's number of unpopped popcorn kernels on the Earth, the alien could cover the United States of America with the kernels to a depth of over nine miles.

SEE ALSO Charles' Gas Law (1787), Atomic Theory (1808), Kinetic Theory (1859).

Place 24 numbered golden balls (numbered 1 through 24) in a bowl. If you randomly drew them out one at a time, the probability of removing them in numerical order is about 1 chance in Avogadro's number—a very small chance!

1814

Fraunhofer Lines

Joseph von Fraunhofer (1787–1826)

A *spectrum* often shows the variation in the intensity of an object's radiation at different wavelengths. Bright lines in atomic spectra occur when electrons fall from higher energy levels to lower energy levels. The color of the lines depends on the energy difference between the energy levels, and the particular values for the energy levels are identical for atoms of the same type. Dark absorption lines in spectra can occur when an atom absorbs light and the electron jumps to a higher energy level.

By examining absorption or emission spectra, we can determine which chemical elements produced the spectra. In the 1800s, various scientists noticed that the spectrum of the Sun's electromagnetic radiation was not a smooth curve from one color to the next; rather, it contained numerous dark lines, suggesting that light was being absorbed at certain wavelengths. These dark lines are called Fraunhofer lines after the Bavarian physicist Joseph von Fraunhofer, who recorded them.

Some readers may find it easy to imagine how the Sun can produce a radiation spectrum but not how it can also produce dark lines. How can the Sun absorb its own light?

You can think of stars as fiery gas balls that contain many different atoms emitting light in a range of colors. Light from the surface of a star—the *photosphere*—has a continuous spectrum of colors, but as the light travels through the outer atmosphere of a star, some of the colors (i.e., light at different wavelengths) are absorbed. This absorption is what produces the dark lines. In stars, the missing colors, or dark absorption lines, tell us exactly which chemical elements are in the outer atmosphere of stars.

Scientists have catalogued numerous missing wavelengths in the spectrum of the Sun. By comparing the dark lines with spectral lines produced by chemical elements on the Earth, astronomers have found over seventy elements in the Sun. Note that, decades later, scientists Robert Bunsen and Gustav Kirchhoff studied emission spectra of heated elements and discovered cesium in 1860.

SEE ALSO Newton's Prism (1672), Electromagnetic Spectrum (1864), Mass Spectrometer (1898), Bremsstrahlung (1909), Stellar Nucleosynthesis (1946).

Visible solar spectrum with Fraunhofer lines. The y-axis represents the wavelength of light, starting from 380 nm at top and ending at 710 nm at bottom.

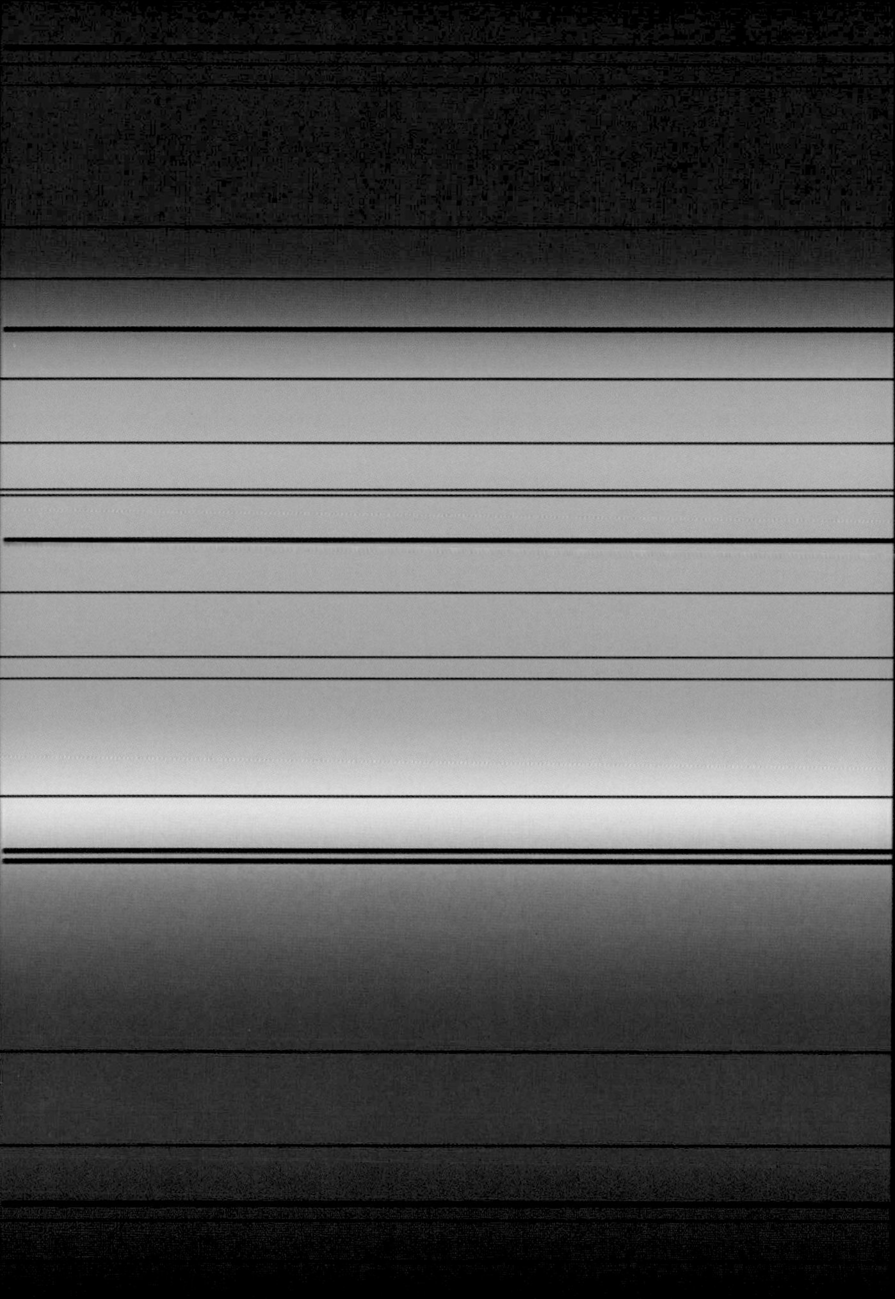

1814

Laplace's Demon

Pierre-Simon, Marquis de Laplace (1749–1827)

In 1814, French mathematician Pierre-Simon Laplace described an entity, later called *Laplace's Demon*, that was capable of calculating and determining all future events, provided that the demon was given the positions, masses, and velocities of every atom in the universe and the various known formulae of motion. "It follows from Laplace's thinking," writes scientist Mario Markus, "that if we were to include the particles in our brains, free will would become an illusion. . . . Indeed, Laplace's God simply turns the pages of a book that is already written."

During Laplace's time, the idea made a certain sense. After all, if one could predict the position of billiard balls bouncing around a table, why not entities composed of atoms? In fact, Laplace has no need of God at all in his universe.

Laplace wrote, "We may regard the present state of the universe as the effect of its past and the cause of its future. An intellect which at a certain moment would know all forces that set nature in motion, and all positions of all items of which nature is composed, if this intellect were also vast enough to submit these data to analysis, [it would embrace in] a single formula the movements of the greatest bodies of the universe and those of the tiniest atom; for such an intellect nothing would be uncertain and the future just like the past would be present before its eyes."

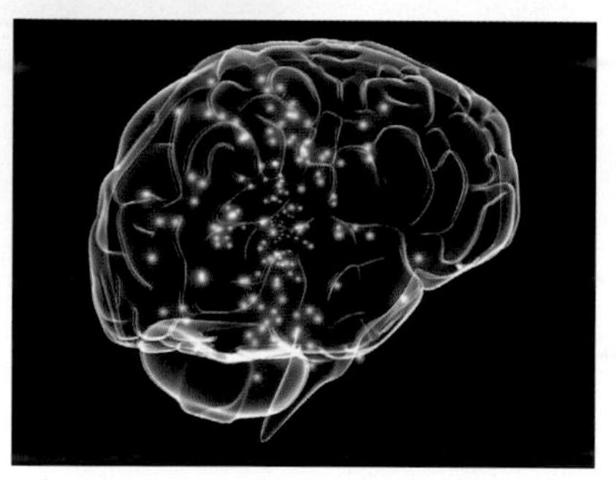

Later, developments such as **Heisenberg's Uncertainty Principle** (HUP) and **Chaos Theory** appear to make Laplace's demon an impossibility. According to chaos theory, even minuscule inaccuracies in measurement at some initial time may lead to vast differences between a predicted outcome and an actual outcome. This means that Laplace's demon would have to know the position and motion of every particle to infinite precision, thus making the demon more complex than the universe itself. Even if this demon existed outside the universe, the HUP tells us that infinitely precise measurements of the type required are impossible.

SEE ALSO Maxwell's Demon (1867), Heisenberg Uncertainty Principle (1927), Chaos Theory (1963).

UPPER LEFT: *Pierre-Simon Laplace (posthumous portrait by Madame Feytaud [1842]).* LOWER LEFT: *In a universe with Laplace's Demon, would free will be an illusion?* RIGHT: *Artistic rendition of Laplace's Demon observing the positions, masses, and velocities of every particle (represented here as bright specks) at a particular time.*

1815

Brewster's Optics

Sir David Brewster (1781–1868)

Light has fascinated scientists for centuries, but who would think that creepy cuttlefish might have something to teach us about the nature of light? A lightwave consists of an electric field and a magnetic field that oscillate perpendicular to each other and to the direction of travel. It is possible, however, to restrict the vibrations of the electric field to a particular plane by *plane-polarizing* the light beam. For example, one approach for obtaining plane-polarized light is via the reflection of light from a surface between two media, such as air and glass. The component of the electric field parallel to the surface is most strongly reflected. At one particular angle of incidence on the surface, called the Brewster's angle after Scottish physicist David Brewster, the reflected beam consists entirely of light whose electric vector is parallel to the surface.

Polarization by light scattering in our atmosphere sometimes produces a glare in the skies. Photographers can reduce this partial polarization using special materials to prevent the glare from producing an image of a washed-out sky. Many animals, such as bees and cuttlefish, are quite capable of perceiving the polarization of light, and bees use polarization for navigation because the linear polarization of sunlight is perpendicular to the direction of the Sun.

Brewster's experiments with light polarization led him to his 1816 invention of the kaleidoscope, which has often fascinated physics students and teachers who attempt to create ray diagrams in order to understand the kaleidoscope's multiple reflections. Cozy Baker, founder of the Brewster Kaleidoscope Society, writes, "His kaleidoscope created unprecedented clamor. . . . A universal mania for the instrument seized all classes, from the lowest to the highest, from the most ignorant to the most learned, and every person not only felt, but expressed the feeling that a new pleasure had been added to their existence." American inventor Edwin H. Land wrote "The kaleidoscope was the television of the 1850s. . . ."

SEE ALSO Snell's Law of Refraction (1621), Newton's Prism (1672), Fiber Optics (1841), Electromagnetic Spectrum (1864), Laser (1960), Unilluminable Rooms (1969).

LEFT: *Using skin patterns that involve polarized light, cuttlefish can produce intricate "designs" as a means of communication. Such patterns are invisible to human eyes.* RIGHT: *Brewster's experiments with light polarization led him to his 1816 invention of the kaleidoscope.*

1816

Stethoscope

René-Théophile-Hyacinthe Laennec (1781–1826)

Social historian Roy Porter writes, "By giving access to body noises—the sound of breathing, the blood gurgling around the heart—the stethoscope changed approaches to internal disease and hence doctor-patient relations. As last, the living body was no longer a closed book: pathology could now be done on the living."

In 1816, French physician René Laennec invented the stethoscope, which consisted of a wooded tube with a trumpet-like end that made contact with the chest. The air-filled cavity transmitted sounds from the patient's body to the physician's ear. In the 1940s, stethoscopes with two-sided chest-pieces became standard. One side of the chest-piece is a diaphragm (e.g. a plastic disc that covers the opening), which vibrates when detecting body sounds and produces acoustic pressure waves that travel through the air cavity of the stethoscope. The other side contains a bell-shaped endpiece (e.g. a hollow cup) that is better at transmitting low-frequency sounds. The diaphragm side actually tunes out the low frequencies associated with heart sounds and is used to listen to the respiratory system. When using the bell side, the physician can vary the pressure of the bell on the skin and "tune" the skin vibration frequency in order to best reveal the heartbeat. Many other refinements occurred over the years that involved improved amplification, noise reduction, and other characteristics that were optimized by application of simple physical principles (see "Notes and Further Reading").

In Laennec's day, a physician often placed his ear directly on the patient's chest or back. However, Laennec complained that this technique "is always inconvenient, both to the physician and the patient; in the case of females, it is not only indelicate but often impracticable." Later, an extra-long stethoscope was used to treat the very poor when physicians wanted to be farther away from their flea-ridden patients. Aside from inventing the device, Laennec carefully recorded how specific physical diseases (e.g. pneumonia, tuberculosis, and bronchitis) corresponded to the sounds heard. Ironically, Laennec himself died at age 45 of tuberculosis, which his nephew diagnosed using a stethoscope.

SEE ALSO Tuning Fork (1711), Poiseuille's Law of Fluid Flow (1840), Doppler Effect (1842), War Tubas (1880).

Modern stethoscope. Various acoustic experiments have been conducted to determine the effect of chest-piece size and material on sound collection.

1822

Fourier's Law of Heat Conduction

Jean Baptiste Joseph Fourier (1768–1830)

"Heat cannot be separated from fire, or beauty from the Eternal," wrote Dante Alighieri. The nature of heat had also fascinated the French mathematician Joseph Fourier, well known for his formulas on the conduction of heat in solid materials. His Law of Heat Conduction suggests that the rate of heat flow between two points in a material is proportional to the difference in the temperatures of the points and inversely proportional to the distance between the two points.

If we place one end of an all-metal knife into a hot cup of cocoa, the temperature of the other end of the knife begins to rise. This heat transfer is caused by molecules at the hot end exchanging their kinetic and vibrational energies with adjacent regions of the knife through random motions. The rate of flow of energy, which might be thought of as a "heat current," is proportional to the difference in temperatures at locations A and B, and inversely proportional to the distance between A and B. This means that the heat current is doubled if the temperature difference is doubled or if the length of the knife is halved.

If we let U be the conductance of the material—that is, the measure of the ability of a material to conduct heat—we may incorporate this variable into Fourier's Law. Among the best thermal conductors, in order of thermal conductivity values, are diamond, carbon nanotubes, silver, copper, and gold. With the use of simple instruments, the high thermal conductivity of diamonds is sometimes employed to help experts distinguish real diamonds from fakes. Diamonds of any size are cool to the touch because of their high thermal conductivity, which may help explain why the word "ice" is often used when referring to diamonds.

Even though Fourier conducted foundational work on heat transfer, he was never good at regulating his own heat. He was always so cold, even in the summer, that he wore several large overcoats. During his last months, Fourier often spent his time in a box to support his weak body.

SEE ALSO Fourier Analysis (1807), Carnot Engine (1824), Joule's Law of Electric Heating (1840), Thermos (1892).

LEFT: *Raw copper ore. Copper is both an excellent thermal and electrical conductor.* RIGHT: *The transfer of heat by various means plays a crucial role in the development of heat-sinks for computer chips. In this photo, the central object with a rectangular base is used to transfer heat away from the chip.*

BAT
CLR_CMOS
CLR_CMOS
SHORT CLEAR CMOS
OPEN NORMAL
SB_HEATSINK
ICH
RU1
RU2
RQ1
REC2

1823

Olbers' Paradox

Heinrich Wilhelm Matthäus Olbers (1758–1840)

"Why is the sky dark at night?" In 1823, the German astronomer Heinrich Wilhelm Olbers presented a paper that discussed this question, and the problem subsequently became known as *Olbers' Paradox*. Here is the puzzle. If the universe is infinite, as you follow a line of sight in *any* direction, that line must eventually intercept a star. This characteristic appears to imply that the night sky should be dazzlingly bright with starlight. Your first thought might be that the stars are far away and that their light dissipates as it travels such great distances. Starlight does dim as it travels, but by the square of the distance from the observer. However, the *volume* of the universe—and hence the total number of stars—would grow as the cube of the distance. Thus, even though the stars become dimmer the further away they are, this dimming is compensated by the increased number of stars. If we lived in an infinite visible universe, the night sky should indeed be very bright.

Here's the solution to Olbers' Paradox. We do not live in an infinite and static visible universe. The universe has a finite age and is expanding. Because only about 13.7 billion years have elapsed since the **Big Bang**, we can only observe stars out to a finite distance. This means that the number of stars that we can observe is finite. Because of the speed of light, there are portions of the universe we never see, and light from very distant stars has not had time to reach the Earth. Interestingly, the first person to suggest this resolution to Olbers' Paradox was the writer Edgar Allan Poe.

Another factor to consider is that the expansion of the universe also acts to darken the night sky because starlight expands into a space that is ever more vast. Also, the **Doppler Effect** causes a red shift in the wavelengths of light emitted from the rapidly receding stars. Life as we know it would not have evolved without these factors because the night sky would have been extremely bright and hot.

SEE ALSO Big Bang (13.7 Billion B.C.), Doppler Effect (1842), Hubble's Law of Cosmic Expansion (1929).

If the universe is infinite, as you follow a line of sight in any *direction, that line must eventually intercept a star. This characteristic appears to imply that the night sky should be dazzlingly bright with starlight.*

Greenhouse Effect

Joseph Fourier (1768–1830), **Svante August Arrhenius** (1859–1927), **John Tyndall** (1820–1893)

"Despite all its bad press," write authors Joseph Gonzalez and Thomas Sherer, "the process known as the *greenhouse effect* is a very natural and necessary phenomenon. . . . The atmosphere contains gases that enable sunlight to pass through to the earth's surface but hinder the escape of reradiated heat energy. Without this natural greenhouse effect, the earth would be much too cold to sustain life." Or, as Carl Sagan once wrote, "A little greenhouse effect is a good thing."

Generally speaking, the greenhouse effect is the heating of the surface of a planet as a result of atmospheric gases that absorb and emit infrared radiation, or heat energy. Some of the energy reradiated by the gases escapes into outer space; another portion is reradiated back toward the planet. Around 1824, mathematician Joseph Fourier wondered how the Earth stays sufficiently warm to support life. He proposed that although some heat does escape into space, the atmosphere acts a little like a translucent dome—a glass lid of a pot, perhaps—that absorbs some of the heat of the Sun and reradiates it downward to the Earth.

In 1863, British physicist and mountaineer John Tyndall reported on experiments that demonstrated that water vapor and carbon dioxide absorbed substantial amounts of heat. He concluded that water vapor and carbon dioxide must therefore play an important role in regulating the temperature at the Earth's surface. In 1896, Swedish chemist Svante Arrhenius showed that carbon dioxide acts as a very strong "heat trap" and that halving the amount in the atmosphere might trigger an ice age. Today we use the term *anthropogenic global warming* to denote an enhanced greenhouse effect due to human contributions to greenhouse gases, such as the burning of fossil fuels.

Aside from water vapor and carbon dioxide, methane from cattle belching can also contribute to the greenhouse effect. "Cattle belching?" Thomas Friedman writes. "That's right—the striking thing about greenhouse gases is the diversity of sources that emit them. A herd of cattle belching can be worse than a highway full of Hummers."

SEE ALSO Aurora Borealis (1621), Fourier's Law of Heat Conduction (1822), Rayleigh Scattering (1871).

LEFT: *"Coalbrookdale by Night" (1801), by Philip James de Loutherbourg (1740–1812), showing the Madeley Wood Furnaces, a common symbol of the early Industrial Revolution.* RIGHT: *Large changes in manufacturing, mining, and other activities since the Industrial Revolution have increased the amount of greenhouse gases in the air. For example, steam engines, fuelled primarily by coal, helped to drive the Industrial Revolution.*

1824

Carnot Engine

Nicolas Léonard Sadi Carnot (1796–1832)

Much of the initial work in thermodynamics—the study of the conversion of energy between work and heat—focused on the operation of engines and how fuel, such as coal, could be efficiently converted to useful work by an engine. Sadi Carnot is probably most often considered the "father" of thermodynamics, thanks to his 1824 work *Réflexions sur la puissance motrice du feu* (*Reflections on the Motive Power of Fire*).

Carnot worked tirelessly to understand heat flow in machines partly because he was disturbed that British steam engines seemed to be more efficient than French engines. During his day, steam engines usually burned wood or coal in order to convert water into steam. The high-pressure steam moved the pistons of the engine. When the steam was released through an exhaust port, the pistons returned to their original positions. A cool radiator converted the exhaust steam to water, so it could be heated again to steam in order to drive the pistons.

Carnot imagined an ideal engine, known today as the *Carnot engine*, that would theoretically have a work output equal to that of its heat input and not lose even a small amount of energy during the conversion. After experiments, Carnot realized that no device could perform in this ideal matter—some energy had to be lost to the environment. Energy in the form of heat could not be converted completely into mechanical energy. However, Carnot did help engine designers improve their engines so that the engines could work close to their peak efficiencies.

Carnot was interested in "cyclical devices" in which, at various parts of their cycles, the device absorbs or rejects heat; it is impossible to make such an engine that is 100 percent efficient. This impossibility is yet another way of stating the **Second Law of Thermodynamics**. Sadly, in 1832 Carnot contracted cholera and, by order of the health office, nearly all his books, papers, and other personal belongings had to be burned!

SEE ALSO Perpetual Motion Machines (1150), Fourier's Law of Heat Conduction (1822), Second Law of Thermodynamics (1850), Drinking Bird (1945).

LEFT: *An 1813 portrait of Sadi Carnot.* RIGHT: *Locomotive steam engine. Carnot worked to understand heat flow in machines, and his theories have relevance to this day. During his time, steam engines usually burned wood or coal.*

1825

Ampère's Law of Electromagnetism

André-Marie Ampère (1775–1836), **Hans Christian Ørsted** (1777–1851)

By 1825, French physicist André-Marie Ampère had established the foundation of electromagnetic theory. The connection between electricity and magnetism was largely unknown until 1820, when Danish physicist Hans Christian Ørsted discovered that a compass needle moves when an electric current is switched on or off in a nearby wire. Although not fully understood at the time, this simple demonstration suggested that electricity and magnetism were related phenomena, a finding that led to various applications of electromagnetism and eventually culminated in telegraphs, radios, TVs, and computers.

Subsequent experiments during a period from 1820 to 1825 by Ampère and others showed that any conductor that carries an electric current I produces a magnetic field around it. This basic finding, and its various consequences for conducting wires, is sometimes referred to as *Ampère's Law of Electromagnetism*. For example, a current-carrying wire produces a magnetic field **B** that circles the wire. (The use of bold signifies a vector quantity.) **B** has a magnitude that is proportional to I, and points along the circumference of an imaginary circle of radius r centered on the axis of the long, straight wire. Ampère and others showed that electric currents attract small bits of iron, and Ampère proposed a theory that electric currents are the source of magnetism.

Readers who have experimented with electromagnets, which can be created by wrapping an insulated wire around a nail and connecting the ends of the wire to a battery, have experienced Ampère's Law first hand. In short, Ampère's Law expresses the relationship between the magnetic field and the electric current that produces it.

Additional connections between magnetism and electricity were demonstrated by the experiments of American scientist Joseph Henry (1797–1878), British scientist Michael Faraday (1791–1867), and James Clerk Maxwell. French physicists Jean-Baptiste Biot (1774–1862) and Félix Savart (1791–1841) also studied the relationship between electrical current in wires and magnetism. A religious man, Ampère believed that he had proven the existence of the soul and of God.

SEE ALSO Faraday's Laws of Induction (1831), Maxwell's Equations (1861), Galvanometer (1882).

LEFT: *Engraving of André-Marie Ampère by A. Tardieu (1788–1841).* RIGHT: *Electric motor with exposed rotor and coil. Electromagnets are widely used in motors, generators, loudspeakers, particle accelerators, and industrial lifting magnets.*

1826

Rogue Waves

Jules Sébastien César Dumont d'Urville (1790–1842)

"From the time of the earliest civilizations," writes marine physicist Susanne Lehner, "humankind has been fascinated with stories of giant waves—the 'monsters' of the seas... towers of water pounding down on a helpless boat. You can observe the wall of water coming... but you cannot run away and you cannot fight it.... Can we cope with [this nightmare] in the future? Predict extreme waves? Control them? Ride giant waves like surfers?"

It may seem surprising that in the twenty-first century, physicists do not have a complete understanding of the ocean surface, but the origin of rogue waves is not completely clear. In 1826, when French explorer and naval officer Captain Dumont d'Urville reported waves up to 98 feet (30 meters) in height—approximately the height of a 10-story building—he was ridiculed. However, after using satellite monitoring and many models that incorporate the relevant probability theory of wave distributions, we now know that waves this high are much more common than expected. Imagine the horrors of such a wall of water appearing without warning in mid-ocean, sometimes in clear weather, preceded by a trough so deep as to form a frightening "hole" in the ocean.

One theory is that ocean currents and seabed shapes act almost like optical lenses and focus wave actions. Perhaps high waves are generated by the superposition of crossing waves from two different storms. However, additional factors seem to play a role in creating such nonlinear wave effects that can produce the tall wall of water in a relatively calm sea. Before breaking, the rogue wave can have a crest four times higher than crests of neighbor waves. Many papers have been written that attempt to model formation of rogue waves using nonlinear Schrödinger equations. The effect of wind on the nonlinear evolution of waves has also been a productive area of research. Because rogue waves are responsible for the loss of ships and lives, scientists continue to search for ways to predict and avoid such waves.

SEE ALSO Fourier Analysis (1807), Soliton (1834), Fastest Tornado Speed (1999).

Rogue waves can be terrifying, appearing without warning in mid-ocean, sometimes in clear weather, preceded by a trough so deep as to form a "hole" in the ocean. Rogue waves are responsible for the loss of ships and lives.

1827

Ohm's Law of Electricity

Georg Ohm (1789–1854)

Although German physicist Georg Ohm discovered one of the most fundamental laws in the field of electricity, his work was ignored by his colleagues, and he lived in poverty for much of his life. His harsh critics called his work a "web of naked fancies." Ohm's Law of electricity states that the steady electric current I in a circuit is proportional to the constant voltage V (or total electromotive force) across a resistance and inversely proportional to the value R of the resistance: $I = V/R$.

Ohm's experimental discovery of the law in 1827 suggested that the law held for a number of different materials. As made obvious from the equation, if the potential difference V (in units of volts) between the two ends of a wire is doubled, then the current I in amperes also doubles. For a given voltage, if the resistance doubles, the current is decreased by a factor of two.

Ohm's Law has relevance in determining the dangers of electrical shocks on the human body. Generally, the higher the current flow, the more dangerous the shock is. The amount of current is equal to the voltage applied between two points on the body, divided by the electrical resistance of the body. Precisely how much voltage a person can experience and survive depends on the total resistance of the body, which varies from person to person and may depend on such parameters as body fat, fluid intake, skin sweatiness, and how and where contact is made with the skin.

Electrical resistance is used today to monitor corrosion and material loss in pipelines. For example, a net change in the resistance in a metal wall may be attributable to metal loss. A corrosion-detection device may be permanently installed to provide continuous information, or the device may be portable to gather information as needed. Note that without resistance, electric blankets, certain kettles, and **Incandescent Light Bulbs** would be useless.

SEE ALSO Joule's Law of Electric Heating (1840), Kirchhoff's Circuit Laws (1845), Incandescent Light Bulb (1878).

LEFT: *Electric kettles rely on electrical resistance to create heat.* RIGHT: *Circuit board with resistors (cylindrical objects with colored bands). The resistor has a voltage across its terminals that is proportional to the electric current passing through it, as given by Ohm's Law. The colored bands indicate the resistance value.*

2N2905A
RIZ442
RIZ
2N2905A
35 V +

1827

Brownian Motion

Robert Brown (1773–1858), **Jean-Baptiste Perrin** (1870–1942), **Albert Einstein** (1879–1955)

In 1827, Scottish botanist Robert Brown was using a microscope to study pollen grains suspended in water. Particles within the vacuoles of the pollen grains seemed to dance about in a random fashion. In 1905, Albert Einstein predicted the movement of such kinds of small particles by suggesting that they were constantly being buffeted by water molecules. At any instant in time, just by chance, more molecules would strike one side of the particle than another side, thereby causing the particle to momentarily move slightly in a particular direction. Using statistical rules, Einstein demonstrated that this Brownian motion could be explained by random fluctuations in such collisions. Moreover, from this motion, one could determine the dimensions of the hypothetical molecules that were bombarding the macroscopic particles.

In 1908, French physicist Jean-Baptiste Perrin confirmed Einstein's explanation of Brownian motion. As a result of Einstein and Perrin's work, physicists were finally compelled to accept the reality of atoms and molecules, a subject still ripe for debate even at the beginning of the twentieth century. In concluding his 1909 treatise on this subject, Perrin wrote, "I think that it will henceforth be difficult to defend by rational arguments a hostile attitude to molecular hypotheses."

Brownian motion gives rise to diffusion of particles in various media and is so general a concept that it has wide applications in many fields, ranging from the dispersal of pollutants to the understanding of the relative sweetness of syrups on the surface of the tongue. Diffusion concepts help us understand the effect of pheromones on ants or the spread of muskrats in Europe following their accidental release in 1905. Diffusion laws have been used to model the concentration of smokestack contaminants and to simulate the displacement of hunter-gatherers by farmers in Neolithic times. Researchers have also used diffusion laws to study diffusion of radon in the open air and in soils contaminated with petroleum hydrocarbons.

SEE ALSO Perpetual Motion Machines (1150), Atomic Theory (1808), Graham's Law of Effusion (1829), Kinetic Theory (1859), Boltzmann's Entropy Equation (1875), Einstein as Inspiration (1921).

Scientists used Brownian motion and diffusion concepts to model muskrat propagation. In 1905, five muskrats were introduced to Prague from the U.S. By 1914, their descendants had spread 90 miles in all directions. In 1927, they numbered over 100 million.

1829

Graham's Law of Effusion

Thomas Graham (1805–1869)

Whenever I ponder Graham's Law of Effusion, I cannot help but think of death and atomic weaponry. The law, named after Scottish scientist Thomas Graham, states that the rate of effusion of a gas is inversely proportional to the square root of the mass of its particles. This formula can be written as $R_1/R_2 = (M_2/M_1)^{1/2}$, where R_1 is the rate of effusion of gas 1, R_2 is the rate of effusion for gas 2, M_1 is the molar mass of gas 1, and M_2 is the molar mass of gas 2. This law works for both diffusion and effusion, the latter being a process in which individual molecules flow through very small holes without colliding with one another. The rate of effusion depends on the molecular weight of the gas. For example, gases like hydrogen with a low molecular weight effuse more quickly than heavier particles because the low-weight particles are generally moving at higher speeds.

Graham's Law had a particularly ominous application in the 1940s when it was used in nuclear reactor technology to separate radioactive gases that had different diffusion rates due to the molecular weights of the gases. A long diffusion chamber was used to separate two isotopes of uranium, U-235 and U-238. These isotopes were allowed to chemically react with fluorine to produce the gas uranium hexafluoride. The less massive uranium hexafluoride molecules containing fissionable U-235 would travel down the chamber slightly faster than the more massive molecules containing the U-238.

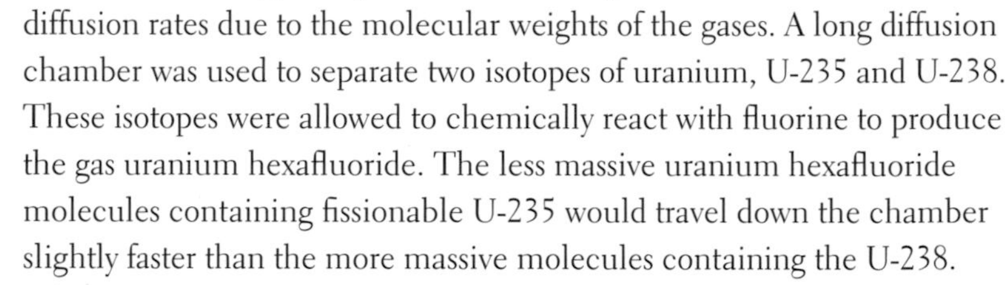

During World War II, this separation process allowed the U.S. to develop an atomic bomb, which required the isolation of U-235 for the nuclear fission chain reaction. To separate U-235 and U-238, the government built a gaseous diffusion plant in Tennessee. The plant used diffusion through porous barriers and processed the uranium for the Manhattan Project, which yielded the atomic bomb dropped on Japan in 1945. In order to perform the isotope separation, the gaseous diffusion plant required 4,000 stages in a space that encompassed 43 acres.

SEE ALSO Brownian Motion (1827), Boltzmann's Entropy Equation (1875), Radioactivity (1896), Little Boy Atomic Bomb (1945).

LEFT: *Uranium ore.* RIGHT: *K-52 gaseous diffusion site in Oak Ridge, Tennessee Manhattan Project. The main building was over half a mile long.* (Photo taken by J. E. Westcott, official government photographer for the Manhattan Project.)

1831

Faraday's Laws of Induction

Michael Faraday (1791–1867)

"Michael Faraday was born in the year that Mozart died," Professor David Goodling writes. "Faraday's achievement is a lot less accessible than Mozart's [but . . .] Faraday's contributions to modern life and culture are just as great. . . . His discoveries of . . . magnetic induction laid the foundations for modern electrical technology . . . and made a framework for unified field theories of electricity, magnetism, and light."

English scientist Michael Faraday's greatest discovery was that of electromagnetic induction. In 1831, he noticed that when he moved a magnet through a stationary coil of wire, he always produced an electric current in the wire. The induced electromotive force was equal to the rate of change of the magnetic flux. American scientist Joseph Henry (1797–1878) carried out similar experiments. Today, this induction phenomenon plays a crucial role in electric power plants.

Faraday also found that if he moved a wire loop near a stationary permanent magnet, a current flowed in the wire whenever it moved. When Faraday experimented with an electromagnet and caused the magnetic field surrounding the electromagnet to change, he then detected electric current flow in a nearby but separate wire.

Scottish physicist James Clerk Maxwell (1831–1879) later suggested that changing the magnetic flux produced an electric field that not only caused electrons to flow in a nearby wire, but that the field also existed in space—even in the absence of electric charges. Maxwell expressed the change in magnetic flux and its relation to the induced electromotive force (ε or emf) in what we call Faraday's Law of Induction. The magnitude of the emf induced in a circuit is proportional to the rate of change of the magnetic flux impinging on the circuit.

Faraday believed that God sustained the universe and that he was doing God's will to reveal truth through careful experiments and through his colleagues, who tested and built upon his results. He accepted every word of the Bible as literal truth, but meticulous experiments were essential in this world before any other kind of assertion could be accepted.

SEE ALSO Ampère's Law of Electromagnetism (1825), Maxwell's Equations (1861), Hall Effect (1879).

LEFT: *Photograph of Michael Faraday (c. 1861) by John Watkins (1823–1874).* RIGHT: *A dynamo, or electrical generator, from G. W. de Tunzelmann's* Electricity in Modern Life, *1889. Power stations usually rely on a generator with rotating elements that convert mechanical energy into electrical energy through relative motions between a magnetic field and an electrical conductor.*

PATENT

1834

Soliton

John Scott Russell (1808–1882)

A soliton is a solitary wave that maintains its shape while it travels for long distances. The discovery of solitons is one of the most enjoyable tales of significant science arising from a casual observation. In August 1834, Scottish engineer John Scott Russell happened to be watching a horse-drawn barge move along a canal. When the cable broke and the barge suddenly stopped, Russell made a startling observation of a humplike water formation, which he described as follows: "a mass of water rolled forward with great velocity, assuming the form of a large solitary elevation, a rounded, smooth and well-defined heap of water, which continued its course along the channel apparently without change of form or diminution of speed. I followed it on horseback, and overtook it still rolling on at a rate of some eight or nine miles an hour, preserving its original figure some 30 feet long and a foot to a foot and a half in height. Its height gradually diminished, and after a chase of one or two miles I lost it in the windings of the channel."

Russell subsequently performed experiments in a wave tank in his home to characterize these mysterious solitons (which he called a *wave of translation*), finding that the speed depends on the size of the soliton. Two solitons of different sizes (and hence velocities) can pass through each other, emerge, and continue their propagation. Soliton behavior has also been observed in other systems, such as plasmas and flowing sand. For example, barchan dunes, which involve arc-shaped sand ridges, have been seen "passing" through each other. The Great Red Spot of Jupiter may also be some kind of soliton.

Today, solitons are considered in a wide range of phenomena, ranging from nerve signal propagation to soliton-based communications in optical fibers. In 2008, the first known unchanging soliton in outer space was reported moving through the ionized gas surrounding the Earth at about 5 miles (8 kilometers) per second.

SEE ALSO Rogue Waves (1826), Fourier Analysis (1807), Self-Organized Criticality (1987).

Barchan dunes on Mars. When two Earthly barchan ridges collide, they may form a compound ridge, and then reform their original shapes. (When one dune "crosses" another dune, sand particles do not actually travel through one another, but the ridge shapes may persist.)

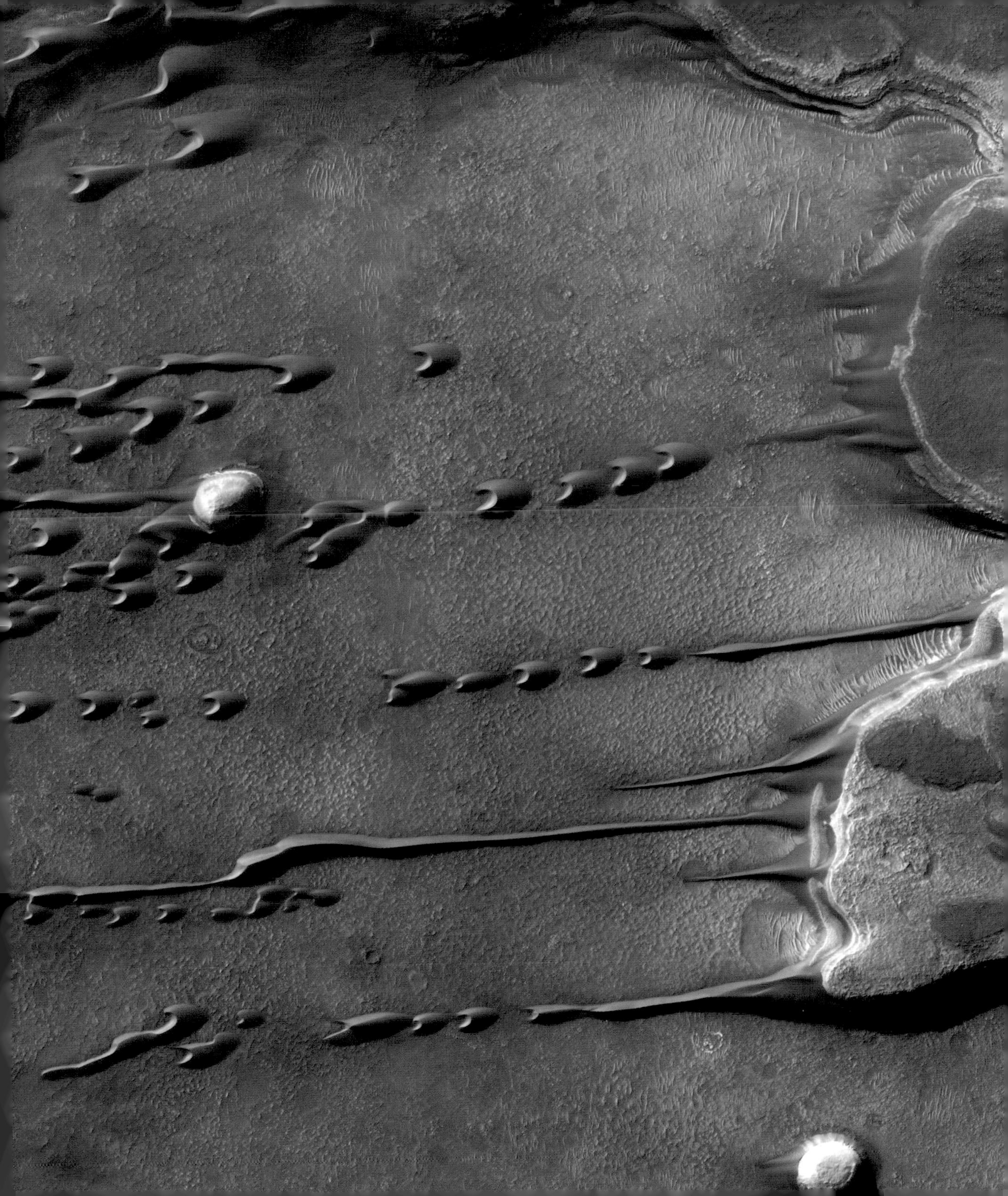

1835

Gauss and the Magnetic Monopole

Carl Friedrich Gauss (1777–1855), **Paul Dirac** (1902–1984)

"One would think that monopoles should exist, because of the prettiness of the mathematics," wrote British theoretical physicist Paul Dirac. Yet no physicist has ever found these strange particles. Gauss' Law for Magnetism, named after German mathematician Carl Gauss, is one of the fundamental equations of electromagnetism and a formal way of stating that isolated magnetic poles (e.g. a magnet with a north pole and no south pole) do not exist. On the other hand, in electrostatics, isolated charges exist, and this lack of symmetry between electric and magnetic fields is a puzzle to scientists. In the 1900s, scientists often wondered precisely why it is possible to isolate positive and negative electric charges but not north and south magnetic poles.

In 1931, Paul Dirac was one of the first scientists to theorize about the possible existence of a magnetic monopole, and a number of efforts through the years have been made to detect magnetic monopole particles. However, thus far, physicists have never discovered an isolated magnetic pole. Note that if you were to cut a traditional magnet (with a north and south pole) in half, the resulting pieces are two magnets—each with its own north pole and south pole.

Some theories that seek to unify the electroweak and strong interactions in particle physics predict the existence of magnetic monopoles. However, if monopoles existed, they would be very difficult to produce using particle accelerators because the monopole would have a huge mass and energy (about 10^{16} giga-electron volts).

Gauss was often extremely secretive about his work. According to mathematical historian Eric Temple Bell, had Gauss published or revealed all of his discoveries when he made them, mathematics would have been advanced by fifty years. After Gauss proved a theorem, he sometimes said that the insight did not come from "painful effort but, so to speak, by the grace of God."

SEE ALSO Olmec Compass (1000 B.C.), *De Magnete* (1600), Maxwell's Equations (1861), Stern-Gerlach Experiment (1922).

LEFT: *Gauss on German postage stamp (1955).* RIGHT: *Bar magnet, with north pole at one end and south pole at the other, along with iron filings showing the magnetic field pattern. Will physicists ever find a magnetic monopole particle?*

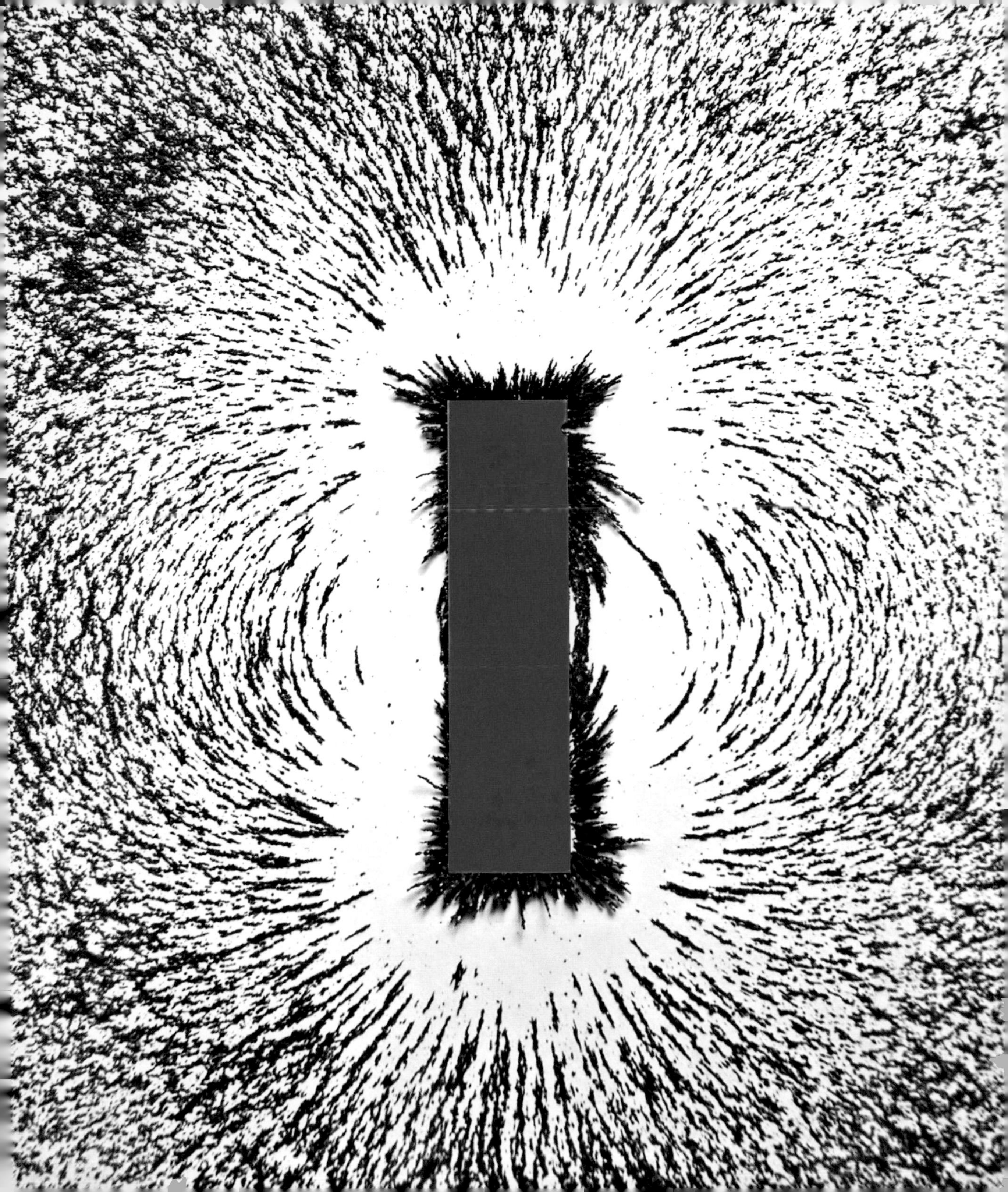

1838

Stellar Parallax

Freidrich Wilhelm Bessel (1784–1846)

Humanity's quest for determining the distance of stars from the Earth has had a long history. The Greek philosopher Aristotle and Polish astronomer Copernicus knew that if the Earth orbited our Sun, one would expect the stars to apparently shift back and forth each year. Unfortunately, Aristotle and Copernicus never observed the tiny *parallaxes* involved, and humans had to wait until the nineteenth century before parallaxes were actually discovered.

Stellar parallax refers to the apparent displacement of a star when viewed along two different lines of sight. Using simple geometry, this displacement angle can be used to determine the distance of the star to the observer. One way to calculate this distance is to determine the position of a star at a particular time of year. Half a year later, when the Earth has traveled halfway around the Sun, we again measure the star's position. A nearby star will have appeared to move against the more distant stars. Stellar parallax is similar to effects you observe when closing one of your eyes. Look at your hand with one eye, then the other. Your hand seems to move. The larger the parallax angle, the closer the object is to your eye.

In the 1830s, there was an impassioned competition between astronomers to be the first person to accurately determine interstellar distances. It was not until 1838 that the first stellar parallax was measured. Using a telescope, German astronomer Freidrich Wilhelm Bessel studied the star 61 Cygni in the constellation of Cygnus the Swan. Cygni displayed significant apparent motion, and Bessel's parallax calculations indicated that the star was 10.4 light-years (3.18 parsecs) away from the Earth. It is awe-inspiring that the early astronomers determined a way to compute vast interstellar distances without leaving the confines of their backyards.

Because the parallax angle is so small for stars, early astronomers could use this approach only for stars that are relatively close to the Earth. In modern times, astronomers have employed the European Hipparcos satellite to measure the distances of over 100,000 stars.

SEE ALSO Eratosthenes Measures the Earth (240 B.C.), Telescope (1608), Measuring the Solar System (1672), Black Drop Effect (1761).

Researchers have measured parallaxes based on observations from NASA's Spitzer Space Telescope and Earth-based telescopes to determine distances to objects that pass in front of stars in the Small Magellanic Cloud (upper left).

1839

Fuel Cell

William Robert Grove (1811–1896)

Some of you may recall the electrolysis of water performed in your high school chemistry class. An electrical current is applied between a pair of metal electrodes, which are immersed in the liquid, and hydrogen and oxygen gas are produced according to the chemical equation: electricity + $2H_2O$ (liquid) → $2H_2$ (gas) + O_2 (gas). (In practice, pure water is a poor conductor of electricity, and one may add dilute sulfuric acid in order to establish a significant current flow.) The energy required to separate the ions is provided by an electrical power supply.

In 1839, lawyer and scientist William Grove created early *fuel cells* (FCs) by using a reverse process to produce electricity from hydrogen and oxygen in a fuel tank. Many combinations of fuels are possible. In a hydrogen FC, chemical reactions remove electrons from the hydrogen atoms, which creates hydrogen protons. The electrons travel through attached wires to provide a useful electrical current. Oxygen then combines with the electrons (which return from the electrical circuit) and hydrogen ions in the FC to produce water as a "waste product." A hydrogen FC resembles a battery, but unlike a battery—which is eventually depleted and discarded or recharged—an FC can work indefinitely, as long as it has a supply of fuel in the form of oxygen from the air and hydrogen. A catalyst such as platinum is used to facilitate the reactions.

Some hope that FCs may one day be used more frequently in vehicles to replace the traditional combustion engine. However, obstacles to widespread use include cost, durability, temperature management, and hydrogen production and distribution. Nevertheless, FCs are very useful in backup systems and spacecraft; they were crucial in helping Americans travel to the moon. Benefits of FCs include zero carbon emissions and reduced dependence on oil.

Note that the hydrogen to power fuel cells is sometimes created by breaking down hydrocarbon-based fuels, which is counter to one desired goal of FCs: reducing greenhouse gases.

SEE ALSO Battery (1800), Greenhouse Effect (1824), Solar Cells (1954).

Photograph of a Direct Methanol Fuel Cell (DMFC), an electrochemical device that creates electricity using a methanol-water solution for fuel. The actual fuel cell is the layered cube shape toward the center of the image.

1840

Poiseuille's Law of Fluid Flow

Jean Louis Marie Poiseuille (1797–1869)

Medical procedures for enlarging an occluded blood vessel can be very helpful because a small increase in the radius of a vessel can cause a dramatic improvement in blood flow. Here's why. Poiseuille's Law, named after French physician Jean Poiseuille, provides a precise mathematical relationship between the flow rate of a fluid in a pipe and the pipe width, fluid viscosity, and pressure change in the pipe. In particular, the law states that $Q = [(\pi r^4)/(8\mu)] \times (\Delta P/L)$. Here, Q is the fluid flow rate in the pipe, r is the internal radius of the pipe, ΔP is the pressure difference between two ends of the pipe, L is the pipe length, and μ is the viscosity of the fluid. The law assumes that the fluid under study is exhibiting laminar (i.e. smooth, non-turbulent) steady flow.

This principle has practical applications in medical fields; in particular, it applies to the study of flow in blood vessels. Note that the r^4 term ensures that the radius of a tube plays a major role in determining the flow rate Q of the liquid. If all other parameters are the same, a doubling of the tube width leads to a sixteenfold increase in Q. Practically speaking, this means that we would need sixteen tubes to pass as much water as one tube twice their diameter. From a medical standpoint, Poiseuille's Law can be used to show the dangers of atherosclerosis: If the radius of a coronary artery decreases twofold, the blood flow through it will decrease 16 times. It also explains why it is so much easier to sip a drink from a wide straw as compared to a slightly thinner straw. For the same amount of sipping effort, if you were to suck on a straw that is twice as wide, you would obtain 16 times as much liquid per unit time of sucking. When an enlarged prostate reduces the urethra's radius, we can blame Poiseuille's Law for why even a small constriction can have dramatic effects on the flow rate of urine.

SEE ALSO Siphon (250 B.C.), Bernoulli's Law of Fluid Dynamics (1738), Stokes' Law of Viscosity (1851).

LEFT: *Poiseuille's Law explains why it so much more difficult to sip a drink using a narrow straw than a wider one.* RIGHT: *Poiseuille's Law can be used to show the dangers of atherosclerosis; for example, if the radius of an artery decreases twofold, the blood flow through it will decrease roughly sixteenfold.*

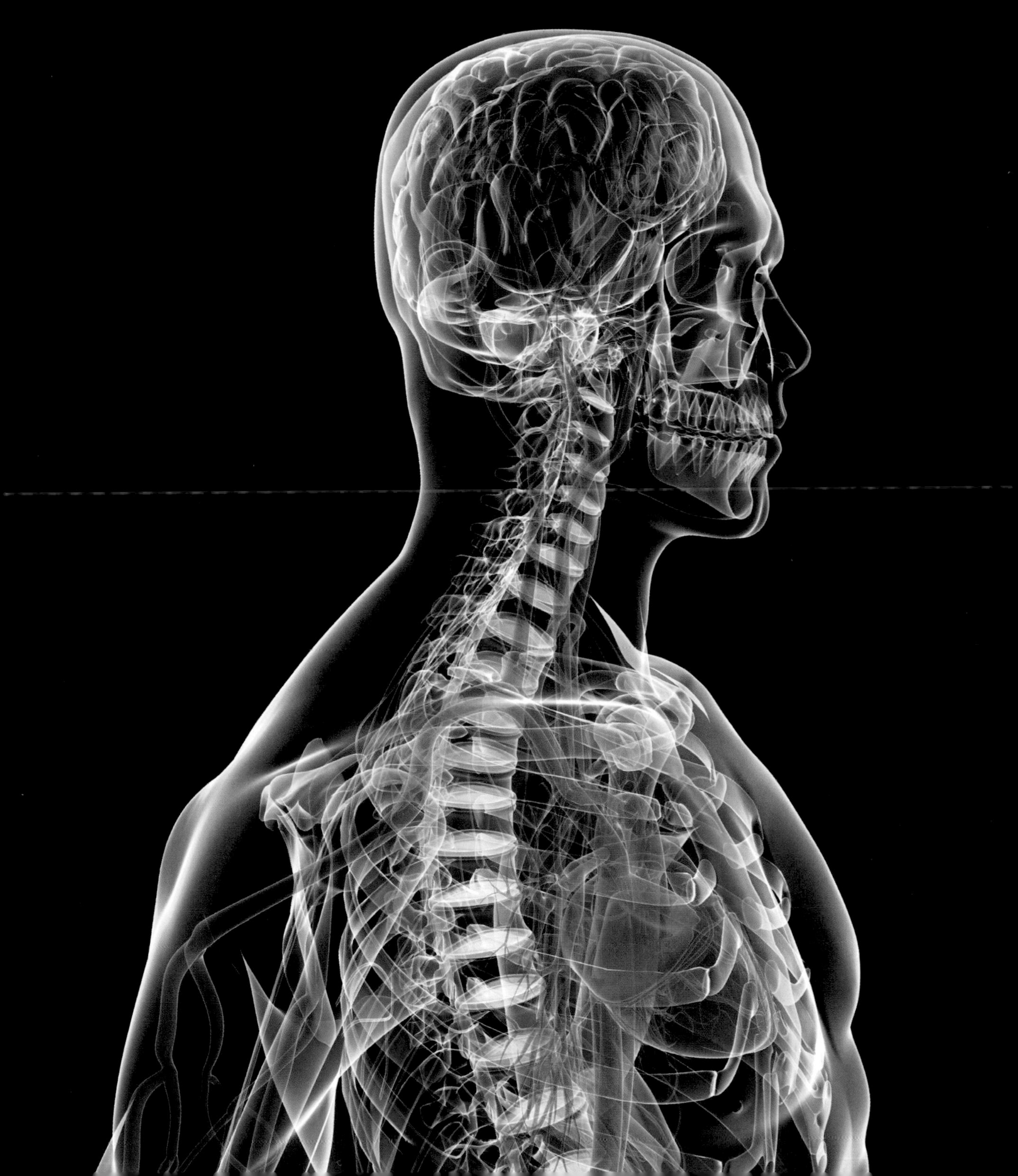

1840

Joule's Law of Electric Heating

James Prescott Joule (1818–1889)

Surgeons often rely on Joule's Law of Electric Heating (named after British physicist James Joule), which states that the amount of heat H generated by a steady electric current flowing in a conductor may be calculated using $H = K \cdot R \cdot I^2 \cdot t$. Here, R is the resistance of the conductor, I is the constant current flowing through the conductor, and t is the duration of current flow.

When electrons travel through a conductor with some resistance R, the electric kinetic energy that the electrons lose is transferred to the resistor as heat. A classical explanation of this heat production involves the lattice of atoms in a conductor. The collisions of the electrons with the lattice cause the amplitude of thermal vibration of the lattice to increase, thereby raising the temperature of the conductor. This process is known as Joule heating.

Joule's Law and Joule heating play a role in modern electrosurgical techniques in which the heat at an electrical probe is determined by Joule's Law. In such devices, current flows from an "active electrode" through the biological tissue to a neutral electrode. The ohmic resistance of the tissue is determined by the resistance of the area in contact with the active electrode (e.g. blood, muscle, or fatty tissue) and the resistance in the total path between the active and neutral electrode. In electrosurgery, the duration (t in Joule's Law) is often controlled by a finger switch or foot pedal. The precise shape of the active electrode can be used to concentrate the heat so that it can be used for cutting (e.g., with a point-shaped electrode), or coagulation, which would result from diffuse heat that is produced by an electrode with a large surface area.

Today, Joule is also remembered for helping to establish that mechanical, electrical, and heat energy are all related and can be converted to one another. Thus, he provided experimental validations for many elements of the law of **Conservation of Energy**, also known as the *First Law of Thermodynamics*.

SEE ALSO Conservation of Energy (1834), Fourier's Law of Heat Conduction (1822), Ohm's Law of Electricity (1827), Incandescent Light Bulb (1878).

LEFT: *Photo of James Joule.* RIGHT: *Joule's Law and Joule heating play a role in modern liquid immersion heaters in which the heat is determined by Joule's Law.*

1841

Anniversary Clock

The first clocks had no minute hands, which gained importance only with the evolution of modern industrial societies. During the Industrial Revolution, trains began to run on schedules, factory work started and stopped at appointed times, and the tempo of life became more precise.

My favorite clocks are called *torsion pendulum clocks*, *400-day clocks*, or *anniversary clocks*, because many versions only had to be wound once a year. I became intrigued with these clocks after reading about eccentric billionaire Howard Hughes, whose favorite room contained, according to biographer Richard Hack, "a world globe on a mahogany stand [and] a large fireplace on whose mantelpiece sat a French bronze 400-day clock that 'was not to be over-wound for any reason'."

The anniversary clock makes use of a weighted disk suspended on a thin wire or ribbon that functions as a torsion spring. The disk rotates back and forth around the vertical axis of the wire—a motion that replaces the swinging action of the traditional pendulum. Ordinary pendulum clocks date back at least to 1656, when Christiaan Huygens, inspired by drawings by Galileo, commissioned their construction. These clocks were more accurate than earlier clocks due to the nearly isochronous motion of the pendulum—the period of swing stays relatively constant, especially if the swings are small.

In the anniversary clock, the rotating disk winds and unwinds the spring slowly and efficiently, allowing the spring to continue powering the clock's gears for long periods after a single initial winding. Early versions of the clock were not very accurate, partly because the spring force was temperature dependent. However, later versions made use of a spring that compensated for changes in temperature. The anniversary clock was patented by American inventor Aaron Crane in 1841. German clockmaker Anton Harder independently invented the clock around 1880. Anniversary clocks became popular wedding gifts at the end of World War II, when American soldiers brought them back to the United States.

SEE ALSO Hourglass (1338), Foucault's Pendulum (1851), Atomic Clocks (1955).

Many versions of the anniversary clock only had to be wound about once a year. The anniversary clock makes use of a weighted disk suspended on a thin wire or ribbon that functions as a torsion spring.

1841

Fiber Optics

Jean-Daniel Colladon (1802–1893), **Charles Kuen Kao** (b. 1933), **George Alfred Hockham** (b. 1938)

The science of fiber optics has a long history, including such wonderful demonstrations as Swiss physicist Jean-Daniel Colladon's light fountains in 1841, in which light traveled within an arcing stream of water from a tank. Modern fiber optics—discovered and independently refined many times through the 1900s—use flexible glass or plastic fibers to transmit light. In 1957, researchers patented the fiberoptic endoscope to allow physicians to view the upper part of the gastrointestinal tract. In 1966, electrical engineers Charles K. Kao and George A. Hockham suggested using fibers to transmit signals, in the form of light pulses, for telecommunications.

Through a process called *total internal reflection* (see entry on **Snell's Law**), light is trapped within the fiber as a result of the higher refractive index of the core material of the fiber relative to that of the thin cladding that surrounds it. Once light enters the fiber's core, it continually reflects off the core walls. The signal propagation can suffer some loss of intensity over very long distances, and thus it may be necessary to boost the light signals using optical regenerators. Today, optical fibers have many advantages over traditional copper wires for communications. Signals travel along relatively inexpensive and lightweight fibers with less attenuation, and they are not affected by electromagnetic interference. Also, fiber optics can be used for illumination or transferring images, thus allowing illumination or viewing of objects that are in tight, difficult-to-reach places.

In optical-fiber communications, each fiber can transmit many independent channels of information via different wavelengths of light. The signal may start as an electronic stream of bits that modulates lights from a tiny source, such as a light-emitting diode or laser diode. The resultant pulses of infrared light are then transmitted. In 1991, technologists developed photonic-crystal fibers that guide light by means of diffraction effects from a periodic structure such as an array of cylindrical holes that run through the fiber.

SEE ALSO Snell's Law of Refraction (1621), Brewster's Optics (1815).

Optical fibers carry light along their lengths. Through a process called total internal reflection, *light is trapped within the fiber until it reaches the end of the fiber.*

1842

Doppler Effect

Christian Andreas Doppler (1803–1853), **Christophorus Henricus Diedericus Buys Ballot** (1817–1890)

"When a police officer zaps a car with a radar gun or a laser beam," writes journalist Charles Seife, "he is really measuring how much the motion of the car is compressing the reflected radiation [via the Doppler Effect]. By measuring that squashing, he can figure out how fast the car is moving, and give the driver a $250 ticket. Isn't science wonderful?"

The Doppler Effect, named after Austrian physicist Christian Doppler, refers to the change in frequency of a wave for an observer as the source of the wave moves relative to the observer. For example, if a car is moving while its horn is sounding, the frequency of the sound you hear is higher (compared to the actual emitted frequency) as the car approaches you, is identical at the instant that it passes you, and is lower as it moves away. Although we often think of the Doppler Effect with respect to sound, it applies to all waves, including light.

In 1845, Dutch meteorologist and physical chemist C. H. D. Buys Ballot performed one of the first experiments verifying Doppler's idea for sound waves. In the experiment, a train carried trumpeters who played a constant note while other musicians listened on the side of the track. Using observers with "perfect pitch," Buys Ballot thus proved the existence of the Doppler Effect, which he then reduced to a formula.

For many galaxies, their velocity away from us can be estimated from the red shift of a galaxy, which is an apparent increase in the wavelength (or decrease in frequency) of electromagnetic radiation received by an observer on the Earth compared to that emitted by the source. Such red shifts occur because galaxies are moving away from our own galaxy at high speeds as space expands. The change in the wavelength of light that results from the relative motion of the light source and the receiver is another example of the Doppler Effect.

SEE ALSO Olbers' Paradox (1823), Hubble's Law of Cosmic Expansion (1929), Quasars (1963), Fastest Tornado Speed (1999).

LEFT: *Portrait of Christian Doppler, in the frontispiece to a reprint of his* Über das farbige Licht der Doppelsterne *("Concerning the Colored Light of Double Stars")*. RIGHT: *Imagine a sound or light source emitting a set of spherical waves. As the source moves right to left, an observer at left sees the waves as compressed. An approaching source is seen as blue-shifted (the wavelengths made shorter).*

1843

Conservation of Energy

James Prescott Joule (1818–1889)

"The law of the conservation of energy offers . . . something to clutch at during those late-night moments of quiet terror, when you think of death and oblivion," writes science-journalist Natalie Angier. "Your private sum of E, the energy in your atoms and the bonds between them, will not be annihilated. . . . The mass and energy of which you're built will change form and location, but they will be here, in this loop of life and light, the permanent party that began with a Bang."

Classically speaking, the principle of the conservation of energy states that the energy of interacting bodies may change forms but remains constant in a closed system. Energy takes many forms, including kinetic energy (energy of motion), potential energy (stored energy), chemical energy, and energy in the form of heat. Consider an archer who deforms, or strains, a bow. This potential energy of the bow is converted into kinetic energy of the arrow when the bow is released. The total energy of the bow and arrow, in principle, is the same before and after release. Similarly, chemical energy stored in a **Battery** can be converted into the kinetic energy of a turning motor. The gravitational potential energy of a falling ball is converted into kinetic energy as it falls. One key moment in the history of the conservation of energy was physicist James Joule's 1843 discovery of how gravitational energy (lost by a falling weight that causes a water paddle to rotate) was equal to the thermal energy gained by water due to friction with the paddle. The First Law of Thermodynamics is often stated as: The increase in internal energy of a system due to heating is equal to the amount of energy added by heating, minus the work performed by the system on its surroundings.

Note that in our bow and arrow example, when the arrow hits the target, the kinetic energy is converted to heat. The **Second Law of Thermodynamics** limits the ways in which heat energy can be converted into work.

SEE ALSO Crossbow (341 B.C.), Perpetual Motion Machines (1150), Conservation of Momentum (1644), Joule's Law of Electric Heating (1840), Second Law of Thermodynamics (1850), Third Law of Thermodynamics (1905), $E = mc^2$ (1905).

This potential energy of the strained bow is converted to kinetic energy of the arrow when the bow is released. When the arrow hits the target, the kinetic energy is converted to heat.

1844

I-Beams

Richard Turner (c. 1798–1881), **Decimus Burton** (1800–1881)

Have you ever wondered why so many steel girders used in construction have a cross-section shaped like the letter I? It turns out that this kind of beam is very efficient for resisting bending in response to a load applied perpendicular to the axis of the beam. For example, imagine a long I-beam supported on both ends, with a heavy elephant balancing in the middle. The upper layers of the beam will be compressed, and the bottom layers will be slightly lengthened, or stretched, by the tension force. Steel is expensive and heavy, so builders try to minimize the material used while preserving the structural strength. The I-beam is efficient and economical because more steel is placed in the top and bottom flanges, where its presence is most effective in resisting the bending. Steel I-beams may be formed by the rolling or extruding of steel, or by the creation of *plate girders*, which are formed by welding plates together. Note that other shapes are more effective than the I-beam if forces are applied side to side; the most efficient and economical shape for resisting bending in any direction is a hollow cylindrical shape.

Historic preservationist Charles Peterson writes of the importance of the I-beam: "The wrought-iron I-beam, perfected in the middle of the nineteenth century, was one of the great structural inventions of all time. The shape, first rolled in wrought-iron, was soon to be rolled in steel. When the Bessemer process made steel cheap, the I-beam came into use universally. It is the stuff of which skyscrapers and great bridges are made."

Among the first-known I-beams introduced into buildings are the ones used in the Kew Gardens Palm House in London, built by Richard Turner and Decimus Burton between 1844 and 1848. In 1853, William Borrow of the Trenton Iron Company (TIC), in New Jersey approximated I-beams by bolting two component pieces back to back. In 1855, Peter Cooper, owner of TIC, rolled out I-beams from a single piece. This came to be known as the *Cooper beam*.

SEE ALSO Truss (2500 B.C.), Arch (1850 B.C.), Tensegrity (1948).

This massive I-beam was once part of the second sub-street level of the World Trade Center. It is now part of a 9/11 memorial at the California State Fair Memorial Plaza. These heavy beams were transported via railway from New York to Sacramento.

1845

Kirchhoff's Circuit Laws

Gustav Robert Kirchhoff (1824–1887)

When Gustav Kirchhoff's wife Clara died, the brilliant physicist was left alone to raise his four children. This task would have been difficult for any man, but it was made especially challenging by a foot injury that forced him to spend his life on crutches or in a wheelchair. Before the death of his wife, Kirchhoff became well known for his electrical circuit laws that focused on the relationships between currents at a circuit junction and the voltages around a circuit loop. *Kirchhoff's Current Law* is a restatement of the principle of conservation of electrical charge in a system. In particular, at any point in an electrical circuit, the sum of currents flowing towards that point is equal to the sum of currents flowing away from that point. This law is often applied to the intersection of several wires to form a junction—i.e., junctions shaped like a + or a T—in which current travels toward the junction for some wires and away in other wires.

Kirchhoff's Voltage Law is a restatement of the **Conversation of Energy** law for a system: The sums of the electrical potential differences around a circuit must be zero. Imagine that we have a circuit with junctions. If we start at any junction and follow a succession of circuit elements that form a closed path back to the starting point, the sum of the changes in potential encountered in the loop is equal to zero. (Elements include conductors, resistors, and batteries.) As an example, voltage rises may occur when we follow the circuit across a **Battery** (traversing from the – to + ends of a typical battery symbol in a circuit drawing). As we continue to trace the circuit in the same direction away from the battery, voltage drops may occur, for example, as a result of the presence of resistors in a circuit.

SEE ALSO Ohm's Law of Electricity (1827), Joule's Law of Electric Heating (1840), Conservation of Energy (1843), Integrated Circuit (1958).

LEFT: *Gustav Kirchhoff.* RIGHT: *Kirchhoff's electrical circuit laws have been used by engineers for many decades to understand the relationships between currents and voltages in circuits, as represented by circuit diagrams such as in this noise reduction circuit diagram (U.S. Patent 3,818,362, 1974).*

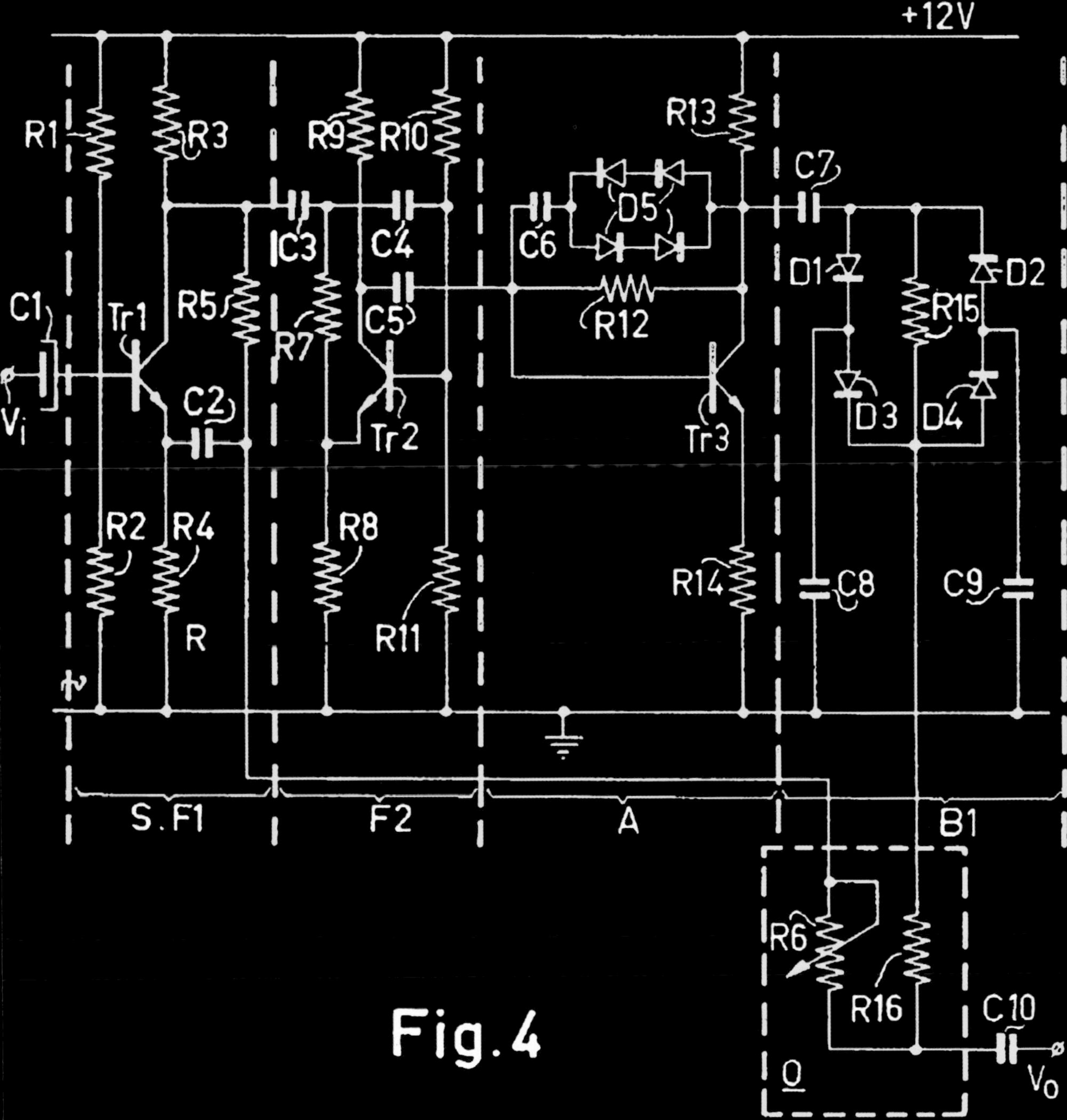
+12V
R1
R3
R9
R10
R13
C7
C3
C4
D5
C6
D1
D2
C1
Tr1
R5
R7
C5
R12
R15
C2
Tr2
Tr3
D3
D4
Vi
R2
R4
R8
R14
C8
C9
R
R11
S.F1
F2
A
B1
R6
R16
C10
O
Vo
Fig. 4

1846

Discovery of Neptune

John Couch Adams (1819–1892), **Urbain Jean Joseph Le Verrier** (1811–1877), **Johann Gottfried Galle** (1812–1910)

"The problem of tracking the planets with the highest precision is an immensely complex one," writes astronomer James Kaler. "For two bodies, we have a beautifully simple set of rules. For just three bodies mutually pulling on one another, it is mathematically proven that no such rules exist. . . . The triumph of this mathematical science [called perturbation theory], and indeed of Newtonian mechanics itself, was in the discovery of Neptune."

Neptune is the only planet in our Solar System whose existence and location was mathematically predicted before the planet was actually observed. Astronomers had noted that Uranus, discovered in 1781, exhibited certain irregularities in its orbit around the Sun. Astronomers wondered if perhaps this meant that Newton's Laws did not apply far out in the Solar System, or that perhaps a large unseen object was perturbing the orbit of Uranus. French astronomer Urbain Le Verrier and British astronomer John Couch Adams both performed calculations to locate a possible new planet. In 1846, Le Verrier told German astronomer Johann Galle where to point Galle's telescope based on Le Verrier's calculations, and in about half an hour Galle found Neptune within one degree of the predicted position—a dramatic confirmation of Newton's Law of Universal Gravitation. Galle wrote to Le Verrier on September 25, "Monsieur, the planet of which you indicated the position *really exists*." Le Verrier replied, "I thank you for the alacrity with which you applied my instructions. We are thereby, thanks to you, definitely in *possession of a new world*."

British scientists argued that Adams had also discovered Neptune at the same time, and a dispute arose as to whom was the true discoverer of the planet. Interestingly, for centuries, a number of astronomers before Adams and Le Verrier had observed Neptune but simply thought it to be a star rather than a planet.

Neptune is invisible to the naked eye. It orbits the Sun once every 164.7 years and has the fastest winds of any planet in our solar system.

SEE ALSO Telescope (1608), Measuring the Solar System (1672), Newton's Laws of Motion and Gravitation (1687), Bode's Law of Planetary Distances (1766), Hubble Telescope (1990).

Neptune, the eighth planet from the Sun, and its moon Proteus. Neptune has 13 known moons, and its equatorial radius is nearly four times that of the Earth's.

1850

Second Law of Thermodynamics

Rudolf Clausius (1822–1888), **Ludwig Boltzmann** (1844–1906)

Whenever I see my sand castles on the beach fall apart, I think of the *Second Law of Thermodynamics* (SLT). The SLT, in one of its early formulations, states that the total entropy, or disorder, of an isolated system tends to increase as it approaches a maximum value. For a closed thermodynamic system, entropy can be thought of as a measure of the amount of thermal energy unavailable to do work. German physicist Rudolf Clausius stated the First and Second Laws of Thermodynamics in the following form: The energy of the universe is constant, and the entropy of the universe tends to a maximum.

Thermodynamics is the study of heat, and more generally the study of transformations of energy. The SLT implies that all energy in the universe tends to evolve toward a state of uniform distribution. We also indirectly invoke the SLT when we consider that a house, body, or car—without maintenance—deteriorates over time. Or, as novelist William Somerset Maugham wrote, "It's no use crying over spilt milk, because all of the forces of the universe were bent on spilling it."

Early in his career, Clausius stated, "Heat does not transfer spontaneously from a cool body to a hotter body." Austrian physicist Ludwig Boltzmann expanded upon the definition for entropy and the SLT when he interpreted entropy as a measure of the disorder of a system due to the thermal motion of molecules.

From another perspective, the SLT says that two adjacent systems in contact with each other tend to equalize their temperatures, pressures, and densities. For example, when a hot piece of metal is lowered into a tank of cool water, the metal cools and the water warms until each is at the same temperature. An isolated system that is finally at equilibrium can do no useful work without energy applied from outside the system, which helps explain how the SLT prevents us from building many classes of **Perpetual Motion Machines**.

SEE ALSO Perpetual Motion Machines (1150), Boltzmann's Entropy Equation (1875), Maxwell's Demon (1867), Carnot Engine (1824), Conservation of Energy (1843), Third Law of Thermodynamics (1905).

LEFT: *Rudolph Clausius.* RIGHT: *Microbes build their "improbable structures" from ambient disordered materials, but they do this at the expense of increasing the entropy around them. The overall entropy of closed systems increases, but the entropy of individual components of a closed system may decrease.*

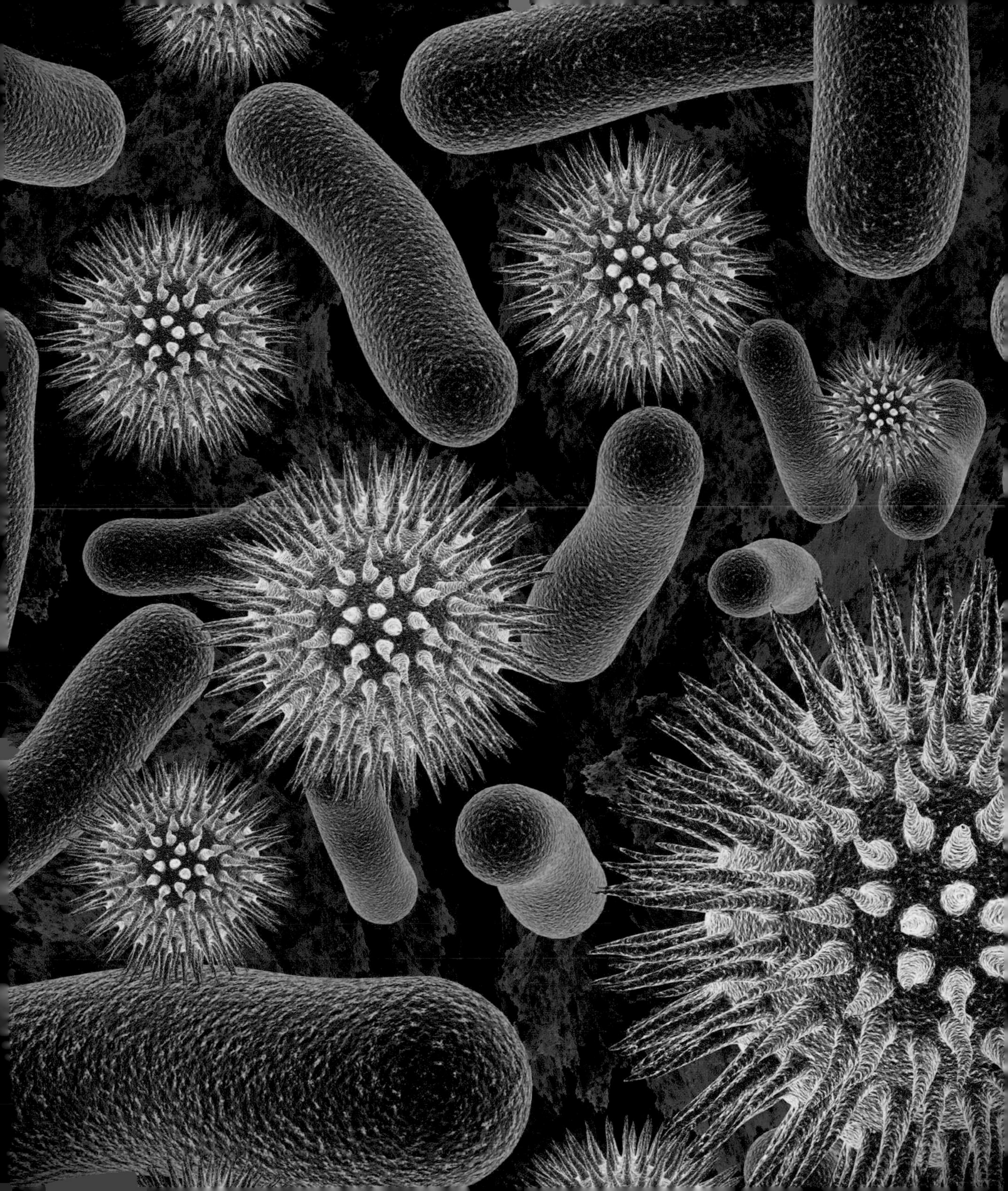

1850

Ice Slipperiness

Michael Faraday (1791–1867), **Gabor A. Somorjai** *(b. 1935)*

"Black ice" generally refers to clear water that has frozen on dark roadways and which poses a particular danger to motorists who are often unable to see it. Interestingly, black ice sometimes forms without the presence of rain, snow, or sleet, because condensation from dew, mist, and fog on roadways can freeze. The frozen water of black ice is transparent because relatively few air bubbles are trapped in the ice.

Over the centuries, scientists have wondered why black ice, or any form of ice, is slippery. On June 7, 1850, English scientist Michael Faraday suggested to the Royal Institution that ice has a hidden layer of liquid water on its surface, which makes the surface slippery. To test his hypotheses, he simply pressed two ice cubes together and they stuck. He then argued that these extremely thin liquid layers froze when they were no longer at the surface.

Why are ice skaters able to skate on ice? For many years, the textbook answer was that the skate blades exerted a pressure that lowers the melting temperature of the ice, thus causing a thin layer of water to form. Although this answer is no longer believed to be valid, the friction between the skate blade and the ice may generate heat and cause some liquid water to temporarily form. Another recent explanation suggests that the water molecules on the surface vibrate more because no water molecules exist on top of them. This creates a very thin layer of liquid water on the surface, even if the temperature is below the freezing point of water. In 1996, chemist Gabor Somorjai used low-energy electron diffraction methods to prove that a thin layer of liquid water exists at the surface of ice. Faraday's 1850 theory seems to be vindicated. Today, scientists are not quite sure whether this intrinsic layer of liquid water or the liquid water caused by friction plays a greater role in the slipperiness of ice.

SEE ALSO Amontons' Friction (1669), Stokes' Law of Viscosity (1851), Superfluids (1937).

LEFT: *Molecular structure of ice crystal.* RIGHT: *Why are ice skaters able to skate on ice? Due to molecular vibrations, a very thin layer of liquid water exists on the surface, even if the temperature is below the freezing point of water.*

1851

Foucault's Pendulum

Jean Bernard Léon Foucault (1819–1868)

"The motion of the pendulum is not due to any transcendental or mysterious forces from without," writes author Harold T. Davis, "but is due simply to the fact that the earth turns under the swinging weight. And yet perhaps the explanation is not so simple since the experiment was performed for the first time in 1851 by Jean Foucault. Simple facts have not usually remained undiscovered for so many years. . . . The principle for which Bruno died and for which Galileo suffered was vindicated. The earth moved!"

In 1851, French physicist Léon Foucault (pronounced "Foo-koh") demonstrated his experiment in the Panthéon, a neoclassical domed building in Paris. An iron ball the size of a pumpkin swung from 220 feet (67 meters) of steel wire. As the pendulum swung, its direction of motions gradually changed, rotating clockwise at a rate of 11 degrees per hour, thus proving that the Earth rotated. To visualize this proof, let us imagine transporting the Panthéon to the North Pole. Once the pendulum is swinging, its plane of oscillation is independent of the Earth's movement, and the Earth simply rotates beneath it. Thus, at the North Pole, the pendulum's plane of oscillation rotates clockwise through 360 degrees every 24 hours. The rate at which the pendulum's plane of oscillation rotates depends on its latitude: At the equator, it does not change at all. At Paris, the pendulum makes a full circle in roughly 32.7 hours.

Of course, by 1851, scientists knew that the Earth rotated, but Foucault's pendulum provided dramatic and dynamic evidence for such rotation in an easily interpreted fashion. Foucault described his pendulum: "The phenomenon develops calmly, but it is inevitable, unstoppable. . . . Any person, brought into the presence of this fact, stops for a few moments and remains pensive and silent; and then generally leaves, carrying with him forever a sharper, keener sense of our incessant motion through space."

Early in his life, Foucault studied medicine, a field that he left for physics when he discovered his fear of blood.

SEE ALSO Tautochrone Ramp (1673), Anniversary Clock (1841), Buys-Ballot's Weather Law (1857), Newton's Cradle (1967).

Foucault's pendulum in the Panthéon, Paris.

AVX ORATEVRS
ET AVX PVBLICISTES
DE LA RESTAVRATION
15
14
13
12
11
10
9
8
7

1851

Stokes' Law of Viscosity

George Gabriel Stokes (1819–1903)

Whenever I think of Stokes' Law, I think of shampoo. Consider a solid sphere of radius r moving with a velocity v through a fluid of viscosity μ. Irish physicist George Stokes determined that the frictional force F that resists the motion of the sphere can be determined with the equation $F = 6\pi r\mu v$. Note that this drag force F is directly proportional to the sphere radius. This was not intuitively obvious because some researchers supposed that the frictional force would be proportional to the cross-section area, which erroneously suggests an r^2 dependence.

Consider a scenario in which a particle in a fluid is subject to the forces of gravity. For example, some older readers may recall the popular Prell shampoo TV commercial that shows a pearl dropping through a container of the green shampoo. The pearl starts off with zero speed and then initially accelerates, but the motion of the pearl quickly generates a frictional resistance counter to the acceleration. Thus, the pearl rapidly reaches a condition of zero acceleration (**Terminal Velocity**) when the force of gravity is balanced by the force of friction.

Stokes' Law is considered in industry when studying sedimentation that occurs during the separation of a suspension of solid particles in a liquid. In these applications, scientists are often interested in the resistance exerted by the liquid to the motion of the descending particle. For example, a sedimentation process is sometimes used in the food industry for separating dirt and debris from useful materials, separating crystals from the liquid in which they are suspended, or separating dust from air streams. Researchers use the law to study aerosol particles in order to optimize drug delivery to the lungs.

In the late 1990s, Stokes' Law was used to provide a possible explanation of how micrometer-sized uranium particles can remain airborne for many hours and traverse great distances—and thus possibly have contaminated Persian Gulf War soldiers. Cannon rounds often contained depleted-uranium penetrators, and this uranium becomes aerosolized when the rounds impact hard targets, such as tanks.

SEE ALSO Archimedes' Principle of Buoyancy (250 B.C.), Amontons' Friction (1669), Ice Slipperiness (1850), Poiseuille's Law of Fluid Flow (1840), Superfluids (1937), Silly Putty (1943).

LEFT: *George Stokes.* RIGHT: *Informally, viscosity is related to the fluid's "thickness" and resistance to flow. Honey, for example, has a higher viscosity than water. Viscosity varies with temperature, and honey flows more readily when heated.*

1852

Gyroscope

Jean Bernard Léon Foucault (1819–1868), **Johann Gottlieb Friedrich von Bohnenberger** (1765–1831)

According to the 1897 *Every Boy's Book of Sport and Pastime*, "The gyroscope has been called the paradox of mechanics: when the disk is not spinning, the apparatus is an inert mass; but with the disc in rapid rotation, it seems to set gravity at defiance, or when held in the hand, a peculiar sensation is felt by the tendency of the thing to move some other way to that in which you wish to turn it, as if it were something alive."

In 1852, the term *gyroscope* was first used by Léon Foucault, the French physicist who conducted many experiments with the device and who is sometimes credited with its invention. In fact, German mathematician Johann Bohnenberger invented the device using a rotating sphere. A mechanical gyroscope is traditionally in the form of a heavy spinning disc suspended within supporting rings called gimbals. When the disc spins, the gyroscope exhibits amazing stability and maintains the direction of the axis of rotation thanks to the principle of conservation of angular momentum. (The direction of the angular momentum vector of a spinning object is parallel to the axis of spin.) As an example, imagine that the gyroscope is pointed toward a particular direction and set spinning within a gimbal mount. The gimbals will reorient, but the axis of the wheel maintains the same position in space, no matter how the frame is moved. Because of this property, gyroscopes are sometime used for navigational purposes when magnetic compasses would be ineffective (in the **Hubble Telescope**, for example) or when they would be insufficiently precise, as in intercontinental ballistic missiles. Airplanes have several gyroscopes associated with navigational systems. The gyroscope's resistance to external motion also makes it useful aboard spacecraft to help them maintain a desired direction. This tendency to continue to point in a particular direction is also found in spinning tops, the wheels of bicycles, and even the rotation of the Earth.

SEE ALSO Boomerang (20,000 B.C.), Conservation of Momentum (1644), Hubble Telescope (1990).

Gyroscope invented by Léon Foucault and built by Dumoulin-Froment, 1852. Photo taken at National Conservatory of Arts and Crafts museum, Paris.

1852

Stokes' Fluorescence

George Gabriel Stokes (1819–1903)

As a child, I collected green fluorescent minerals that reminded me of the Land of Oz. Fluorescence usually refers to the glow of an object caused by visible light that is emitted when the object is stimulated via electromagnetic radiation. In 1852, physicist George Stokes observed phenomena that behaved according to *Stokes' Law of Fluorescence*, which states that the wavelength of emitted fluorescent light is always greater than the wavelength of the exciting radiation. Stokes published his finding in his 1852 treatise "On the Change of Refrangibility of Light." Today, we sometimes refer to the reemission of longer wavelength (lower frequency) photons by an atom that has absorbed photons of shorter wavelengths (higher frequency) as *Stokes' fluorescence*. The precise details of the process depend on the characteristics of a particular atom involved. Light is generally absorbed by atoms in about 10^{-15} seconds, and this absorption causes electrons to become excited and jump to a higher energy state. The electrons remain in the excited state for about 10^{-8} seconds, and then the electron may emit energy as it returns to the ground state. The phrase *Stokes' shift* usually refers to the difference in wavelength or frequency between absorbed and emitted quanta.

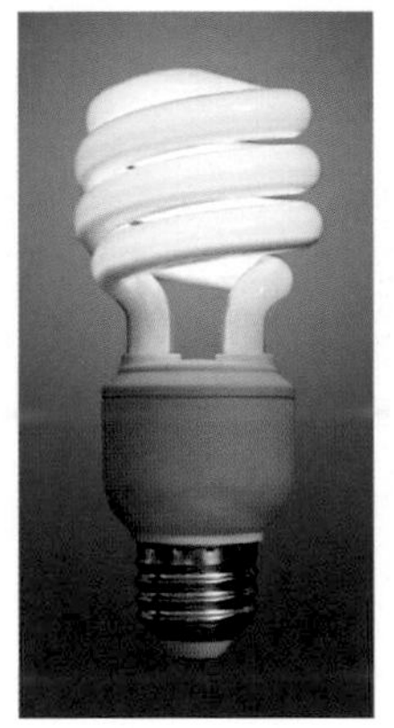

Stokes coined the term *fluorescence* after fluorite, a strongly fluorescent mineral. He was the first to adequately explain the phenomenon in which fluorescence can be induced in some materials through stimulation with ultraviolet (UV) light. Today, we know that these kinds of materials can be made to fluoresce by stimulation with numerous forms of electromagnetic radiation, including visible light, infrared radiation, X-rays, and radio waves.

Applications of fluorescence are many and varied. An electrical discharge in a fluorescent light causes the mercury atoms to emit ultraviolet light, which is then absorbed by a fluorescent material coating the tube that reemits visible light. In biology, fluorescent dyes are used as labels for tracking molecules. *Phosphorescent* materials do not re-emit the absorbed radiation as quickly as fluorescent materials.

SEE ALSO St. Elmo's Fire (78), Black Light (1903), Neon Signs (1923), Jacob's Ladder (1931), Atomic Clocks (1955).

LEFT: *Compact fluorescent light.* RIGHT: *Collection of various fluorescent minerals under UV-A, UV-B, and UV-C light. Photo has been rotated to better fit the space.*

1857

Buys-Ballot's Weather Law

Christophorus Henricus Diedericus Buys Ballot (1817–1890)

You can impress your friends, as I do, by going outside into windy weather with the apparently mystical ability of being able to point toward the direction of lower pressure. Buys-Ballot's Law, named after Dutch meteorologist Christoph Buys Ballot, asserts that in the Northern Hemisphere, if a person stands with his back to the wind, the low pressure area will be to his left. This means that wind travels counterclockwise around low pressure zones in the Northern Hemisphere. (In the Southern Hemisphere, wind travels clockwise.) The law also states that the wind and the pressure gradient are at right angles if measured sufficiently far above the surface of the Earth in order to avoid frictional effects between the air and the Earth's surface.

The weather patterns of the Earth are affected by several planetary features such as the Earth's roughly spherical shape and the Coriolis Effect, which is the tendency for any moving body on or above the surface of the Earth, such as an ocean current, to drift sideways from its course due to the rotation of the Earth. Air that is closer to the equator is generally traveling faster than air farther away because equatorial air is farther from the Earth's axis of rotation. To help visualize this, consider that air farther from the axis must travel faster in a day than air at higher latitudes, which are closer to the axis of the Earth. Thus, if a low pressure system in the north exists, it will draw air from the south that can move faster than the ground below it because the more northerly part of the Earth's surface has a slower eastward motion than the southerly surface. This means that the air from the south will move east as a result of its higher speed. The net effect of air movement from north and south is a counterclockwise swirl around a low pressure area in the Northern Hemisphere.

SEE ALSO Barometer (1643), Boyle's Gas Law (1662), Bernoulli's Law of Fluid Dynamics (1738), Baseball Curveball (1870), Fastest Tornado Speed (1999).

LEFT: *Christophorus Buys Ballot.* RIGHT: *Hurricane Katrina, August 28, 2005. Buys-Ballot's Law can be used on the ground by people trying to determine the approximate location of the center and direction of travel of a hurricane.*

1859

Kinetic Theory

James Clerk Maxwell (1831–1879), **Ludwig Eduard Boltzmann** (1844–1906)

Imagine a thin plastic bag filled with buzzing bees, all bouncing randomly against one another and the surface of the bag. As the bees bounce around with greater velocity, their hard bodies impact the wall with greater force, causing it to expand. The bees are a metaphor for atoms or molecules in a gas. The kinetic theory of gases attempts to explain the macroscopic properties of gases—such as pressure, volume, and temperature—in terms of the constant movements of such particles.

According to kinetic theory, temperature depends on the speed of the particles in a container, and pressure results from the collisions of the particles with the walls of the container. The simplest version of kinetic theory is most accurate when certain assumptions are fulfilled. For example, the gas should be composed of a large number of small, identical particles moving in random directions. The particles should experience elastic collisions with themselves and the container walls but have no other kinds of forces among them. Also, the average separation between particles should be large.

Around 1859, physicist James Clerk Maxwell developed a statistical treatment to express the range of velocities of gas particles in a container as a function of temperature. For example, molecules in a gas will increase speed as the temperature rises. Maxwell also considered how the viscosity and diffusion of a gas depend on the characteristics of the molecules' motion. Physicist Ludwig Boltzmann generalized Maxwell's theory in 1868, resulting in the Maxwell-Boltzmann distribution law, which describes a probability distribution of particle speeds as a function of temperature. Interestingly, scientists still debated the existence of atoms at this time.

We see the kinetic theory in action in our daily lives. For example, when we inflate a tire or balloon, we add more air molecules to the enclosed space, which results in more collisions of the molecules on the inside of the enclosed space than there are on the outside. As a result, the enclosure expands.

SEE ALSO Charles' Gas Law (1787), Atomic Theory (1808), Avogadro's Gas Law (1811), Brownian Motion (1827), Boltzmann's Entropy Equation (1875).

According to kinetic theory, when we blow a soap bubble, we add more air molecules to the enclosed space, leading to more molecular collisions on the bubble's inside than on the outside, causing the bubble to expand.

1861

Maxwell's Equations

James Clerk Maxwell (1831–1879)

"From a long view of the history of mankind," writes physicist Richard Feynman, "seen from, say, ten thousand years from now—there can be no doubt that the most significant event of the 19th century will be judged as Maxwell's discovery of the laws of electrodynamics. The American Civil War will pale into provincial insignificance in comparison with this important scientific event of the same decade."

In general, Maxwell's Equations are the set of four famous formulas that describe the behavior of the electric and magnetic fields. In particular, they express how electric charges produce electric fields and the fact that magnetic charges cannot exist. They also show how currents produce magnetic fields and how changing magnetic fields produce electric fields. If you let **E** represent the electric field, **B** represent the magnetic field, ε_0 represent the electric constant, μ_0 represent the magnetic constant, and **J** represent the current density, you can express Maxwell's equations thus:

$\nabla \cdot \mathbf{E} = \frac{\rho}{\varepsilon_0}$	Gauss' Law for Electricity
$\nabla \cdot \mathbf{B} = 0$	Gauss' Law for Magnetism (no magnetic monopoles)
$\nabla \times \mathbf{E} = -\frac{\partial \mathbf{B}}{\partial t}$	Faraday's Law of Induction
$\nabla \times \mathbf{B} = \mu_0 \mathbf{J} + \mu_0 \varepsilon_0 \frac{\partial \mathbf{E}}{\partial t}$	Ampère's Law with Maxwell's extension

Note the utter compactness of expression, which led Einstein to rate Maxwell's achievement on a par with that of Isaac Newton's. Moreover, the equations predicted the existence of electromagnetic waves.

Philosopher Robert P. Crease writes of the importance of Maxwell's equations: "Although Maxwell's equations are relatively simple, they daringly reorganize our perception of nature, unifying electricity and magnetism and linking geometry, topology and physics. They are essential to understanding the surrounding world. And as the first field equations, they not only showed scientists a new way of approaching physics but also took them on the first step towards a unification of the fundamental forces of nature."

SEE ALSO Ampère's Law of Electromagnetism (1825), Faraday's Laws of Induction (1831), Gauss and the Magnetic Monopole (1835), Theory of Everything (1984).

LEFT: *Mr. and Mrs. James Clerk Maxwell, 1869.* RIGHT: *Computer core memory of the 1960s can be partly understood using Ampere's Law in Maxwell's Equations, which describes how a current-carrying wire produces a magnetic field that circles the wire and, thus, can cause the core (doughnut shape) to change its magnetic polarity.*

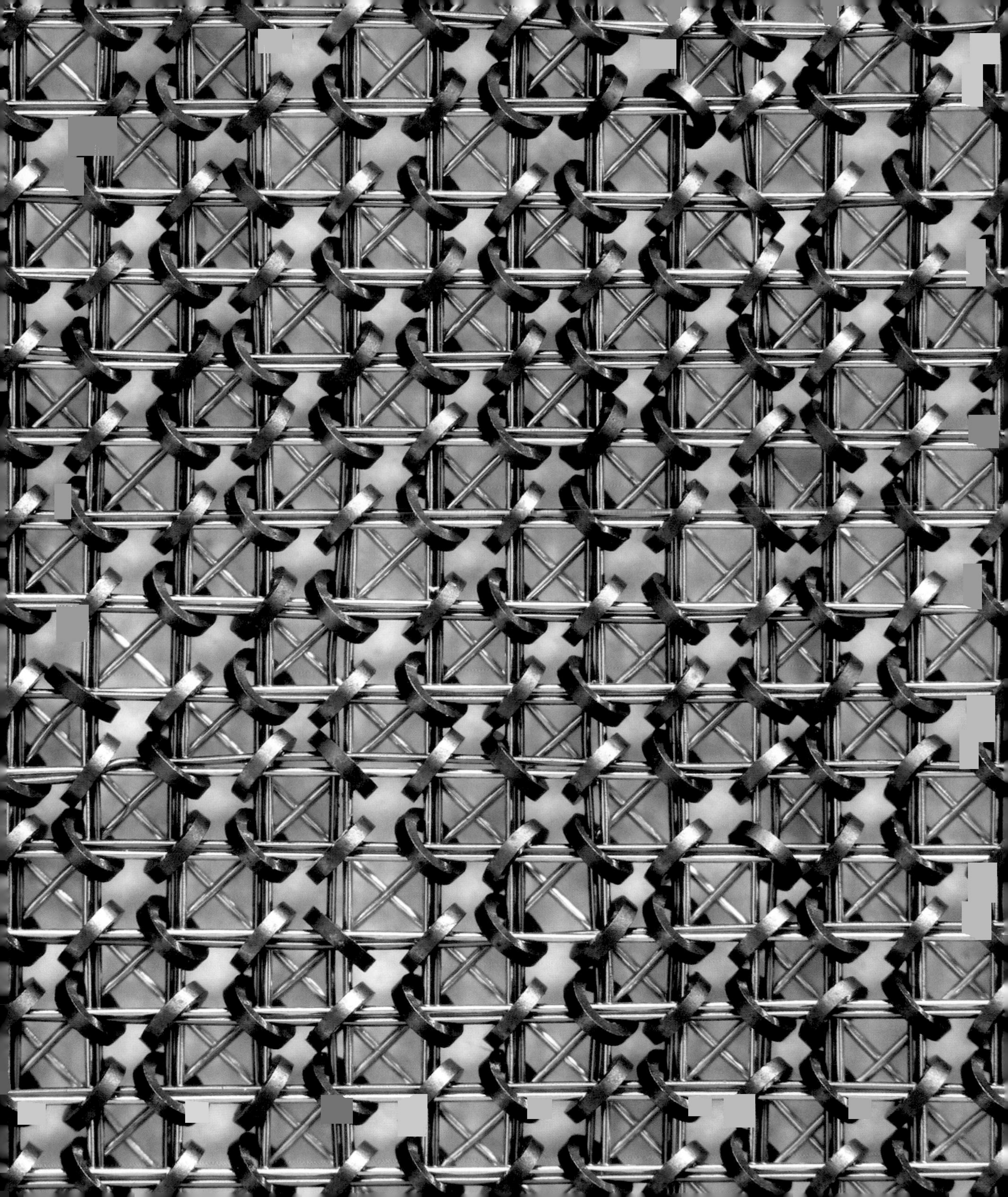

1864

Electromagnetic Spectrum

Frederick William Herschel (1738–1822), **Johann Wilhelm Ritter** (1776–1810), **James Clerk Maxwell** (1831–1879), **Heinrich Rudolf Hertz** (1857–1894)

The electromagnetic spectrum refers to the vast range of frequencies of electromagnetic (EM) radiation. It is composed of waves of energy that can propagate through a vacuum and that contain electric and magnetic field components that oscillate perpendicular to each other. Different portions of the spectrum are identified according to the frequency of the waves. In order of increasing frequency (and decreasing wavelength), we have radio waves, microwaves, infrared radiation, visible light, ultraviolet radiation, X-rays, and gamma rays.

We can see light with wavelengths between 4,000 and 7,000 angstroms, where an angstrom is equal to 10^{-10} meters. Radio waves may be generated by electrons that move back and forth in transmission towers and have wavelengths ranging from several feet to many miles. If we represent the electromagnetic spectrum as a 30-octave piano, in which the wavelength of radiation doubles with each octave, visible light occupies only part of an octave. If we wanted to represent the entire spectrum of radiation that has been detected by our instruments, we would need to add at least 20 octaves to the piano.

Extraterrestrials may have senses beyond our own. Even on the Earth, we find examples of creatures with increased sensitivities. For example, rattlesnakes have infrared detectors that give them "heat pictures" of their surroundings. To our eyes, both the male and female Indian luna moths are light green and indistinguishable from each other, but the luna moths themselves perceive the ultraviolet range of light. Therefore, to them, the female looks quite different from the male. Other creatures have difficulty seeing the moths when they rest on green leaves, but luna moths are not camouflaged to one another; rather, they see each other as brilliantly colored. Bees can also detect ultraviolet light. In fact, many flowers have beautiful patterns that bees can see to guide them to the flower. These attractive and intricate patterns are totally hidden from human perception.

The physicists listed at the top of this entry played key research roles with respect to the electromagnetic spectrum.

SEE ALSO Newton's Prism (1672), Wave Nature of Light (1801), Fraunhofer Lines (1814), Brewster's Optics (1815), Stokes' Fluorescence (1852), X-rays (1895), Black Light (1903), Cosmic Microwave Background (1965), Gamma-Ray Bursts (1967), Blackest Black (2008).

To our eyes, male and female luna moths are light green and indistinguishable from each other. But the luna moths themselves perceive in the ultraviolet range of light, and to them the female looks quite different from the male.

1866

Surface Tension

Loránd von Eötvös (1848–1919)

The physicist Loránd Eötvös once wrote, "Poets can penetrate deeper into the realm of secrets than scientists," yet Eötvös used the tools of science to understand the intricacies of surface tension, which plays a role in numerous aspects of nature. At the surface of a liquid, the molecules are pulled inward by intermolecular forces. Eötvös determined an interesting relationship between the surface tension of a liquid and the temperature of a liquid: $\gamma = k(T_0 - T)/\rho^{3/2}$. Here, the surface tension γ of a liquid is related to its temperature T, the critical temperature of the liquid (T_0), and its density ρ. The constant k is approximately the same for many common liquids, including water. T_0 is the temperature at which the surface tension disappears or becomes zero.

The term surface tension usually refers to a property of liquids that arises from unbalanced molecular forces at or near the surface of a liquid. As a result of these attractive forces, the surface tends to contract and exhibits properties similar to those of a stretched elastic membrane. Interestingly, the surface tension, which may be considered a molecular surface energy, changes in response to temperature essentially in a manner that is independent of the nature of the liquid.

During his experiments, Eötvös had to take special care that the surface of his fluids had no contamination of any kind, so he worked with glass vessels that had been closed by melting. He also used optical methods for determining the surface tension. These sensitive methods were based on optical reflection in order to characterize the local geometry of the liquid surface.

Water strider insects can walk on water because the surface tension causes the surface to behave like an elastic membrane. In 2007, Carnegie Mellon University researchers created robotic striders and found that "optimal" robotic wire legs coated with Teflon were about 2 inches (5 centimeters) long. Furthermore, twelve legs attached to the 0.03-ounce (1-gram) body can support up to 0.3 ounces (9.3 grams).

SEE ALSO Stokes' Law of Viscosity (1851), Superfluids (1937), Lava Lamp (1963).

LEFT: *Water strider.* RIGHT: *Photograph of two red paper clips floating on water, with projected colored stripes indicating water surface contours. Surface tension prevents the paper clips from submerging.*

1866

Dynamite

Alfred Bernhard Nobel (1833–1896)

"Humanity's quest to harness the destructive capacity of fire is a saga that extends back to the dawn of civilization," writes author Stephen Bown. "Although gunpowder did bring about social change, toppling feudalism and ushering in a new military structure . . . , the true great era of explosives, when they radically and irrevocably changed the world, began in the 1860s with the remarkable intuition of a sallow Swedish chemist named Alfred Nobel."

Nitroglycerin had been invented around 1846 and was a powerful explosive that could easily detonate and cause loss of life. In fact, Nobel's own Swedish factory for manufacturing nitroglycerin exploded in 1864, killing five people, including his younger brother Emil. The Swedish government prohibited Nobel from rebuilding his factory. In 1866, Nobel discovered that by mixing nitroglycerin with a kind of mud composed of finely ground rock known as *kieselguhr* (and sometimes referred to as diatomaceous earth), he could create an explosive material that had much greater stability than nitroglycerine. Nobel patented the material a year later, calling it *dynamite*. Dynamite has been used primarily in mining and construction industries, but it has also been used in war. Consider the many British soldiers based in Gallipoli during World War I who made bombs out of jam tins (literally, cans that had contained jam) packed with dynamite and pieces of scrap metal. A fuse was used for detonation.

Nobel never intended for his material to be used in war. In fact, his main goal was to make nitroglycerin safer. A pacifist, he believed that dynamite could end wars quickly, or that the power of dynamite would make war unthinkable—too horrifying to carry out.

Today, Nobel is famous for his founding of the Nobel Prize. The Bookrags staff write, "Many have observed the irony in the fact that he left his multimillion dollar fortune, made by the patenting and manufacture of dynamite and other inventions, to establish prizes awarded 'to those, who during the preceding year, shall have conferred the greatest benefit on mankind'."

SEE ALSO Little Boy Atomic Bomb (1945).

Dynamite is sometimes used in open-pit mining. Open-pit mines that produce building materials and related stones are often referred to as quarries.

1867

Maxwell's Demon

James Clerk Maxwell (1831–1879), **Léon Nicolas Brillouin** (1889–1969)

"Maxwell's demon is no more than a simple idea," write physicists Harvey Leff and Andrew Rex. "Yet it has challenged some of the best scientific minds, and its extensive literature spans thermodynamics, statistical physics, quantum mechanics, information theory, cybernetics, the limits of computing, biological sciences, and the history and philosophy of science."

Maxwell's demon is a hypothetical intelligent entity—first envisioned by Scottish physicist James Clerk Maxwell—that has been used to suggest that the **Second Law of Thermodynamics** might be violated. In one of its early formulations, this law states that the total entropy, or disorder, of an isolated system tends to increase over time as it approaches a maximum value. Moreover, it maintains that heat does not naturally flow from a cool body to a warmer body.

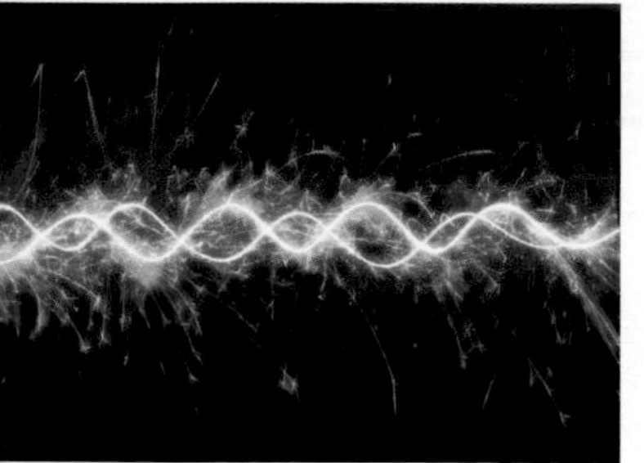

To visualize Maxwell's demon, imagine two vessels, A and B, connected by a small hole and containing gas at equal temperatures. Maxwell's demon could, in principle, open and close this hole to allow individual gas molecules to pass between the vessels. Additionally, the demon allows only fast-moving molecules to pass from vessel A to B, and only slow-moving molecules to go from B to A. In doing so, the demon creates greater kinetic energy (and heat) in B, which could be used as a source of power for devices. This appears to provide a loophole in the Second Law of Thermodynamics. The little creature—whether alive or mere machine—exploits the random, statistical features of molecular motions to decrease entropy. If some mad scientist could create such an entity, the world would have an endless source of energy.

One "solution" to the problem of Maxwell's demon came from French physicist Léon Brillouin around 1950. Brillouin and others banished the demon by showing that the decrease in entropy resulting from the demon's careful observations and actions would be exceeded by the increase in entropy needed to actually choose between the slow and fast molecules. The demon requires energy to operate.

SEE ALSO Perpetual Motion Machines (1150), Laplace's Demon (1814), Second Law of Thermodynamics (1850).

LEFT: *Maxwell's Demon is able to segregate collections of hot and cold particles, depicted here in red and blue colors. Could Maxwell's Demon provide us with an endless source of energy?* RIGHT: *Artistic depiction of Maxwell's Demon, allowing fast-moving molecules (orange) to accumulate in one region and slow-moving molecules (blue-green) in another.*

1868

Discovery of Helium

Pierre Jules César Janssen (1824–1907), **Joseph Norman Lockyer** (1836–1920), **William Ramsay** (1852–1916)

"It may seem surprising now," write authors David and Richard Garfinkle, "when helium-filled balloons are at every child's birthday party, but [in 1868] helium was a mystery in much the same way dark matter is today. Here was a substance that had never been seen on the Earth, was only seen in the Sun, and was only known indirectly through the presence of its spectral line."

Indeed, the discovery of helium is notable because it represents the first example of a chemical element discovered on an extraterrestrial body before being found on the Earth. Even though helium is abundant in the universe, it was completely unknown for most of human history.

Helium is inert, colorless, and odorless and has boiling and melting points that are the lowest among all elements. After hydrogen, it is the second most abundant element in the universe, making up about 24 percent of the galactic stellar mass. Helium was discovered in 1868 by astronomers Pierre Janssen and Norman Lockyer after observing an unknown spectral line signature in sunlight. However, it wasn't until 1895 that British chemist Sir William Ramsay discovered helium on the Earth in a radioactive, uranium-rich mineral. In 1903, large reservoirs of helium were found in U.S. natural gas fields.

Because of its extremely low boiling temperature, liquid helium is the standard coolant for superconducting magnets used in MRI (magnetic resonance imaging) devices and particle accelerators. At very low temperatures, liquid helium exhibits the peculiar properties of a **Superfluid**. Helium is also important for deep-sea divers (to prevent too much oxygen from entering the brain) and for welders (to reduce oxidation during the application of high temperatures). It is also used for rocket launches, lasers, weather balloons, and leak detection.

Most helium in the universe is the helium-4 isotope—containing two protons, two neutrons, and two electrons—formed during the **Big Bang**. A smaller portion is formed in stars through nuclear fusion of hydrogen. Helium is relatively rare above ground on the Earth because, as demonstrated by an untethered helium-filled balloon that rises into the atmosphere, it is so light that most has escaped into space.

SEE ALSO Big Bang (13.7 Billion B.C.), Thermos (1892), Superconductivity (1911), Superfluids (1937), Nuclear Magnetic Resonance (1938).

USS Shenandoah (ZR-1), *flying in the vicinity of New York City, c. 1923. The* Shenandoah *is notable in that it was the first rigid airship to use helium gas rather than flammable hydrogen gas.*

U.S. NAVY

1870

Baseball Curveball

Fredrick Ernest Goldsmith (1856–1939), **Heinrich Gustav Magnus** (1802–1870)

Robert Adair, author of *The Physics of Baseball*, writes, "The pitcher's action up to the release of the ball is part of the art of pitching; the action of the ball after release . . . is addressed by physics." For years, arguments raged in popular magazines as to whether the curveballs pitched in baseball really curved, or whether they were simply some kind of optical illusion.

While it may be impossible to definitively say which baseball player first developed the curveball, professional pitcher Fred Goldsmith is often credited with giving the first publicly recorded demonstration of a curveball on August 16, 1870, in Brooklyn, New York. Many years later, research into the physics of curveballs showed that, for example, when a topspin was placed on the ball—so that the top of the ball rotated in the direction of the pitch—a significant deviation from the ball's ordinary course of travel took place. In particular, a layer of air rotates with the ball, like a whirlpool, and the air layer near the bottom of the ball travels faster than near the top—the top of the whirlpool travels against the direction of ball travel. According to Bernoulli's principle, when air or liquid flows, it creates a region of low air pressure that is related to the flow velocity (see **Bernoulli's Law of Fluid Dynamics**). This pressure difference between the top and the bottom of the ball causes it to have a curving trajectory, dropping as it nears the batter. This drop or "break" may be a deviation of as many as 20 inches off the course of a ball that has no spin. German physicist Heinrich Magnus described this effect in 1852.

In 1949, engineer Ralph Lightfoot used a wind tunnel to prove that the curveball actually curves. However, an optical illusion enhances the effect of the curveball because, as the ball nears the plate and shifts from the batter's direct vision to his peripheral vision, the spinning motion distorts the batter's perception of the ball's trajectory so that it appears to drop suddenly.

SEE ALSO Cannon (1132), Bernoulli's Law of Fluid Dynamics (1738), Golf Ball Dimples (1905), Terminal Velocity (1960).

A layer of air rotates with the curveball, creating a pressure difference between the top and the bottom of the ball. This difference can cause the ball to have a curving trajectory so that it drops as it nears the batter.

1871

Rayleigh Scattering

John William Strutt, 3rd Baron Rayleigh (1842–1919)

In 1868, the Scottish poet George MacDonald wrote, "When I look into the blue sky, it seems so deep, so peaceful, so full of a mysterious tenderness that I could lie for centuries, and wait for the dawning of the face of God out of the awful loving-kindness." For many years, both scientists and laypeople have wondered what made the sky blue and the sunsets a fiery red. Finally, in 1871, Lord Rayleigh published a paper providing the answers. Recall that the "white light" from the Sun is actually composed of a range of hidden colors, which you can reveal with a simple glass prism. *Rayleigh scattering* refers to the scattering of sunlight by the gas molecules and microscopic density fluctuations in the atmosphere. In particular, the angle through which sunlight scatters varies inversely as the fourth power of the color wavelength. This means that blue light scatters much more than other colors, such as red, because the wavelength of blue light is shorter than the wavelength of red light. Blue light scatters strongly across much of the sky, and thus someone on the Earth observes a blue sky. Interestingly, the sky does not look violet (even though this color has a shorter wavelength than blue light), partly because there is more blue light than violet light in the spectrum of sunlight and because our eyes are more sensitive to blue than to violet light.

When the Sun is near the horizon, such as during sunset, the amount of air through which sunlight passes to an observer is greater than when the Sun is higher in the sky. Thus, more of the blue light is scattered away from the observer, leaving longer wavelength colors to dominate the view of the sunset.

Note that Rayleigh scattering applies to particles in the air having a radius of less than about a tenth of the wavelength of radiation, such as gas molecules. Other laws of physics apply when a significant amount of larger particles is in the air.

SEE ALSO Explaining the Rainbow (1304), Aurora Borealis (1621), Newton's Prism (1672), Greenhouse Effect (1824), Green Flash (1882).

For centuries, both scientists and laypeople have wondered what made the sky blue and the sunsets a fiery red. Finally, in 1871 Lord Rayleigh published a paper providing the answers.

1873

Crookes Radiometer

William Crookes (1832–1919)

As a child, I had three "light mills" lined up on the window sill, with their paddles all spinning, as if by magic. The explanation of the movement has generated much debate through the decades, and even the brilliant physicist James Maxwell was initially confounded by the mill's mode of operation.

The Crookes radiometer, also called the light mill, was invented in 1873 by English physicist William Crookes. It consists of a glass bulb that is partially evacuated and, within it, four vanes mounted on a spindle. Each vane is black on one side and shiny or white on the other. When exposed to light, the black sides of the paddles absorb photons and become warmer than the light sides, causing the paddles to turn so that the black sides move away from the light source, as explained below. The brighter the light, the faster the rotation speed. The paddles will not turn if the vacuum is too extreme within the bulb, and this suggests that movement of the gas molecules inside the bulb is the cause of the motion. Additionally, if the bulb is not evacuated at all, too much air resistance prevents rotation of the paddles.

At first, Crookes suggested that the force turning the paddles resulted from the actual pressure of the light upon the blades, and Maxwell originally agreed with this hypothesis. However, it became clear that this theory was insufficient, given that the paddles did not turn in a strong vacuum. Also, light pressure would be expected to cause the shiny, reflective side of the paddles to move away from the light. In fact, the light mill's rotation can be attributed to the motions of gas molecules as a result of the difference in temperature between the sides of the paddle. The precise mechanism appears to make use of a process called *thermal transpiration*, which involves the motion of gas molecules from the cooler sides to the warmer sides near the edges of the blades, creating a pressure difference.

SEE ALSO Perpetual Motion Machines (1150), Drinking Bird (1945).

LEFT: *Sir William Crookes, from J. Arthur Thomson's* The Outline of Science, *1922.* RIGHT: *The Crookes radiometer, also called the light mill, consists of a glass bulb that is partially evacuated. Inside are four vanes, mounted on a spindle. When light is shined on the radiometer, the blades rotate.*

1875

Boltzmann's Entropy Equation

Ludwig Eduard Boltzmann (1844–1906)

"One drop of ink may make a million think," says an old proverb. Austrian physicist Ludwig Boltzmann was fascinated by statistical thermodynamics, which focuses on the mathematical properties of large numbers of particles in a system, including ink molecules in water. In 1875, he formulated a mathematical relationship between entropy S (roughly, the disorder of a system) and the number of possible states of the system W in a compact expression: $S = k \cdot \log W$. Here, k is Boltzmann's constant.

Consider a drop of ink in water. According to **Kinetic Theory**, the molecules are in constant random motion and always rearranging themselves. We assume that all possible arrangements are equally probable. Because most of the arrangements of ink molecules do not correspond to a drop of clustered ink molecules, most of the time we will not observe a drop. Mixing occurs spontaneously simply because so many more arrangements exist that are mixed than that are not. A spontaneous process occurs because it produces the most probable final state. Using the formula $S = k \cdot \log W$, we can calculate the entropy and can understand why the more states that exist, the greater the entropy. A state with a high probability (e.g. a mixed ink state) has a large value for the entropy, and a spontaneous process produces the final state of greatest entropy, which is another way of stating the **Second Law of Thermodynamics**. Using the terminology of thermodynamics, we can say that there are a number of ways W (number of *microstates*) that exist to create a particular *macrostate*—in our case, a mixed state of ink in a glass of water.

Although Boltzmann's idea of deriving thermodynamics by visualizing molecules in a system seems obvious to us today, many physicists of his time criticized the concept of atoms. Repeated clashes with other physicists, combined with an apparent lifelong struggle with bipolar disorder, may have contributed to the physicist's suicide in 1906 while on vacation with his wife and daughter. His famous entropy equation is engraved on his tombstone in Vienna.

SEE ALSO Brownian Motion (1827), Second Law of Thermodynamics (1850), Kinetic Theory (1859).

LEFT: *Ludwig Eduard Boltzmann. RIGHT: Imagine that all possible arrangements of ink and water molecules are equally probable. Because most of the arrangements do not correspond to a drop of clustered ink molecules, most of the time we will not observe a droplet once the ink drop is added.*

1878

Incandescent Light Bulb

Joseph Wilson Swan (1828–1914), **Thomas Alva Edison** (1847–1931)

The American inventor Thomas Edison, best known for his development of the light bulb, once wrote, "To invent you need a good imagination and a pile of junk." Edison was not the only person to have invented a version of the *incandescent* light bulb—that is, a light source that makes use of heat-driven light emissions. Other equally notable inventors include Joseph Swan of England. However Edison is best remembered because of the combination of factors he helped to promote—a long-lasting filament, the use of a higher vacuum within the bulb than others were able to produce, and a power distribution system that would make the light bulb of practical value in buildings, streets, and communities.

In an incandescent light bulb, an electric current passes through the filament, heating it to produce light. A glass enclosure prevents oxygen in the air from oxidizing and destroying the hot filament. One of the greatest challenges was to find the most effective material for the filament. Edison's carbonized filament of bamboo could emit light for more than 1,200 hours. Today, a filament made of tungsten wire is often used, and the bulb is filled with an inert gas such as argon to reduce evaporation of material from the filament. Coiled wires increase the efficiency, and the filament within a typical 60-watt, 120-volt bulb is actually 22.8 inches (580 millimeters) in length.

If a bulb is operated at low voltages, it can be surprisingly long lasting. For example, the "Centennial Light" in a California fire station has been burning almost continually since 1901. Generally, incandescent lights are inefficient in the sense that about 90% of the power consumed is converted to heat rather than visible light. Although today more efficient forms of light bulbs (e.g. compact fluorescent lamps) are starting to replace the incandescent light bulbs, the simple incandescent bulb once replaced the soot-producing and more dangerous lamps and candles, changing the world forever.

SEE ALSO Joule's Law of Electric Heating (1840), Stokes' Fluorescence (1852), Ohm's Law of Electricity (1827), Black Light (1903), Vacuum Tube (1906).

Edison light bulb with a looping carbon filament.

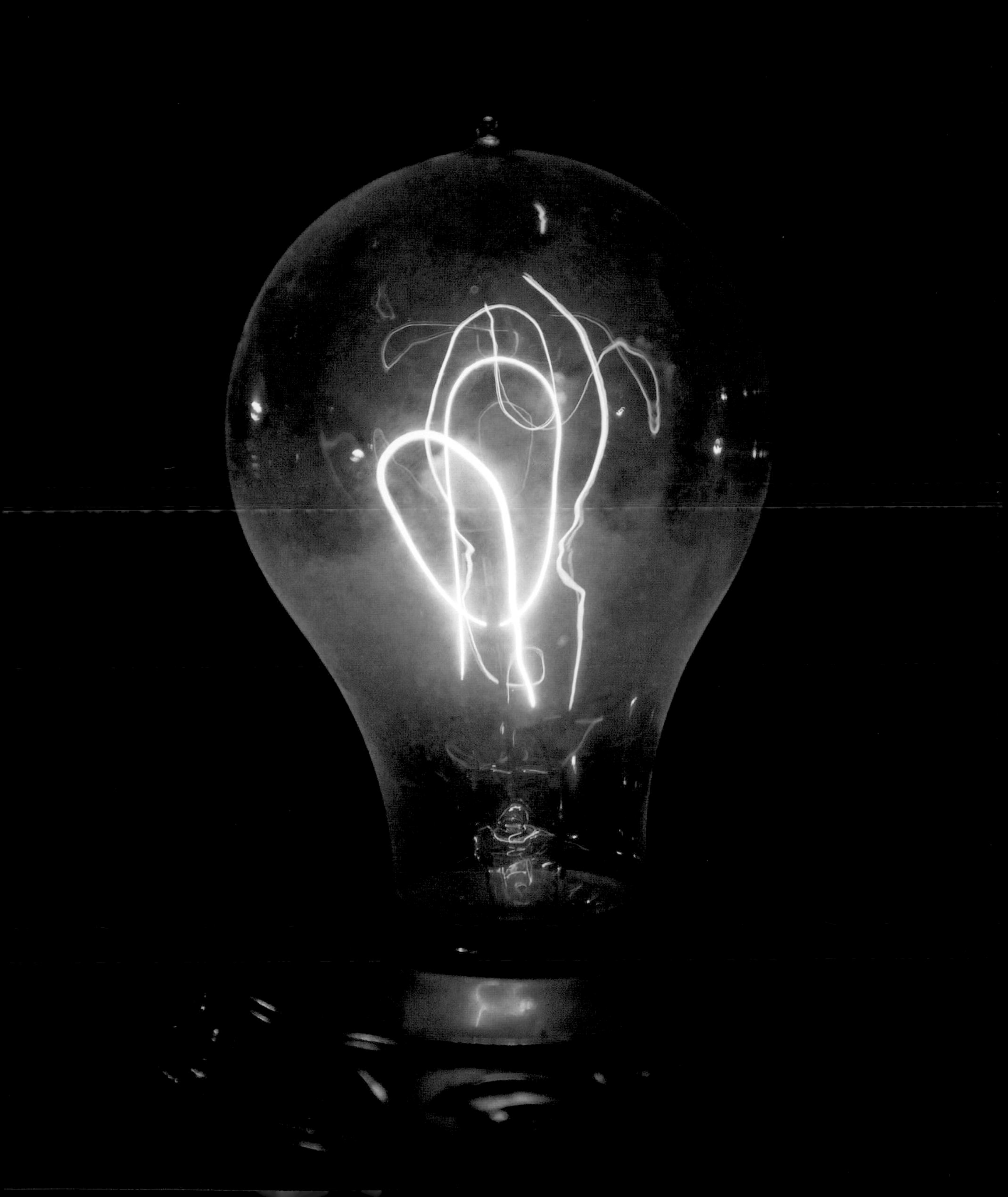

1879

Plasma

William Crookes (1832–1919)

A plasma is a gas that has become ionized, which means that the gas contains a collection of free-moving electrons and ions (atoms that have lost electrons). The creation of a plasma requires energy, which may be supplied in a variety of forms, including thermal, radiant, and electrical. For example, when a gas is sufficiently heated so that the atoms collide with one another and knock their electrons off, a plasma can be formed. Like a gas, a plasma does not have a definite shape unless enclosed in a container. Unlike ordinary gases, magnetic fields may cause a plasma to form a tapestry of unusual structures, such as filaments, cells, layers, and other patterns of startling complexity. Plasmas can exhibit a rich variety of waves not present in ordinary gases.

British physicist William Crookes first identified plasmas in 1879 while experimenting with a partially evacuated electrical discharge tube called a Crookes tube. Interestingly, a plasma is the most common state of matter—far more common than solids, liquids, and gases. Shining stars are made of this "fourth state of matter." On the Earth, common examples of plasma producers include fluorescent lights, plasma TVs, neon signs, and lightning. The ionosphere—the upper atmosphere of the Earth—is a plasma produced by solar radiation that has a practical importance because it influences radio communication around the world.

Plasma studies employ a large range of plasma gases, temperatures, and densities in fields that range from astrophysics to fusion power. The charged particles in plasmas are sufficiently close so that each particle influences many nearby charged particles. In plasma TVs, xenon and neon atoms release photons of light when they are excited. Some of these photons are ultraviolet photons (which we can't see) that interact with phosphor materials, causing them, in turn, to emit visible light. Each pixel in the display is made up of smaller subpixels that have different phosphors for green, red, and blue colors.

SEE ALSO St. Elmo's Fire (78), Neon Signs (1923), Jacob's Ladder (1931), Sonoluminescence (1934), Tokamak (1956), HAARP (2007).

Plasma lamp, exhibiting complex phenomena such as filamentation. The beautiful colors result from the relaxation of electrons in excited states to lower energy states.

1879

Hall Effect

Edwin Herbert Hall (1855–1938), **Klaus von Klitzing** (b. 1943)

In 1879, American physicist Edwin Hall placed a thin gold rectangle in a strong magnetic field that was perpendicular to the rectangle. Let us imagine that x and x' are two parallel sides of the rectangle and that y and y' are the other parallel sides. Hall then connected battery terminals to sides x and x' to produce a current flow in the x direction along the rectangle. He discovered that this produced a tiny voltage difference from y to y', which was proportional to the strength of an applied magnetic field, B_z, multiplied by the current. For many years, the voltage produced by the Hall Effect was not used in practical applications because it was small. However, in the second half of the twentieth century, the Hall Effect blossomed into countless areas of research and development. Note that Hall discovered his tiny voltage 18 years before the electron was actually discovered.

The Hall coefficient R_H is the ratio of the induced electric field E_y to the current density j_x multiplied by B_z: $R_H = E_y/(j_x B_z)$. The ratio of the created voltage in the y direction to the amount of current is known as the *Hall resistance*. Both the Hall coefficient and resistance are characteristics of the material under study. The Hall Effect turned out to be very useful for measuring either the magnetic field or the *carrier* density. Here, we use the term *carrier* instead of the more familiar term *electron* because, in principle, an electrical current may be carried by charged particles other than electrons (e.g. positively charged carriers referred to as *holes*).

Today, the Hall Effect is used in many kinds of magnetic field sensors within applications that range from fluid flow sensors, to pressure sensors, to automobile ignition timing systems. In 1980, German physicist Klaus von Klitzing discovered the quantum Hall Effect when, using a large magnetic field strength and low temperatures, he noticed discrete steps in the Hall resistance.

SEE ALSO Faraday's Laws of Induction (1831), Piezoelectric Effect (1880), Curie's Magnetism Law (1895).

High-end paintball guns make use of Hall Effect sensors to provide a very short throw, *allowing a high rate of fire. The throw is the distance that the trigger travels before actuating.*

1880

Piezoelectric Effect

Paul-Jacques Curie (1856–1941), **Pierre Curie** (1859–1906)

"Every [scientific] discovery, however small, is a permanent gain," French physicist Pierre Curie wrote to Marie, a year before they married, urging her to join him in "our scientific dream." As a young teenager, Pierre Curie had a love of mathematics—spatial geometry, in particular—which would later be of value to him in his work on crystallography. In 1880, Pierre and his brother Paul-Jacques demonstrated that electricity was produced when certain crystals were compressed—a phenomenon now called *piezoelectricity*. Their demonstrations involved crystals such as tourmaline, quartz, and topaz. In 1881, the brothers demonstrated the reverse effect: that electric fields could cause some crystals to deform. Although this deformation is small, it was later found to have applications in the production and detection of sound and the focusing of optical components. Piezoelectric applications have been used in the design of phonograph cartridges, microphones, and ultrasonic submarine detectors. Today, electric cigarette lighters use a piezoelectric crystal to produce a voltage in order to ignite the gas from the lighter. The U.S. military has explored the possible use of piezoelectric materials in soldiers' boots for generating power in the battlefield. In piezoelectric microphones, sound waves impinge on the piezoelectric material and create a change in voltage.

Science-journalist Wil McCarthy explains the molecular mechanism of the piezoelectric effect, which "occurs when pressure on a material creates slight dipoles within it, by deforming neutral molecules or particles so that they become positively charged on one side and negatively charged on the other, which in turn increase an electrical voltage across the material." In an old-fashioned phonograph, the needle glides within the wiggly grooves of the record, which deform the needle tip made of Rochelle salt and create voltages that are converted to sound.

Interestingly, the material of bone exhibits the piezoelectric effect, and piezoelectric voltages may play a role in bone formation, nourishment, and the effect that mechanical loads have on bone.

SEE ALSO Triboluminescence (1620), Hall Effect (1879).

An electric cigarette lighter employs a piezoelectric crystal. Pressing a button causes a hammer to hit the crystal, which produces an electric current that flows across a spark gap and ignites the gas.

1880

War Tubas

War tuba is the informal name given to a large variety of huge acoustic locators—many of which had an almost-comical appearance—that played a crucial role in the history of warfare. These devices were primarily used to locate aircraft and guns from World War I to the early years of World War II. Once radar (detection systems using electromagnetic waves) was introduced in the 1930s, the amazing war tubas were mostly rendered obsolete, but they were sometimes used either for misinformation (e.g. to mislead the Germans into thinking that radar was not being used) or if radar-jamming methods were deployed. Even as late as 1941, Americans used acoustic locators to detect the first Japanese attack on Corregidor in the Philippines.

Through the years, the acoustic locators have taken many forms, ranging from personal devices that resemble horns strapped to the shoulders, such as the topophone of the 1880s, to huge multi-horn protuberances mounted on carriages and operated by multiple users. The German version—a *Ringtrichterrichtungshoerer* ("ring-horn acoustic direction detector")—was used in World War II to assist with the initial nighttime aiming of searchlights at airplanes.

The December 1918 *Popular Science* described how 63 German guns were detected by acoustic location in a single day. Microphones concealed beneath rocks across a landscape were connected by wires to a central location. The central station recorded the precise moment at which every sound was received. When used to locate guns, the station recorded the sound of the flying shell passing overhead, the boom of the gun, and shell explosion sounds. Corrections were made to adjust for variations in the speed of sound waves due to atmospheric conditions. Finally, the difference in times at which the same sound was recorded from the receiving stations was compared with the known distances among stations. British and French military observers then directed airplane bombers to destroy the German guns, some of which were camouflaged and nearly undiscoverable without the acoustic locators.

SEE ALSO Tuning Fork (1711), Stethoscope (1816).

LEFT: *A huge two-horn system at Bolling Field, Washington DC (1921).* RIGHT: *Photograph (mirrored) of Japanese Emperor Hirohito inspecting an array of acoustic locators, also known today as* war tubas, *that were used to locate aircraft before the invention of radar.*

1882

Galvanometer

Hans Christian Oersted (1777–1851), **Johann Carl Friedrich Gauss** (1777–1855), **Jacques-Arsène d'Arsonval** (1851–1940)

A galvanometer is a device for measuring electric currents using a needle, or pointer, that rotates in response to electrical current. In the mid-1800s, the Scottish scientist George Wilson spoke with awe of the galvanometer's dancing needle, writing that a similar compass needle was the "guide of Columbus to the New World [and] was the precursor and pioneer of the telegraph. Silently . . . it led the explorers across the waste of waters to the new homes of the world; but when these were largely filled, and houses . . . longed to exchange affectionate greetings, it . . . broke silence. The quivering magnetic needle which lies in the coil of the galvanometer is the tongue of the electric telegraph, and already engineers talk of it as speaking."

One of the earliest forms of galvanometers emerged from the 1820 work of Hans Christian Oersted, who discovered that electric current flowing through a wire created a surrounding magnetic field that deflected a magnetized needle. In 1832, Carl Friedrich Gauss built a telegraph that made use of signals that deflected a magnetic needle. This older kind of galvanometer made use of a moving magnet, which had the disadvantage of being affected by any nearby magnets or iron masses, and its deflection was not linearly proportional to the current. In 1882, Jacques-Arsène d'Arsonval developed a galvanometer that used a stationary permanent magnet. Mounted between the magnet's poles was a coil of wire that generated a magnetic field and rotated when electric current flowed through the coil. The coil was attached to a pointer, and the angle of deflection was proportional to the current flow. A small torsion spring returned the coil and pointer to the zero position when no current was present.

Today, galvanometer needles are typically replaced by digital readouts. Nevertheless, modern galvanometer-like mechanisms have had a variety of uses, ranging from positioning pens in analog strip chart recorders to positioning heads in hard-disk drives.

SEE ALSO Ampère's Law of Electromagnetism (1825), Faraday's Laws of Induction (1831).

LEFT: *Hans Christian Oersted.* RIGHT: *An antique ammeter with binding posts and a dial with scale for DC milliamperes. (This is original physics laboratory equipment at the State University of New York College at Brockport.) The D'Arsonval galvanometer is a moving-coil ammeter.*

HIGH
LOW
MILLIAMPERES
D.C.

1882

Green Flash

Jules Gabriel Verne (1828–1905), **Daniel Joseph Kelly O'Connell** (1896–1982)

Interest in the mysterious green flashes that are sometimes glimpsed above the setting or rising sun was kindled in the West by Jules Verne's romantic novel *The Green Ray* (1882). The novel describes the quest for the eerie green flash, "a green which no artist could ever obtain on his palette, a green of which neither the varied tints of vegetation nor the shades of the most limpid sea could ever produce the like! If there is a green in Paradise, it cannot be but of this shade, which most surely is the true green of Hope. . . . He who has been fortunate enough once to behold it is enabled to see closely into his own heart and to read the thoughts of others."

Green flashes result from several optical phenomena and are usually easier to see over an unobstructed ocean horizon. Consider a setting sun. The Earth's atmosphere, with its layers of varying density, acts like a prism, causing different colors of light to bend at different angles. Higher-frequency light, such as green and blue light, bends more than lower-frequency light, such as red and orange. As the Sun dips below the horizon, the lower-frequency reddish image of the Sun is obstructed by the Earth, but the higher-frequency green portion can be briefly seen. Green flashes are enhanced by mirage effects, which can create distorted images (including magnified images) of distant objects because of differences in air densities. For example, cold air is denser than warm air and thus has a greater refractive index than warm air. Note that the color blue is usually not seen during the green flash because blue light is scattered beyond view (see **Rayleigh Scattering**).

For years, scientists often believed that reports of green flashes were optical illusions induced by staring too long at the setting sun. However, in 1954, the Vatican priest Daniel O'Connell took color photographs of a green flash as the Sun was setting over the Mediterranean Sea, thus "proving" the existence of the unusual phenomenon.

SEE ALSO Aurora Borealis (1621), Snell's Law of Refraction (1621), Black Drop Effect (1761), Rayleigh Scattering (1871), HAARP (2007).

Green flash, photographed in San Francisco, 2006.

1887

Michelson-Morley Experiment

Albert Abraham Michelson (1852–1931) **and Edward Williams Morley** (1838–1923)

"It is hard to imagine nothing," physicist James Trefil writes. "The human mind seems to want to fill empty space with some kind of material, and for most of history that material was called the aether. The idea was that the emptiness between celestial objects was filled with a kind of tenuous Jell-O."

In 1887, physicists Albert Michelson and Edward Morley conducted pioneering experiments in order to detect the luminiferous aether thought to be pervading space. The aether idea was not too crazy—after all, water waves travel through water and sound travels through air. Didn't light also require a medium through which to propagate, even in an apparent vacuum? In order to detect aether, the researchers split a light beam into two beams that traveled at right angles to each other. Both beams were reflected back and recombined to produce a striped interference pattern that depended on the time spent traveling in both directions. If the Earth moved through an aether, this should be detectable as a change in the interference pattern produced when one of the light beams (which had to travel *into* the aether "wind") was slowed relative to the other beam. Michelson explained the idea to his daughter, "Two beams of light race against each other, like two swimmers, one struggling upstream and back, while the other, covering the same distance just crosses and returns. The second swimmer will always win if there is any *current* in the river."

In order to make such fine measurements, vibrations were minimized by floating the apparatus on a pool of mercury, and the apparatus could be rotated relative to the motion of the Earth. No significant change in the interference patterns was found, suggesting that the Earth did not move through an "aether wind"—making the experiment the most famous "failed" experiment in physics. This finding helped to persuade other physicists to accept Einstein's **Special Theory of Relativity**.

SEE ALSO Electromagnetic Spectrum (1864), Lorentz Transformation (1904), Special Theory of Relativity (1905).

The Michelson-Morley Experiment demonstrated that the earth did not move through an aether *wind. In the late 1800s, a* luminiferous aether *(the light-bearing substance artistically depicted here) was thought to be a medium for the propagation of light.*

1889

Birth of the Kilogram

Louis Lefèvre-Gineau (1751–1829)

Since 1889, the year the Eiffel Tower opened, a kilogram has been defined by a platinum-iridium cylinder the size of a salt shaker, carefully sequestered from the world within a jar within a jar within a jar in a temperature- and humidity-controlled vault in the basement of the International Bureau of Weights and Measures near Paris. Three keys are required to open the vault. Various nations have official copies of this mass to serve as their national standards. The physicist Richard Steiner once remarked, partly in humor, "If somebody sneezed on that kilogram standard, all the weights in the world would be instantly wrong."

Today, the kilogram is the only base unit of measurement that is still defined by a physical artifact. For example, the meter is now equal to the distance traveled by light in vacuum during a time interval of 1/299,792,458 of a second—rather than by marks on a physical bar. The kilogram is a unit of mass, which is a fundamental measure of the amount of matter in the object. As can be understood from Newton's $\mathbf{F} = m\mathbf{a}$ (where $\mathbf{F}$ is force, m is mass, and $\mathbf{a}$ is acceleration), the greater an object's mass, the less the object accelerates under a given force.

Researchers are so nervous about scratching or contaminating the Paris cylinder that it has been removed from its secure location only in 1889, 1946, and 1989. Scientists have discovered that the masses of the entire worldwide collection of kilogram prototypes have been mysteriously diverging from the Paris cylinder. Perhaps the copies grew heavier by absorbing air molecules, or perhaps the Paris cylinder has become lighter. These deviations have led physicists on the quest to redefine the kilogram in terms of a fundamental unchanging constant, independent of a specific hunk of metal. In 1799, French chemist Louis Lefèvre-Gineau defined the kilogram with respect to a mass of 1000 cubic centimeters of water, but measurements of the mass and volume were inconvenient and imprecise.

SEE ALSO Acceleration of Falling Objects (1638), Birth of the Meter (1889).

Industries have often used reference masses for standards. Accuracies varied as the references became scraped and suffered other deteriorations. Here, the label dkg stands for decagram (or dekagram)—a metric unit of mass equal to 10 grams.

20dkg
10dkg
1kg

1889

Birth of the Meter

In 1889, the basic unit of length known as the meter was in the form of a special one-meter bar made of platinum and iridium, measured at the melting point of ice. Physicist and historian Peter Galison writes, "When gloved hands lowered this polished standard meter **M** into the vaults of Paris, the French, literally, held the keys to a universal system of weights and measures. Diplomacy and science, nationalism and internationalism, specificity and universality converged in the secular sanctity of that vault."

Standardized lengths were probably among the earliest "tools" invented by humans for constructing dwellings or for bartering. The word *meter* comes from the Greek *métron*, meaning "a measure," and the French *mètre*. In 1791, the French Academy of Sciences suggested that the meter be set equal to one ten-millionth of the distance from the Equator to the North Pole, passing through Paris. In fact, a multi-year French expedition took place in an effort to determine this distance.

The history of the meter is both long and fascinating. In 1799, the French created a bar made from platinum with the appropriate length. In 1889, a more definitive platinum-iridium bar was accepted as the international standard. In 1960, the meter was defined as being equal to the impressive-sounding 1,650,763.73 wavelengths in vacuum of the radiation corresponding to the transition between the $2p10$ and $5d5$ quantum levels of the krypton-86 atom! No longer did the meter have a direct correspondence to a measurement of the Earth. Finally, in 1983, the world agreed that the meter was equal to the distance traveled by light in vacuum during a time interval of 1/299,792,458 of a second.

Interestingly, the first prototype bar was short by a fifth of a millimeter because the French did not take into account that the Earth is not precisely spherical but flattened closer to the poles. However, despite this error, the actual length has not changed; rather, the *definition* has changed in order to increase the possible precision of the measurement.

SEE ALSO Stellar Parallax (1838), Birth of the Kilogram (1889).

For centuries, engineers have been interested in making increasingly fine measurements of length. For example, calipers can be used to measure and compare the distances between two points on an object.

Tyrrhenian
Sea
Gibr. to Naples
Hamburg to Yokohama
Porto d'Anzio
C. Circello
G. of Gaeta
Ischia
Capri
Amalfi
G. of Salerno
Manfredonia
Bari
Lecce
Gallipoli
Taranto
Corigliano
Rossano
Catanzaro
Cotrone
Paola
Squillace
C. Stilo
Stromboli
Lipari I.
Ustica
Palermo
Messina
Str. of Messina
C. Spartivento
Catania
C. Passaro
Malta to Corfu
Marsala
Mazzara
Sciacca
Trieste to Brazil
G. of Cagliari
C. Spartivento
C. Blanco
C. Bon
Tunis
Pantellaria
Ras al Mahmour
G. of Hammamet
Kuriat I.
Mahadiah
Malta (B.)
Gozo
Comino
Philippeville
Collo
Bona
Constantine
Batna
Biskra
G. of Gabes

Eötvös' Gravitational Gradiometry

Loránd von Eötvös (1848–1919)

Hungarian physicist and world-renowned mountain-climber Loránd Eötvös was not the first to use a torsion balance (an apparatus that twists in order to measure very weak forces) to study the gravitational attraction between masses, but Eötvös refined his balance to gain added sensitivity. In fact, the Eötvös balance became one of the best instruments for measuring gravitational fields at the surface of the Earth and for predicting the existence of certain structures beneath the surface. Although Eötvös focused on basic theory and research, his instruments later proved important for prospecting for oil and natural gas.

This device was essentially the first instrument useful for gravitational gradiometry—that is, for the measurement of very local gravitational properties. For example, Eötvös' early measurements involved his mapping changes of the gravitational potential at different locations in his office and, shortly after, the entire building. Local masses in the rooms influence the values he obtained. The Eötvös balance could also be used to study the gravitational changes due to slow motions of massive bodies or fluids. According to physicist Péter Király, "changes in the water level of the Danube could allegedly be detected from a cellar 100 meters away with a centimeter precision, but that measurement was not well documented."

Eötvös' measurements also showed that *gravitational mass* (the mass m in Newton's Law of Gravitation, $F = Gm_1m_2/r^2$) and *inertial mass* (the constant mass m responsible for inertia in Newton's Second Law, which we often write as $\mathbf{F} = m\mathbf{a}$) were the same—at least to an accuracy of about five parts in 10^9. In other words, Eötvös showed that the inertial mass (the measure of the resistance of an object to acceleration by an applied force) is the same as gravitational mass (the factor determining the weight of an object) to within a great level of accuracy. This information later proved useful for Einstein when he formulated the General Theory of Relativity. Einstein cited Eötvös' work in Einstein's 1916 paper, "The Foundation of the **General Theory of Relativity**."

SEE ALSO Newton's Laws of Motion and Gravitation (1687), Cavendish Weighs the Earth (1798), General Theory of Relativity (1915).

LEFT: *Loránd Eötvös, 1889.* RIGHT: *Visualization of gravity, created with data from NASA's Gravity Recovery and Climate Experiment (GRACE). Variations in the gravity field are shown across the Americas. Red shows the areas where gravity is stronger.*

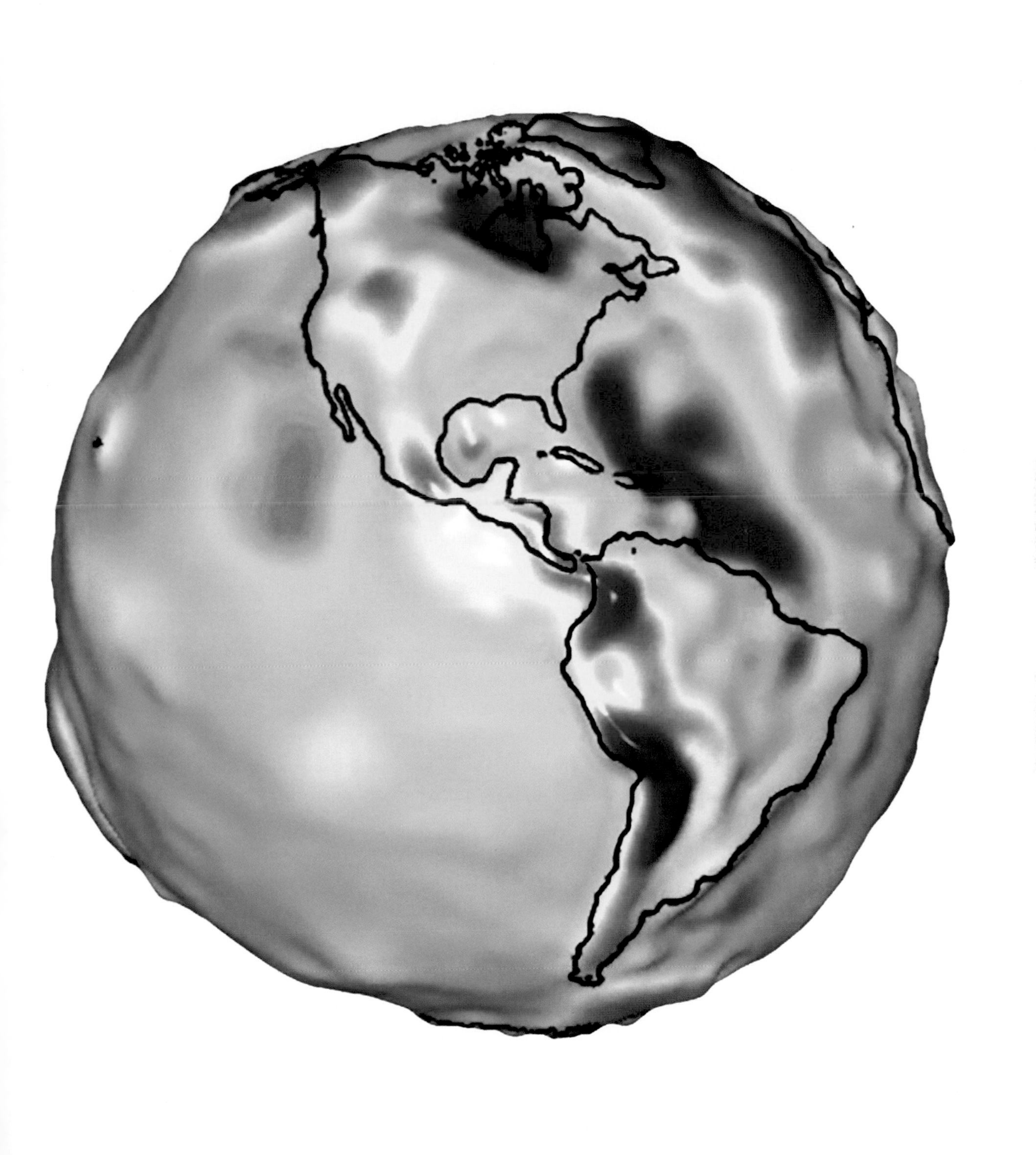

1891

Tesla Coil

Nikola Tesla (1856–1943)

The Tesla coil (TC) has played a significant role in stimulating generations of students to become interested in the wonders of science and electrical phenomena. On the wilder side, it is sometimes used in horror movies by mad scientists to create impressive lightning effects, and paranormal researchers creatively suggest that "heightened supernatural activity has been reported when they are in use!"

Developed around 1891 by the inventor Nikola Tesla, the TC can be used to produce high-voltage, low-current, high-frequency, alternating-current electricity. Tesla used the TCs to conduct experiments involving the transmission of electrical energy without wires to extend the limits of our understanding of electrical phenomena. "None of the circuit's typical components were unknown at the time," writes the Public Broadcasting Service (PBS), "but its design and operation together achieved unique results—not the least because of Tesla's masterful refinements in construction of key elements, most particularly of a special transformer, or coil, which is at the heart of the circuit's performance."

Generally speaking, an electrical transformer transfers electrical energy from one circuit to another through the transformer's coils. A varying current in the primary coil winding creates a varying magnetic flux in the transformer's core, which then creates a varying magnetic field through the secondary winding that induces a voltage in the secondary winding. In the TC, a high-voltage capacitor and spark gap are used to periodically excite a primary coil with bursts of current. The secondary coil is excited through resonant inductive coupling. The more turns the secondary winding has in relation to the primary winding, the larger the increase in voltage. Millions of volts can be produced in this fashion.

Often, the TC has a large metal ball (or other shape) on top from which streams of electricity chaotically shoot. In practice, Tesla had constructed a powerful radio transmitter, and he also used the device to investigate phosphorescence (a process in which energy absorbed by an object is released in the form of light) and X-rays.

SEE ALSO Von Guericke's Electrostatic Generator (1660), Leyden Jar (1744), Ben Franklin's Kite (1752), Lichtenberg Figures (1777), Jacob's Ladder (1931).

High-voltage arcs of a Tesla coil discharging to a piece of copper wire. The voltage is approximately 100,000 volts.

1892

Thermos

James Dewar (1842–1923), **Reinhold Burger** (1866–1954)

Invented by Scottish physicist James Dewar in 1892, the thermos (also called the Dewar flask or vacuum flask) is a double-walled container with a vacuum space between the walls that enables the flask to keep its contents hotter or colder than the surroundings for a significant period of time. After it was commercialized by German glass-blower Reinhold Burger, the thermos "was an instant success and a worldwide bestseller," writes author Joel Levy, "thanks in part to the free publicity it gained from the day's leading explorers and pioneers. Thermos vacuum flasks were carried to the South Pole by Ernest Shackleton, the North Pole by William Parry, the Congo by Colonel Roosevelt and Richard Harding Davis, Mount Everest by Sir Edmund Hillary, and into the skies by both the Wright brothers and Count von Zeppelin."

The flask works by reducing the three principal ways in which objects exchange heat with the environment: *conduction* (e.g. the way heat spreads from the hot end of an iron bar to the colder end), *radiation* (e.g. the heat one feels radiating from bricks in a fireplace after the fires have burned out), and *convection* (e.g. circulation of soup in a pot that is heated from below). The narrow, hollow region between the inner and outer walls of the thermos, which is evacuated of air, reduces loss by conduction and convection, while the reflective coating on the glass reduces loss by infrared radiation.

The thermos has important uses beyond keeping a beverage hot or cold; its insulating properties have allowed for the transport of vaccines, blood plasma, insulin, rare tropical fish, and much more. During World War II, the British military produced around 10,000 thermoses that were taken by bomber crews on their nighttime raids over Europe. Today, in laboratories all over the world, thermoses are used to store ultra-cold liquids such as liquid nitrogen or liquid oxygen.

In 2009, researchers at Stanford University showed how a stack of photonic crystals (periodic structures known for blocking narrow-frequency ranges of light) layered within a vacuum could provide significantly better suppression of thermal radiation than a vacuum alone.

SEE ALSO Fourier's Law of Heat Conduction (1822), Discovery of Helium (1868).

Beyond keeping a beverage hot or cold, thermoses have been used to carry vaccines, blood plasma, rare tropical fish, and more. In laboratories, vacuum flasks are used to store ultracold liquids, such as liquid nitrogen or liquid oxygen.

1895

X-rays

Wilhelm Conrad Röntgen (1845–1923), **Max von Laue** (1879–1960)

Upon seeing her husband's X-ray image of her hand, Wilhelm Röntgen's wife "shrieked in terror and thought that the rays were evil harbingers of death," writes author Kendall Haven. "Within a month, Wilhelm Röntgen's X-rays were the talk of the world. Skeptics called them death rays that would destroy the human race. Eager dreamers called them miracle rays that could make the blind see again and could beam . . . diagrams straight into a student's brains." However, for physicians, X-rays marked a turning point in the treatment of the sick and wounded.

On November 8, 1895, the German physicist Wilhelm Röntgen was experimenting with a cathode-ray tube when he found that a discarded fluorescent screen lit up over a meter away when he switched on the tube, even though the tube was covered with a heavy cardboard. He realized that some form of invisible ray was coming from the tube, and he soon found that they could penetrate various materials, including wood, glass, and rubber. When he placed his hand in the path of the invisible rays, he saw a shadowy image of his bones. He called the rays *X-rays* because they were unknown and mysterious at that time, and he continued his experiments in secrecy in order to better understand the phenomena before discussing them with other professionals. For his systematic study of X-rays, Röntgen would win the first Nobel Prize.

Physicians quickly made use of X-rays for diagnoses, but the precise nature of X-rays was not fully elucidated until around 1912, when Max von Laue used X-rays to create a diffraction pattern of a crystal, which verified that X-rays were electromagnetic waves, like light, but of a higher energy and a shorter wavelength that was comparable to the distance between atoms in molecules. Today, X-rays are used in countless fields, ranging from X-ray crystallography (to reveal the structure of molecules) to X-ray astronomy (e.g., the use of X-ray detectors on satellites to study X-ray emissions from sources in outer space).

SEE ALSO Telescope (1608), Triboluminescence (1620), Radioactivity (1896), Electromagnetic Spectrum (1864), Bremsstrahlung (1909), Bragg's Law of Crystal Diffraction (1912), Compton Effect (1923).

X-ray of the side view of a human head, showing screws used to reconstruct the jaw bones.

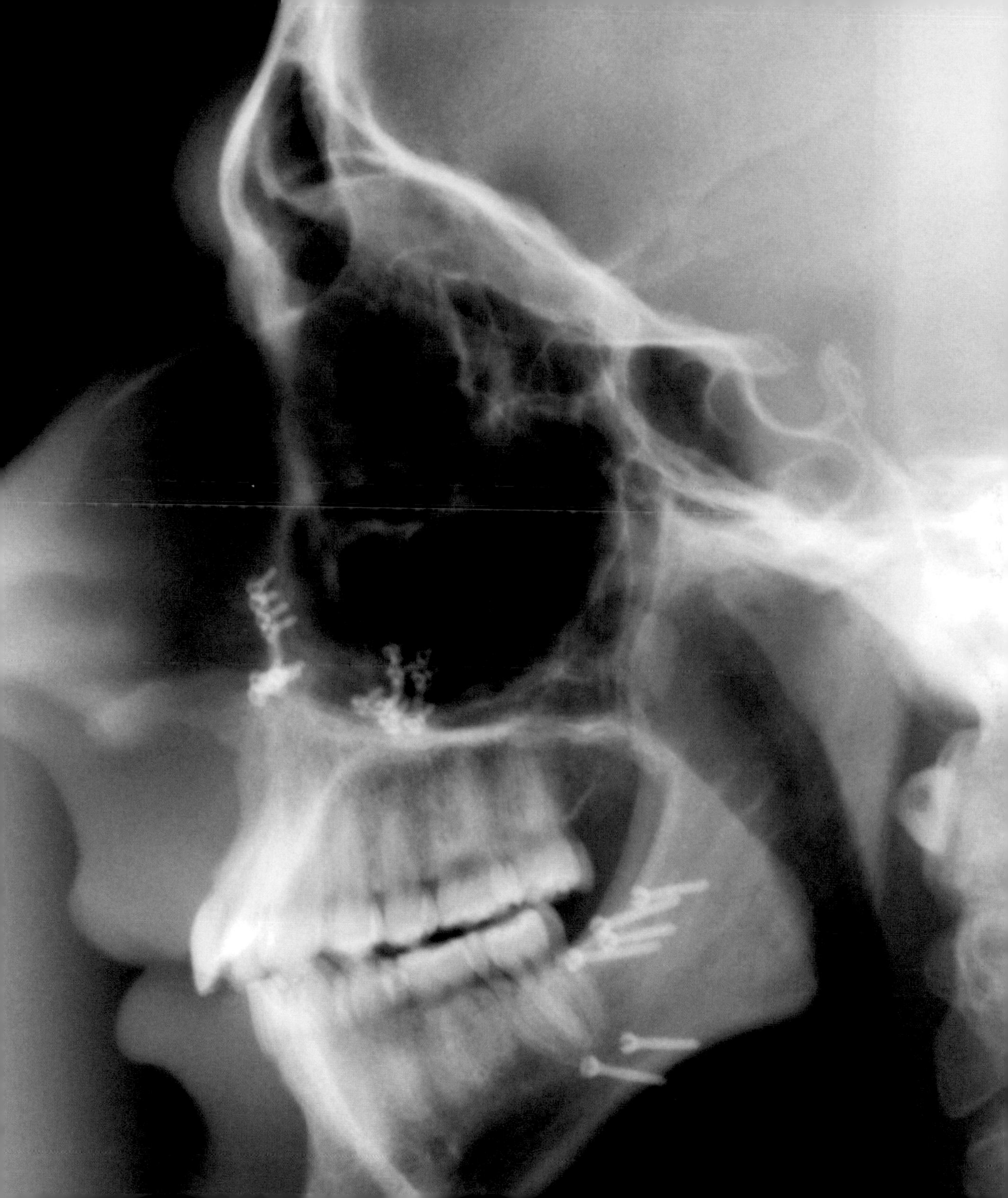

1895

Curie's Magnetism Law

Pierre Curie (1859–1906)

French physical chemist Pierre Curie considered himself to have a feeble mind and never went to elementary school. Ironically, he later shared the Nobel Prize with his wife Marie for their work on radioactivity. In 1895, he illuminated an interesting relationship between the magnetization of certain kinds of materials and the applied magnetic field and temperature T: $M = C \times (B_{ext}/T)$. Here, M is the resulting magnetization, and B_{ext} is the magnetic flux density of the applied (external) field. C is the Curie point, a constant that depends on the material. According to Curie's Law, if one increases the applied magnetic field, one tends to increase the magnetization of a material in that field. As one increases the temperature while holding the magnetic field constant, the magnetization decreases.

Curie's Law is applicable to *paramagnetic* materials, such as aluminum and copper, whose tiny atomic magnetic dipoles have a tendency to align with an external magnetic field. These materials can become very weak magnets. In particular, when subject to a magnetic field, paramagnetic materials suddenly attract and repel like standard magnets. When there is no external magnetic field, the magnetic moments of particles in a paramagnetic material are randomly oriented and the paramagnet no longer behaves as a magnet. When placed in a magnetic field, the moments generally align parallel to the field, but this alignment may be counteracted by the tendency for the moments to be randomly oriented due to thermal motion.

Paramagnetic behavior can also be observed in ferromagnetic materials—e.g., iron and nickel—that are above their Curie temperatures, T_c. The Curie temperature is a temperature above which the materials lose their ferromagnetic ability—that is, the ability to possess a net (*spontaneous*) magnetization even when no external magnetic field is nearby. Ferromagnetism is responsible for most of the magnets you encounter at home, such as permanent magnets that may be sticking to your refrigerator door or the horseshoe magnet you played with as a child.

SEE ALSO *De Magnete* (1600), Hall Effect (1879), Piezoelectric Effect (1880).

LEFT: *Photograph of Pierre Curie and his wife Marie, with whom he shared the Nobel Prize.* RIGHT: *Platinum is an example of a paramagnetic material at room temperature. This platinum nugget is from Konder mine, Yakutia, Russia.*

1896

Radioactivity

Abel Niépce de Saint-Victor (1805–1870), **Antoine Henri Becquerel** (1852–1908), **Pierre Curie** (1859–1906), **Marie Skłodowska Curie** (1867–1934), **Ernest Rutherford** (1871–1937), **Frederick Soddy** (1877–1956)

To understand the behavior of radioactive nuclei (the central regions of atoms), picture popcorn popping on your stove. Kernels appear to pop at random over several minutes, and a few don't seem to pop at all. Similarly, most familiar nuclei are stable and are essentially the same now as they were centuries ago. However, other kinds of nuclei are unstable and spew fragments as the nuclei disintegrate. Radioactivity is the emission of such particles.

The discovery of radioactivity is usually associated with French scientist Henri Becquerel's 1896 observations of phosphorescence in uranium salts. Roughly a year before Becquerel's discovery, German physicist Wilhelm Röntgen discovered X-rays while experimenting with electrical discharge tubes, and Becquerel was curious to see if phosphorescent compounds (compounds that emit visible light after being stimulated by sunlight or other excitation waves) might also produce X-rays. Becquerel placed uranium potassium sulfate on a photographic plate that was wrapped in black paper. He wanted to see if this compound would phosphoresce and produce X-rays when stimulated by light.

To Becquerel's surprise, the uranium compound darkened the photographic plate even when the packet was in a drawer. Uranium seemed to be emitting some kind of penetrating "rays." In 1898, physicists Marie and Pierre Curie discovered two new radioactive elements, polonium and radium. Sadly, the dangers of radioactivity were not immediately recognized, and some physicians began to provide radium enema treatments among other dangerous remedies. Later, Ernest Rutherford and Frederick Soddy discovered that these kinds of elements were actually transforming into other elements in the radioactive process.

Scientists were able to identify three common forms of radioactivity: alpha particles (bare helium nuclei), beta rays (high-energy electrons), and gamma rays (high-energy electromagnetic rays). Author Stephen Battersby notes that, today, radioactivity is used for medical imaging, killing tumors, dating ancient artifacts, and preserving food.

SEE ALSO Prehistoric Nuclear Reactor (2 Billion B.C.), X-rays (1895), Graham's Law of Effusion (1829), $E = mc^2$ (1905), Geiger Counter (1908), Quantum Tunneling (1928), Cyclotron (1929), Neutron (1932), Energy from the Nucleus (1942), Little Boy Atomic Bomb (1945), Radiocarbon Dating (1949), CP Violation (1964).

During the late 1950s, fallout shelters grew in number across the US. These spaces were designed to protect people from radioactive debris from a nuclear explosion. In principle, people might remain in the shelters until the radioactivity had decayed to a safer level outside.

CAPACITY
FALLOUT SHELTER

1897

Electron

Joseph John "J. J." Thomson (1856–1940)

"The physicist J. J. Thomson loved to laugh," writes author Josepha Sherman. "But he was also clumsy. Test tubes broke in his hands, and experiments refused to work." Nevertheless, we are lucky that Thomson persisted and revealed what Benjamin Franklin and other physicists had suspected—that electrical effects were produced by minuscule units of electrical charge. In 1897, J. J. Thomson identified the electron as a distinct particle with a mass much smaller than the atom. His experiments employed a cathode ray tube: an evacuated tube in which a beam of energy travels between a positive and negative terminal. Although no one was sure what cathode rays actually were at the time, Thompson was able to bend the rays using a magnetic field. By observing how the cathode rays moved through electric and magnetic fields, he determined that the particles were identical and did not depend on the metal that emitted them. Also, the particles all had the same ratio of electric charge to mass. Others had made similar observations, but Thomson was among the first to suggest that these "corpuscles" were the carriers of all forms of electricity and a basic component of matter.

Discussions of the various properties of electrons are presented in many sections of this book. Today, we know that the electron is a subatomic particle with negative electric charge and a mass that is 1/1,836 of the mass of a proton. An electron in motion generates a magnetic field. An attractive force, known as the Coulomb force, between the positive proton and the negative electron causes electrons to be bound to atoms. Chemical bonds between atoms may result when two or more electrons are shared between atoms.

According to the American Institute of Physics, "Modern ideas and technologies based on the electron, leading to the television and the computer and much else, evolved through many difficult steps. Thomson's careful experiments and adventurous hypotheses were followed by crucial experimental and theoretical work by many others [who] opened for us new perspective—a view from *inside* the atom."

SEE ALSO Atomic Theory (1808), Millikan Oil Drop Experiment (1913), Photoelectric Effect (1905), De Broglie Relation (1924), Bohr Atom (1913), Stern-Gerlach Experiment (1922), Pauli Exclusion Principle (1925), Schrödinger's Wave Equation (1926), Dirac Equation (1928), Wave Nature of Light (1801), Quantum Electrodynamics (1948).

A lightning discharge involves a flow of electrons. The leading edge of a bolt of lightning can travel at speeds of 130,000 miles per hour (60,000 meters/second) and can reach temperatures approaching 54,000 °F (30,000 °C).

1898

Mass Spectrometer

Wilhelm Wien (1864–1928), **Joseph John "J. J." Thomson** (1856–1940)

"One of the pieces of equipment that has contributed most to the advancement in scientific knowledge in the twentieth century is undoubtedly the mass spectrometer," writes author Simon Davies. The mass spectrometer (MS) is used to measure the masses and relative concentrations of atoms and molecules in a sample. The basic principle involves the generation of ions from a chemical compound, then separating these ions according to their mass-to-charge ratio (m/z), and finally detecting the ions and characterizing them according to their m/z and abundance in the sample. The specimen may be ionized by many methods, including by bombarding it with energetic electrons. The resulting ions can be charged atoms, molecules, or molecular fragments. For example, the specimen may be bombarded with an electron beam, which forms positively charged ions when it strikes a molecule and knocks an electron out of the molecule. Sometimes, molecular bonds are broken, creating charged fragments. In a MS, the speed of a charged particle may change while passing through an electric field, and its travel direction altered by a magnetic field. The amount of ion deflection is affected by m/z (e.g., the magnetic force deflects lighter ions more than heavier ions). The detector records the relative abundance of each ion type.

When identifying the fragments detected in a sample, the resulting mass spectrum is often compared against spectra of known chemicals. MS can be used for many applications, including determining different isotopes in a sample (i.e. elements that have a different number of neutrons), protein characterization (e.g. using an ionization method called *electrospray ionization*), and exploring outer space. For example, MS devices were taken on space probes used to study atmospheres of other planets and moons.

Physicist Wilhelm Wien established the foundation of mass spectrometry in 1898, when he found that beams of charged particles were deflected by electric and magnetic fields according to their m/z. J. J. Thomson and others refined the spectrometry apparatus through the years.

SEE ALSO Fraunhofer Lines (1814), Electron (1897), Cyclotron (1929), Radiocarbon Dating (1949).

A mass spectrometer on the Cassini–Huygens spacecraft was used to analyze particles in the atmospheres of Saturn and its moons and rings. Launched in 1997, the craft was part of a joint mission of NASA, the European Space Agency, and the Italian Space Agency.

Blackbody Radiation Law

Max Karl Ernst Ludwig Planck (1858–1947), **Gustav Robert Kirchhoff** (1824–1887)

"Quantum mechanics is magic," writes quantum physicist Daniel Greenberger. Quantum theory, which suggests that matter and energy have the properties of both particles and waves, had its origin in pioneering research concerning hot objects that emit radiation. For example, imagine the coil on an electric heater that glows brown and then red as it gets hotter. The Blackbody Radiation Law, proposed by German physicist Max Planck in 1900, quantifies the amount of energy emitted by blackbodies at a particular wavelength. Blackbodies are objects that emit and absorb the maximum possible amount of radiation at any given wavelength and at any given temperature.

The amount of thermal radiation emitted by a blackbody changes with frequency and temperature, and many of the objects that we encounter in our daily lives emit a large portion of their radiation spectrum in the infrared, or far-infrared, portion of the spectrum, which is not visible to our eyes. However, as the temperature of a body increases, the dominant portion of its spectrum shifts so that we can see a glow from the object.

In the laboratory, a blackbody can be approximated using a large, hollow, rigid object such as a sphere, with a hole poked in its side. Radiation entering the hole reflects off the inner walls, dissipating with each reflection as the walls absorb the radiation. By the time the radiation exits through the same hole, its intensity is negligible. Thus, this hole acts as a blackbody. Plank modeled the cavity walls of blackbodies as a collection of tiny electromagnetic oscillators. He posited that the energy of oscillators is discrete and could assume only certain values. These oscillators both *emit* energy into the cavity and *absorb* energy from it via discrete jumps, or in packages called quanta. Planck's quantum approach involving discrete oscillator energies for theoretically deriving his Radiation Law led to his 1918 Nobel Prize. Today, we know that the universe was a near-perfect blackbody right after the **Big Bang**. German physicist Gustav Kirchhoff introduced the actual term *blackbody* in 1860.

SEE ALSO Big Bang (13.7 Billion B.C.), Photoelectric Effect (1905).

LEFT: *Max Planck, 1878.* RIGHT: *Molten glowing lava is an approximation of blackbody radiation, and the temperature of the lava can be estimated from the color.*

1901

Clothoid Loop

Edwin C. Prescott (1841–1931)

The next time you see a vertical loop in a roller coaster track—aside from wondering why people would allow themselves to be turned upside down and careened through insane curves—notice that the loop is not circular, but instead has an inverted teardrop shape. Mathematically speaking, the loop is called a clothoid, which is a section of a cornu spiral. The curve is employed in such loops for safety reasons.

On traditional roller coasters, the potential energy provided by gravity is often introduced in a large initial climb that is converted to kinetic energy as the roller coaster begins to plummet. Circular loops would require that the cars enter the loop with a greater entry speed than cars in a clothoid in order to complete the loop. The larger entry speed required by a circular loop would subject riders to greater centripetal acceleration through the lower half of the loop and dangerous g-forces (where 1 g is the g-force acting on a stationary object resting on the Earth's surface).

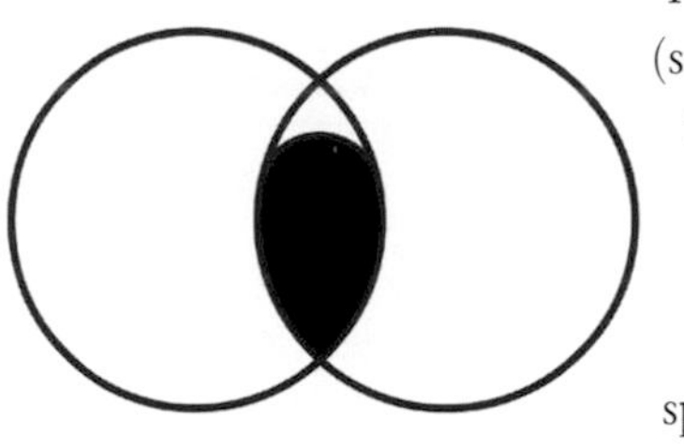

For example, if you were to superimpose a clothoid curve on a circle (see diagram), the top of the clothoid loop is lower than the top of the circle. Because a roller coaster car converts its kinetic energy to potential energy as it rolls around the top of the loop, the decreased height of the clothoid can be reached at lower speeds and decreased g-force. Also, the short arc at the top of the clothoid allows the car to spend less time at a location where the car is both upside-down and traveling slowly.

One of the earliest uses of the teardrop shape was observed in 1901, when inventor Edwin Prescott built the Loop-the-Loop at Coney Island, New York. Prescott was largely self-taught in mechanics. His 1898 version of the "centrifugal railway" called for a circular loop, but the sudden centripetal acceleration placed too much strain on the riders' bodies when the cars rounded the loop at high speed. Prescott patented the teardrop-like design in 1901.

SEE ALSO Acceleration of Falling Objects (1638), Tautochrone Ramp (1673), Newton's Laws of Motion and Gravitation (1687).

Roller-coaster loops are usually not circular but instead have an inverted teardrop shape. Clothoid loops are employed partly for safety reasons.

1903

Black Light

Robert Williams Wood (1868–1955)

Those of you old enough to remember the ubiquitous black-light posters of the psychedelic Sixties will appreciate author Edward J. Rielly's reminiscence: "Individuals tripping on LSD were fond of 'black lights,' which could be purchased in 'headshops' along with fluorescent paints and dyes. Fluorescent clothes or posters in the presence of black-light bulbs created a visual counterpart to the effect of LSD." Even restaurants sometimes installed black lights to provide a mysterious ambience. Author Laren Stover writes, "It would not be unusual to find black-light posters of Jimi Hendrix in the Gypsy/ Psychedelic Bohemian home, as well as a Lava lamp bubbling away in a corner."

A black light (also called a Wood's lamp) emits electromagnetic radiation that is mostly in the near-ultraviolet range (see entry on **Stokes' Fluorescence**). To create a fluorescent black light, technologists often use *Wood's glass*, a purple glass containing nickel oxide, which blocks most visible light but lets ultraviolet light (UV) pass through. The phosphor inside the light bulb has an emission peak below 400 nanometers. Although the first black lights were incandescent bulbs made of Wood's glass, these black lights were inefficient and hot. Today, one version of a black light is used to attract insects in outdoor bug zappers; however, in order to keep costs down, these lights omit the Wood's glass and therefore produce more visible light.

The human eye cannot see UV, but we can see the effects of fluorescence and phosphorescence when the light is shone upon psychedelic posters. Black lights have countless applications ranging from crime investigations, where the lights can be used to reveal trace amounts of blood and semen, to dermatological use, where they are used to detect various skin conditions and infections.

Although the black light was invented by chemist William H. Byler, black lights are most often associated with American physicist Robert Wood, the "father of ultraviolet photography" and inventor of Wood's glass in 1903.

SEE ALSO Stokes' Fluorescence (1852), Electromagnetic Spectrum (1864), Incandescent Light Bulb (1878), Neon Signs (1923), Lava Lamp (1963).

Homeowners in the American southwest use black lights to locate nocturnal scorpions that may have invaded homes. The scorpion's body brightly fluoresces in response to black light.

1903

Tsiolkovsky Rocket Equation

Konstantin Eduardovich Tsiolkovsky (1857–1935)

"The Space Shuttle launch is perhaps the best known image of modern space travel," writes author Douglas Kitson, "but the leap from fireworks to space ships was a big one, and would have been impossible without the ideas of Konstantin Tsiolkovsky." The Russian schoolteacher Tsiolkovsky was aware of descriptions of fictitious journeys to the Moon with birds or guns as the propelling force, but by solving the equation of motion of the rocket, he was able to show that actual space travel was possible from the standpoint of physics. Knowing that a rocket that can apply acceleration to itself by expelling part of its mass at high speed in the opposite direction, Tsiolkovsky's rocket equation describes the change in velocity Δv of the rocket with respect to the initial total mass m_0 of the rocket (which includes the propellant), the final total mass m (e.g. after all the propellant has been burned), and the constant effective exhaust velocity v_e of the propellant: $\Delta v = v_e \cdot \ln (m_0/m_1)$.

His rocket equation, derived in 1898 and published in 1903, also led him to another important conclusion. It would be unrealistic for a single-stage rocket to carry humans to outer space using rocket fuel because the weight of the fuel would exceed, by several times, the weight of the entire rocket, including the people and devices on board. He was then able to analyze, in depth, how a multi-stage rocket would make the journey feasible. The multi-stage rocket has more than one rocket joined together. When the first one has burned all its fuel, it falls away, and a second stage starts to fire.

"Mankind will not forever remain on Earth," wrote Tsiolkovsky in 1911, "but in the pursuit of light and space, mankind will at first timidly penetrate beyond the limits of the atmosphere, and then conquer all the space around the Sun. Earth is the cradle of humanity, but one cannot remain in the cradle forever."

SEE ALSO Hero's Jet Engine (50), Cannon (1132), Conservation of Momentum (1644), Newton's Laws of Motion and Gravitation (1687), Conservation of Energy (1843), Dynamite (1867), Fermi Paradox (1950).

Soyuz spacecraft and launch vehicle on the launchpad at the Baikonur complex in Kazakhstan. Baikonur is the world's largest space center. The launch was part of a cooperative space mission in 1975 between the United States and the USSR.

1904

Lorentz Transformation

Hendrik Antoon Lorentz (1853–1928), **Walter Kaufmann** (1871–1947), **Albert Einstein** (1879–1955)

Imagine two spatial frames of reference, represented by two shoeboxes. One frame is stationary in a laboratory. The other is moving at constant velocity v along the x-axis of the first box—visualized by sliding one shoebox inside the other. We'll use variables with a prime sign to denote the moving frame. Next, we place a clock at the origin of each frame. When the origins are on top of one another, we set both clocks to time zero—that is, $t = t' = 0$. We can now write down four equations that describe how the primed system moves: $y' = y$; $z' = z$; $x' = (x - vt)/[1 - (v/c)^2]^{1/2}$; and $t' = [t - (vx)/c^2]/[1 - (v/c)^2]^{1/2}$. Here, c is the speed of light in a vacuum.

Play with the equations. Notice that as you increase the velocity in the third equation, x' shrinks. In fact, if something travels 0.999 times the speed of light, it shrinks 22 times and slows down by a factor of 22 relative to the laboratory observer. This means that the Statue of Liberty (151 feet, or 46 meters, tall) would appear to shrink to 6.8 feet (2 meters) in its direction of travel if we were to launch it at 0.999 times the speed of light. Someone inside the statue's inner chambers would contract by the same amount and also age 22 times slower than you in the laboratory frame.

The formulas are called the Lorentz transformations after the physicist Hendrik Lorentz, who developed them in 1904. A similar-looking equation shows that the mass of a moving body depends on speed $m = m_0/[1 - (v/c)^2]^{1/2}$, where m_0 is the mass when velocity $v = 0$. In 1901, the velocity-dependent change of an electron's mass was first experimentally observed by the physicist Walter Kaufmann. Notice how this formula appears to preclude increasing the velocity of a starship close to the speed of light, because the value of its mass would approach infinity within the lab reference frame. Several scientists had been discussing the physics behind these equations since 1887. Albert Einstein reinterpreted the Lorentz transformation as it relates to the fundamental nature of both space and time.

SEE ALSO Michelson-Morley Experiment (1887), Special Theory of Relativity (1905), Tachyons (1967).

Statue of Liberty contracting in accordance with the Lorentz transformation. Candles indicate the approximate height of the statue when it is stationary (first candle), at 0.9 × c (second candle), 0.99 × c (third candle), and 0.999 × c (fourth candle).

1905

Special Theory of Relativity

Albert Einstein (1879–1955)

Albert Einstein's Special Theory of Relativity (STR) is one of humankind's greatest intellectual triumphs. When Albert Einstein was only 26, he made use of one of the key foundations of STR—namely, that the speed of light in a vacuum is independent of the motion of the light source and the same for all observers, no matter how they are moving. This is in contrast to the speed of sound, which changes for an observer who moves, for example, with respect to the sound source. This property of light led Einstein to deduce the *relativity of simultaneity*: Two events occurring at the same time as measured by one observer sitting in a laboratory frame of reference occur at different times for an observer moving relative to this frame.

Because time is relative to the speed one is traveling at, there can never be a clock at the center of the Universe to which everyone can set their watches. Your entire life can be the blink of an eye to an alien who leaves the Earth traveling close to the speed of light and then returns an hour later to find that you have been dead for centuries. (The word "relativity" partly derives from the fact that the appearance of the world depends on our relative states of motion—the appearance is "relative.")

Although the strange consequences of STR have been understood for over a century, students still learn of them with a sense of awe and bewilderment. Nevertheless, from tiny subatomic particles to galaxies, STR appears to accurately describe nature.

To help understand another aspect of STR, imagine yourself in an airplane traveling at constant speed relative to the ground. This may be called the *moving frame of reference*. The principle of relativity makes us realize that without looking out the window, you cannot tell how fast you are moving. Because you cannot see the scenery moving by, for all you know, you could be in an airplane on the ground in a stationary frame of reference with respect to the ground.

SEE ALSO Michelson-Morley Experiment (1887), General Theory of Relativity (1915), Time Travel (1949), Lorentz Transformation (1904), $E = mc^2$ (1905), Einstein as Inspiration (1921), Dirac Equation (1928), Tachyons (1967).

There can never be a clock at the center of the Universe to which everyone can set their watches. Your entire life can be the blink of an eye to an alien who leaves earth traveling at high speed and then returns.

QUARTZ

1905

$E = mc^2$

Albert Einstein (1879–1955)

"Generations have grown up knowing that the equation $E = mc^2$ changed the shape of our world," writes author David Bodanis, ". . . governing, as it does, everything from the atomic bomb to a television's cathode ray tube to the carbon dating of prehistoric paintings." Of course, part of the equation's appeal, independent from its meaning, is its simplicity. Physicist Graham Farmelo writes, "Great equations also share with the finest poetry an extraordinary power—poetry is the most concise and highly charged form of language, just as the great equations of science are the most succinct form of understanding of the aspect of physical reality they describe."

In a short article published in 1905, Einstein derived his famous $E = mc^2$, sometimes called the law of mass-energy equivalence, from the principles of **Special Relativity**. In essence, the formula indicates that the mass of a body is a "measure" of its energy content. c is the speed of light in vacuum, which is about 186,000 miles per second (299,792,468 meters per second).

Radioactive elements are constantly converting part of their masses to energy as governed by $E = mc^2$, and the formula was also used in the development of the atomic bomb to better understand the nuclear binding energy that holds the atomic nucleus together, which can be used to determine the energy released in a nuclear reaction.

$E = mc^2$ explains why the Sun shines. In the Sun, four hydrogen nuclei (four protons) are fused into a helium nucleus, which is less massive than the hydrogen nuclei that combine to make it. The fusion reaction converts the missing mass into energy that allows the Sun to heat the Earth and permits the formation of life. The mass loss, m, during the fusion supplies an energy, E, according to $E = mc^2$. Every second, fusion reactions convert about 700 million metric tons of hydrogen into helium within the Sun's core, thereby releasing tremendous energy.

SEE ALSO Radioactivity (1896), Special Theory of Relativity (1905), Atomic Nucleus (1911), Energy from the Nucleus (1942), Conservation of Energy (1843), Stellar Nucleosynthesis (1946), Tokamak (1956).

A USSR stamp from 1979, dedicated to Albert Einstein and $E = mc^2$.

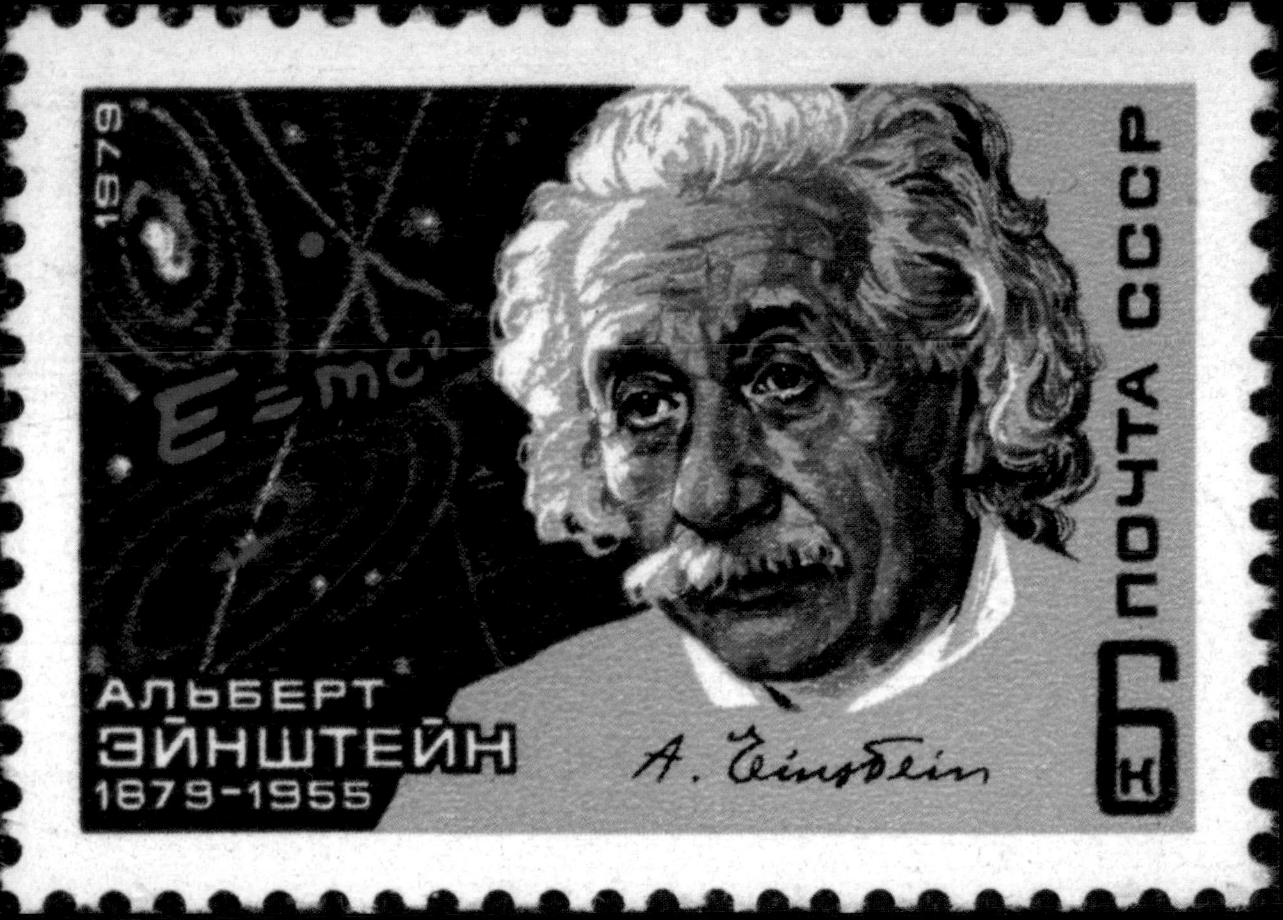

1979

$E=mc^2$

АЛЬБЕРТ
ЭЙНШТЕЙН
1879-1955

A. Einstein

ПОЧТА СССР

6 К

1905

Photoelectric Effect

Albert Einstein (1879–1955)

Of all of Albert Einstein's masterful achievements, including the **Special Theory of Relativity** and the **General Theory of Relativity**, the achievement for which he won the Nobel Prize, was his explanation of the workings the photoelectric effect (PE), in which certain frequencies of light shined on a copper plate cause the plate to eject electrons. In particular, he suggested that packets of light (now called photons) could explain the PE. For example, it had been noted that high-frequency light, such as blue or ultraviolet light, could cause electrons to be ejected—but not low-frequency red light. Surprisingly, even intense red light did not lead to electron ejection. In fact, the energy of individual emitted electrons increases with the frequency (and, hence, the color) of the light.

How could the frequency of light be the key to the PE? Rather than light exerting its effect as a classical wave, Einstein suggested that the energy of light came in the form of packets, or *quanta*, and that this energy was equal to the light frequency multiplied by a constant (later called *Planck's constant*). If the photon was below a threshold frequency, it just did not have the energy to kick out an electron. As a very rough metaphor for low-energy red quanta, imagine the impossibility of chipping a fragment from a bowling ball by tossing peas at it. It just won't work, even if you toss lots of peas! Einstein's explanation for the energy of the photons seemed to account for many observations, such as for a given metal, there exists a certain minimum frequency of incident radiation below which no photoelectrons can be emitted. Today, numerous devices, such as solar cells, rely on conversion of light to electric current in order to generate power.

In 1969, American physicists suggested that one could actually account for the PE without the concept of photons; thus, the PE did not provide definitive *proof* of photon existence. However, studies of the statistical properties of the photons in the 1970s provided experimental verification of the manifestly quantum (nonclassical) nature of the electromagnetic field.

SEE ALSO Atomic Theory (1808), Wave Nature of Light (1801), Electron (1897), Special Theory of Relativity (1905), General Theory of Relativity (1915), Compton Effect (1923), Solar Cells (1954), Quantum Electrodynamics (1948).

Photograph through night-vision device. U.S. Army paratroopers train using infrared lasers and night-vision optics in Camp Ramadi, Iraq. Night-vision goggles make use of the ejection of photoelectrons due to the photoelectric effect to amplify the presence of individual photons.

1905

Golf Ball Dimples

Robert Adams Paterson (1829–1904)

A golfer's worst nightmare may be the sand trap, but the golfer's best friends are the dimples on the golf ball that help send it sailing down the fairway. In 1848, Rev. Dr. Robert Adams Paterson invented the gutta-percha golf ball using the dried sap of a sapodilla tree. Golfers began to notice that small scrapes or nicks in the ball actually increased the golf ball's flight distance, and soon ball makers used hammers to add defects. By 1905, virtually all golf balls were manufactured with dimples.

Today we know that dimpled balls fly farther than smooth ones due to a combination of effects. First, the dimples delay separation of the boundary layer of air that surrounds the ball as it travels. Because the clinging air stays attached longer, it produces a narrower low-pressure wake (disturbed air) that trails behind the ball, reducing drag as compared to a smooth ball. Second, when the golf club strikes the ball, it usually creates a backspin that generates lift via the Magnus effect, in which an increase in the velocity of the air flowing over the ball generates a lower-pressure area at the top of the spinning ball relative to the bottom. The presence of dimples enhances the Magnus effect.

Today, most golf balls have between 250–500 dimples that can reduce the amount of drag by as much as half. Research has shown that polygonal shapes with sharp edges, such a hexagons, reduce drag more than smooth dimples. Various ongoing studies employ supercomputers to model air flow in the quest for the perfect dimple design and arrangement.

In the 1970s, Polara golf balls had an asymmetric arrangement of dimples that helped counteract the side spin that causes hooks and slices. However, the United States Golf Association banned it from tournament play, ruling it would "reduce the skill required to play golf." The association also added a symmetry rule, requiring that a ball perform essentially the same regardless of where on the surface it is struck. Polara sued, received $1.4 million, and removed the ball from the market.

SEE ALSO Cannon (1132), Baseball Curveball (1870), Kármán Vortex Street (1911).

Today, most golf balls have between 250–500 dimples that can reduce the amount of drag "felt" by the golf ball by as much as half.

1905

Third Law of Thermodynamics

Walther Nernst (1864–1941)

The humorist Mark Twain once told the crazy tale of weather so cold that a sailor's shadow froze to the deck! How cold could the environment really become?

From a classical physics perspective, the Third Law of Thermodynamics states that as a system approaches absolute zero temperature (0 K, −459.67 °F, or −273.15 °C), all processes cease and the entropy of the system approaches a minimum value. Developed by German chemist Walther Nernst around 1905, the law can be stated as follows: As the temperature of a system approaches absolute 0, the entropy, or disorder S, approaches a constant S_0. Classically speaking, the entropy of a pure and perfectly crystalline substance would be 0 if the temperature could actually be reduced to absolute zero.

Using a classical analysis, all motion stops at absolute zero. However, quantum mechanical *zero-point motion* allows systems in their lowest possible energy state (i.e. *ground state*) to have a probability of being found over extended regions of space. Thus, two atoms bonded together are not separated by some unvarying distance but can be thought of as undergoing rapid vibration with respect to each other, even at absolute zero. Instead of stating that the atom is motionless, we say that it is in a state from which no further energy can be removed; the energy remaining is called *zero-point energy*.

The term *zero-point motion* is used by physicists to describe the fact that atoms in a solid—even a super-cold solid—do not remain at exact geometric lattice points; rather, a probability distribution exists for both their positions and moments. Amazingly, scientists have been able to achieve the temperature of 100 picokelvins (0.000,000,000,1 degrees above absolute zero) by cooling a piece of rhodium metal.

It is impossible to cool a body to absolute zero by any finite process. According to physicist James Trefil, "No matter how clever we get, the third law tells us that we can never cross the final barrier separating us from absolute zero."

SEE ALSO Heisenberg Uncertainty Principle (1927), Second Law of Thermodynamics (1850), Conservation of Energy (1843), Casimir Effect (1948).

The rapid expansion of gas blowing from an aging central star in the Boomerang Nebula has cooled molecules in the nebular gas to about one degree above absolute zero, making it the coldest region observed in the distant Universe.

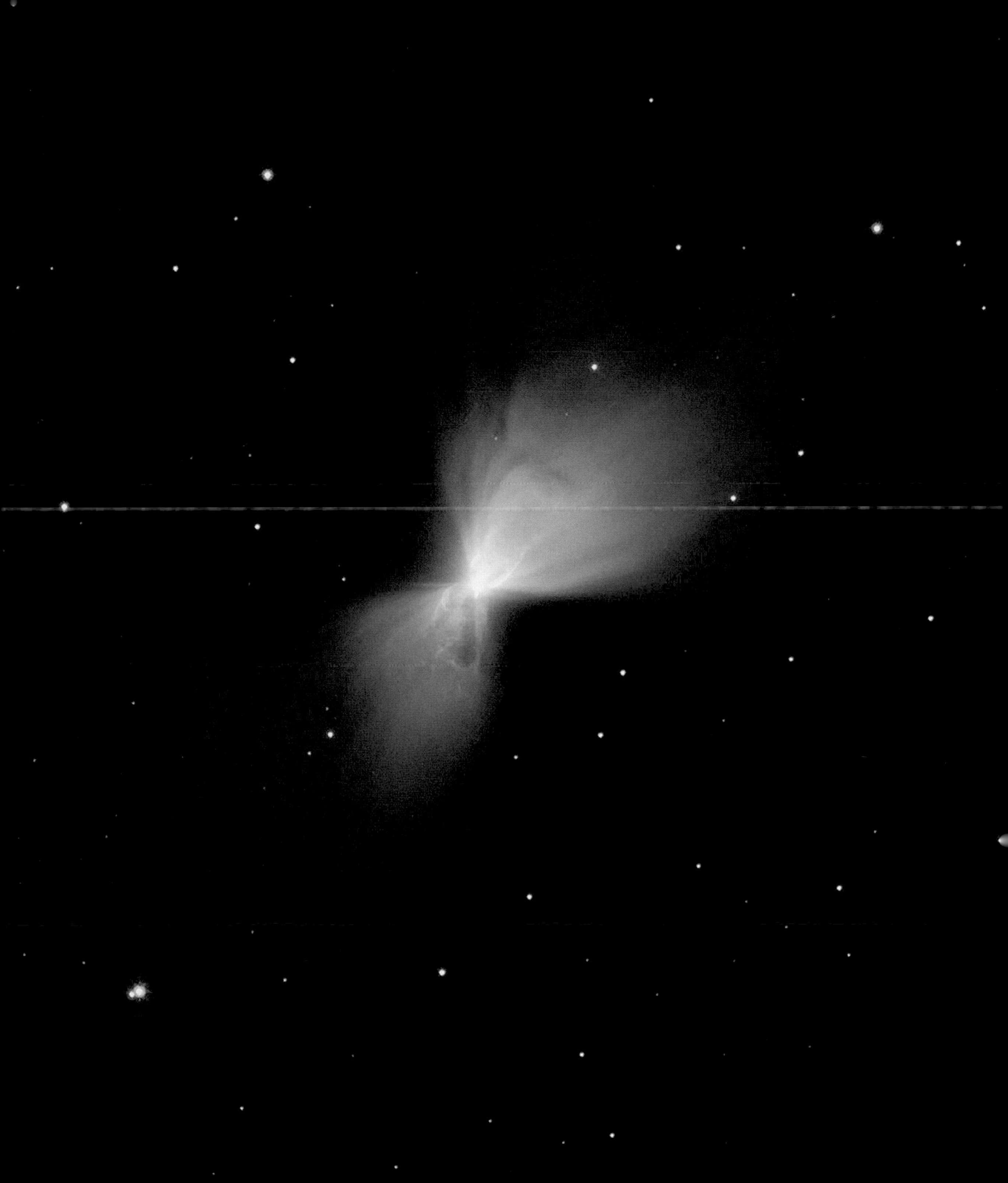

1906

Vacuum Tube

Lee De Forest (1873–1961)

In his Nobel Prize lecture on December 8, 2000, American engineer Jack Kilby noted, "The invention of the vacuum tube launched the electronics industry. . . . These devices controlled the flow of electrons in a vacuum [and] were initially used to amplify signals for audio and other devices. This allowed broadcast radio to reach the masses in the 1920s. Vacuum tubes steadily spread into other devices, and the first tube was used as a switch in calculating machines in 1939."

In 1883, the inventor Thomas Edison noticed that electrical current could "jump" from a hot filament to a metal plate in an experimental **Incandescent Light Bulb**. This primitive version of a vacuum tube eventually led to American inventor Lee De Forest's 1906 creation of the triode, which not only forces current in a single direction but can be used as an amplifier for signals such as audio and radio signals. He had placed a metal grid in his tube, and by using a small current to change the voltage on the grid, he controlled the flow of the second, larger current through the tube. Bell Labs was able to make use of this feature in its "coast to coast" phone system, and the tubes were soon used in other devices, such as radios. Vacuum tubes can also convert AC to DC current and generate oscillating radio-frequency power for radar systems. De Forest's tubes were not evacuated, but the introduction of a strong vacuum was shown to help create a useful amplifying device.

Transistors were invented in 1947, and in the following decade assumed most of the amplifying applications of tubes at much lower cost and reliability. Early computers used tubes. For example, ENIAC, the first electronic, reprogrammable, digital computer that could be used to solve a large range of computing problems, was unveiled in 1946 and contained over 17,000 vacuum tubes. Since tube failures were more likely to occur during warm-up periods, the machine was rarely turned off, reducing tube failures to one tube every two days.

SEE ALSO Incandescent Light Bulb (1878), Transistor (1947).

RCA 808 power vacuum tube. The invention of the vacuum tube launched the electronics industry and allowed broadcast radio to reach the public in the 1920s.

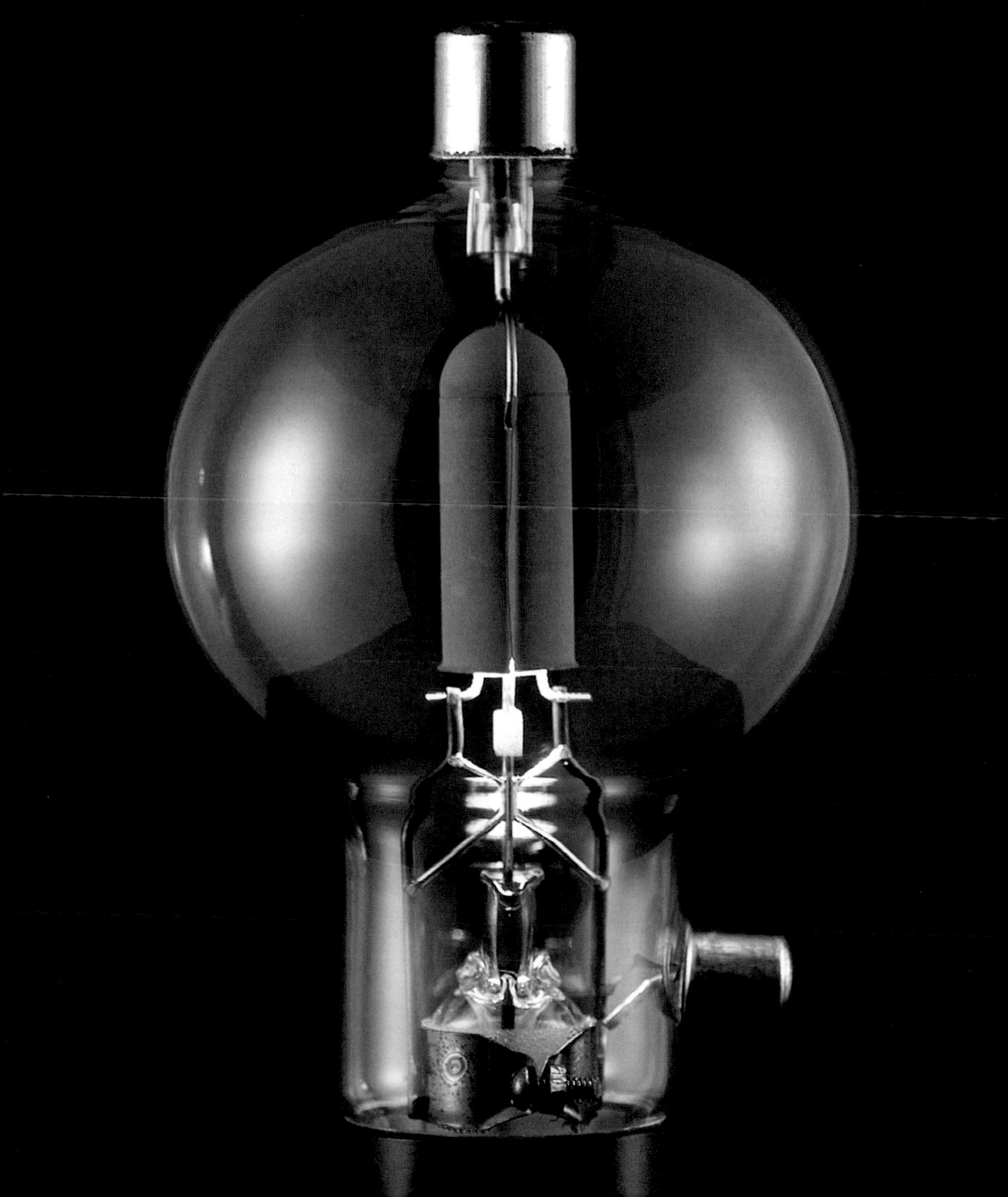

1908

Geiger Counter

Johannes (Hans) Wilhelm (Gengar) Geiger (1882–1945), **Walther Müller** (1905–1979)

During the Cold War in the 1950s, American contractors offered deluxe backyard fallout shelters equipped with bunk beds, a telephone, and a Geiger counter for detecting radiation. Science-fiction movies of the era featured huge radiation-created monsters—and ominously clicking Geiger counters.

Since the early 1900s, scientists have sought technologies for detecting **Radioactivity**, which is caused by particles emitted from unstable **Atomic Nuclei**. One of the most important detection devices was the Geiger counter, first developed by Hans Geiger in 1908 and subsequently improved by Geiger and Walther Müller in 1928. The Geiger counter consists of a central wire within a sealed metal cylinder with a mica or glass window at one end. The wire and metal tube are connected to a high-voltage power source outside the cylinder. When radiation passes through the window, it creates a trail of ion pairs (charged particles) in the gas within the tube. The positive ions of the pairs are attracted to the negatively charged cylinder walls (cathode). The negatively charged electrons are attracted to the central wire, or anode, and a slight detectable voltage drop occurs across the anode and cathode. Most detectors convert these pulses of current to audible clicks. Unfortunately, however, Geiger counters do not provide information about the kind of radiation and energy of the particles detected.

Over the years, improvements in radiation detectors have been made, including ionization counters and proportional counters that can be used to identify the kind of radiation impinging on the device. With the presence of boron trifluoride gas in the cylinder, the Geiger-Müller counter can also be modified so that it is sensitive to non-ionizing radiation, such as neutrons. The reaction of neutrons and boron nuclei generates alpha particles (positively charged helium nuclei), which are detected like other positively charged particles.

Geiger counters are inexpensive, portable, and sturdy. They are frequently used in geophysics, nuclear physics, and medical therapies, as well as in settings in which potential radiation leaks can occur.

SEE ALSO Radioactivity (1896), Wilson Cloud Chamber (1911), Atomic Nucleus (1911), Schrödinger's Cat (1935).

Geiger counters may produce clicking noises to indicate radiation and also contain a dial to show how much radiation is present. Geologists sometimes locate radioactive minerals by using Geiger counters.

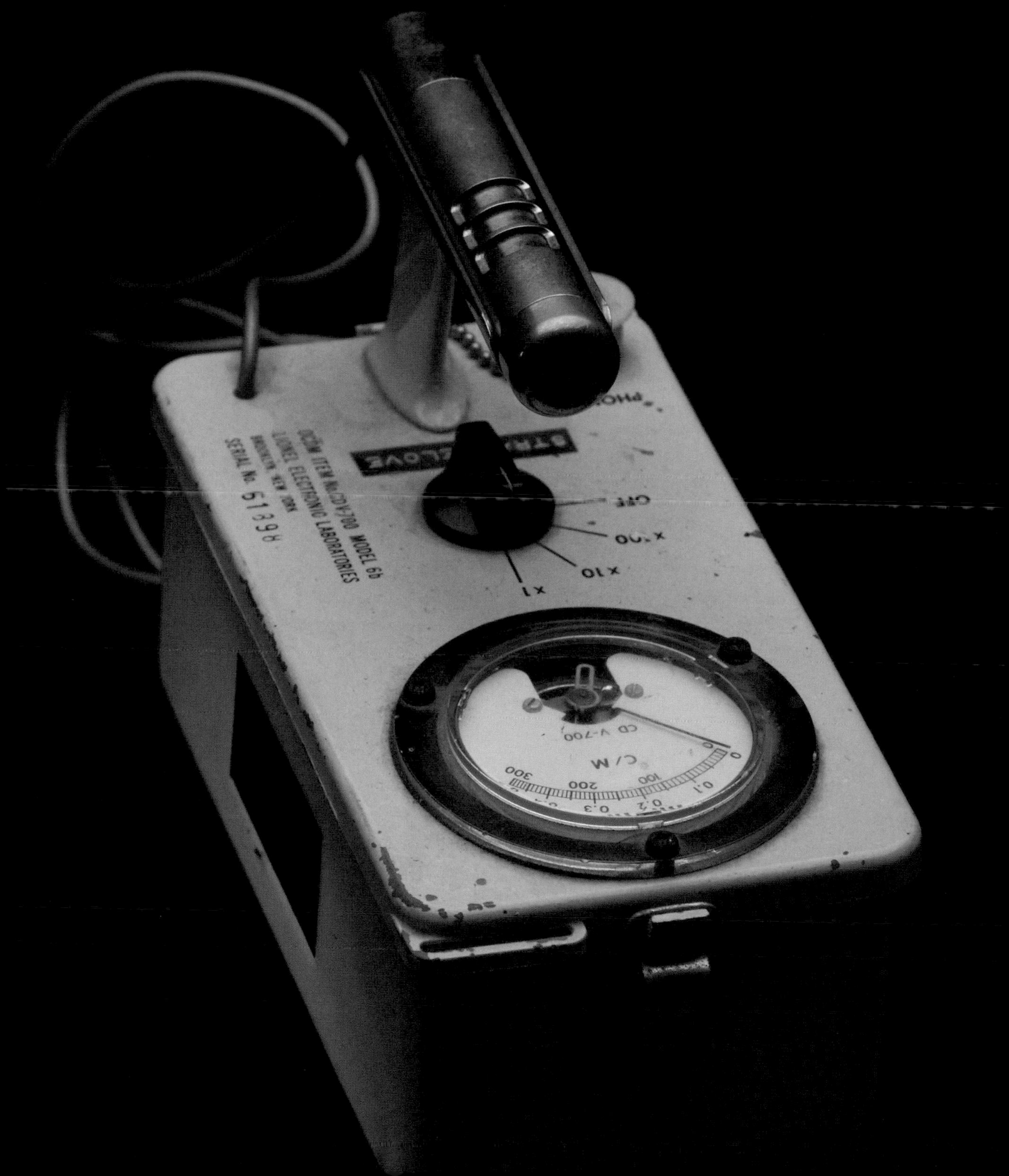
OCDM ITEM No. CDV-700 MODEL 6b
LIONEL ELECTRONIC LABORATORIES
BROOKLYN, NEW YORK
SERIAL No. 61898
OFF
x100
x10
x1
CD V-700
C/M
0
0.1
0.2
0.3
100
200
300

1909

Bremsstrahlung

Wilhelm Conrad Röntgen (1845–1923), **Nikola Tesla** (1856–1943), **Arnold Johannes Wilhelm Sommerfeld** (1868–1951)

Bremsstrahlung, or "braking radiation," refers to X-ray or other electromagnetic radiation that is produced when a charged particle, such as an electron, suddenly slows its velocity in response to the strong electric fields of **Atomic Nuclei**. Bremsstrahlung is observed in many areas of physics, ranging from materials science to astrophysics.

Consider the example of X-rays emitted from a metal target that is bombarded by high-energy electrons (HEEs) in an X-ray tube. When the HEEs collide with the target, electrons from the target are knocked out of the inner energy levels of the target atoms. Other electrons may fall into these vacancies, and X-ray photons are then emitted with wavelengths characteristic of the energy differences between the various levels within the target atoms. This radiation is called the *characteristic X-rays*.

Another type of X-ray emitted from this metal target is *bremsstrahlung*, as the electrons are suddenly slowed upon impact with the target. In fact, any accelerating or decelerating charge emits bremsstrahlung radiation. Since the rate of deceleration may be very large, the emitted radiation may have short wavelengths in the X-ray spectrum. Unlike the characteristic X-rays, bremsstrahlung radiation has a continuous range of wavelengths because the decelerations can occur in many ways—from near head-on impacts with nuclei to multiple deflections by positively charged nuclei.

Even though physicist Wilhelm Röntgen discovered X-rays in 1895 and Nikola Tesla began observing them even earlier, the separate study of the *characteristic* line spectrum and the superimposed *continuous* bremsstrahlung spectrum did not start until years later. The physicist Arnold Sommerfeld coined the term *bremsstrahlung* in 1909.

Bremsstrahlung is everywhere in the universe. Cosmic rays lose some energy in the Earth's atmosphere after colliding with atomic nuclei, slowing, and producing bremsstrahlung. Solar X-rays result from the deceleration of fast electrons in the Sun as they pass through the Sun's atmosphere. Additionally, when beta decay (a kind of radioactive decay that emits electrons or positrons, referred to as beta particles) occurs, the beta particles can be deflected by one of their own nuclei and emit *internal bremsstrahlung*.

SEE ALSO Fraunhofer Lines (1814), X-rays (1895), Cosmic Rays (1910), Atomic Nucleus (1911), Compton Effect (1923), Cherenkov Radiation (1934).

Large solar flares produce an X-ray and gamma-ray continuum of radiation partly due to the bremsstrahlung effect. Shown here is the NASA RHESSI spacecraft, launched in 2002, as it watches the Sun for X-rays and gamma rays.

1910

Cosmic Rays

Theodor Wulf (1868–1946), **Victor Francis Hess** (1883–1964)

"The history of cosmic ray research is a story of scientific adventure," write the scientists at the Pierre Auger Cosmic Ray Observatory. "For nearly a century, cosmic ray researchers have climbed mountains, ridden hot air balloons, and traveled to the far corners of the earth in the quest to understand these fast-moving particles from space."

Nearly 90 percent of the energetic cosmic-ray particles bombarding the Earth are protons, with the remainder being helium nuclei (alpha particles), electrons, and a small amount of heavier nuclei. The variety of particle energies suggests that cosmic rays have sources ranging from the Sun's solar flares to galactic cosmic rays that stream to the Earth from beyond the solar system. When cosmic ray particles enter the atmosphere of the Earth, they collide with oxygen and nitrogen molecules to produce a "shower" of numerous lighter particles.

Cosmic rays were discovered in 1910 by German physicist and Jesuit priest Theodor Wulf, who used an electrometer (a device for detecting energetic charged particles) to monitor radiation near the bottom and top of the Eiffel Tower. If radiation sources were on the ground, he would have detected less radiation as he moved away from the Earth. To his surprise, he found that the level of radiation at the top was larger than what would be expected if it were from terrestrial radioactivity. In 1912, Austrian-American physicist Victor Hess carried detectors in a balloon to an altitude of 17,300 feet (5,300 meters) and found that the radiation increased to a level four times that at the ground.

Cosmic rays have been creatively referred to as "killer rays from outer space" because they may contribute to more than 100,000 deaths from cancer each year. They also have sufficient energy to damage electronic **Integrated Circuits** and may alter data stored in the memory of a computer. The highest-energy cosmic-ray events are quite curious, because their arrival directions do not always suggest a particular source; however, supernovae (exploding stars) and stellar winds from massive stars are probably accelerators of cosmic-ray particles.

SEE ALSO Bremsstrahlung (1909), Integrated Circuit (1958), Gamma Ray Bursts (1967), Tachyons (1967).

Cosmic rays may have deleterious effects on components of electronic integrated circuits. For example, studies in the 1990s suggested that roughly one cosmic-ray-induced error occurred for 256 megabytes of RAM (computer memory) every month.

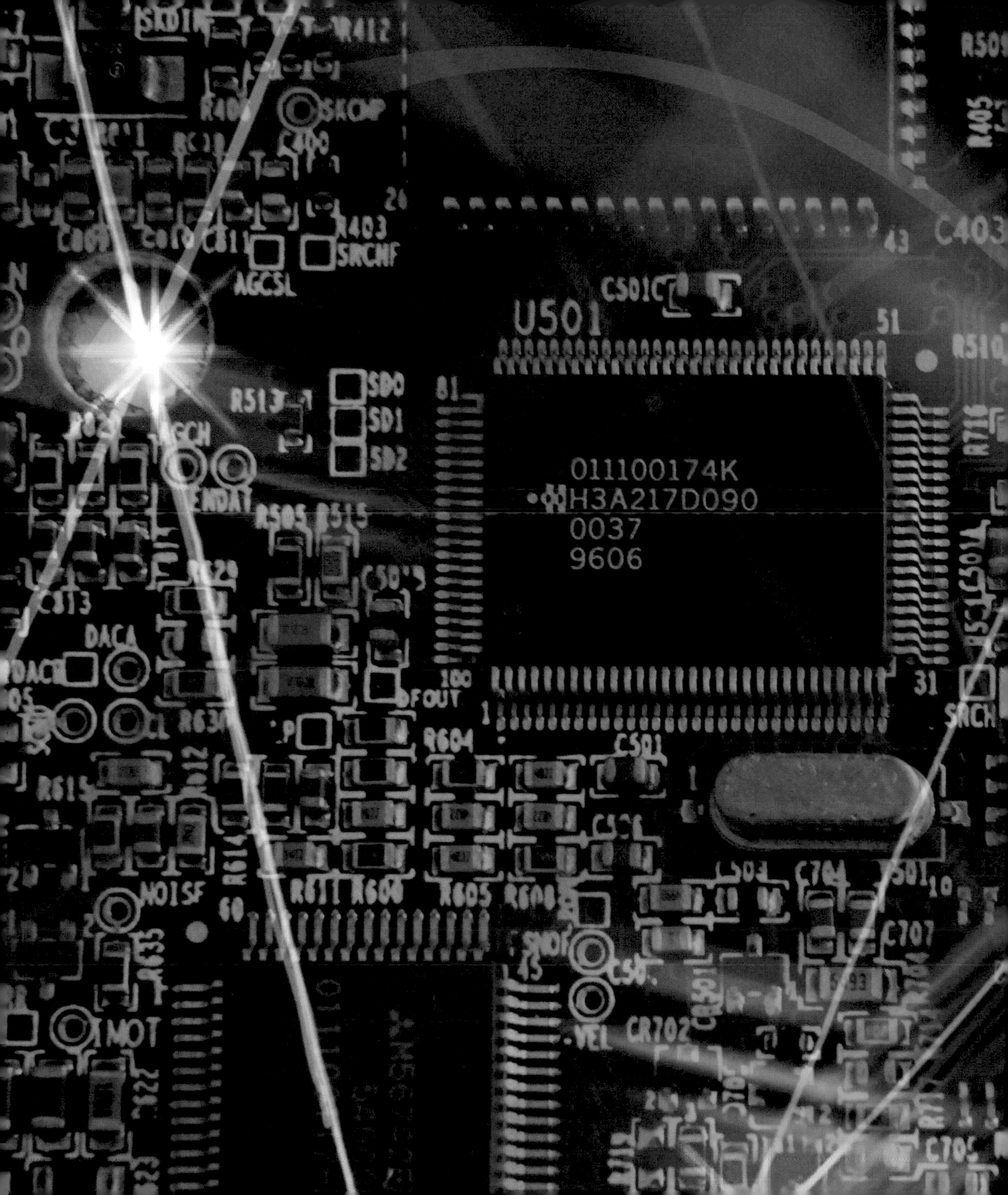
U501
011100174K
H3A217D090
0037
9606
SD0
SD1
SD2
AGCSL
C403

1911

Superconductivity

Heike Kamerlingh Onnes (1853–1926), **John Bardeen** (1908–1991), **Karl Alexander Müller** (b. 1927), **Leon N. Cooper** (b. 1930), **John Robert Schrieffer** (b. 1931), **Johannes Georg Bednorz** (b. 1950)

"At very low temperatures," writes science-journalist Joanne Baker, "some metals and alloys conduct electricity without any resistance. The current in these superconductors can flow for billions of years without losing any energy. As electrons become coupled and all move together, avoiding the collisions that cause electrical resistance, they approach a state of perpetual motion."

In fact, many metals exist for which the resistivity is zero when they are cooled below a critical temperature. This phenomenon, called superconductivity, was discovered in 1911 by Dutch physicist Heike Onnes, who observed that when he cooled a sample of mercury to 4.2 degrees above absolute zero (−452.1 °F), its electrical resistance plunged to zero. In principle, this means that an electrical current can flow around a loop of superconducting wire forever, with no external power source. In 1957, American physicists John Bardeen, Leon Cooper, and Robert Schrieffer determined how electrons could form pairs and appear to ignore the metal around them: Consider a metal window screen as a metaphor for the arrangement of positively charged atomic nuclei in a metal lattice. Next, imagine a negatively charged electron zipping between the atoms, creating a distortion by pulling on them. This distortion attracts a second electron to follow the first; they travel together in a pair, and encounter less resistance overall.

In 1986, Georg Bednorz and Alex Müller discovered a material that operated at the higher temperature of roughly −396 °F (35 kelvins), and in 1987 a different material was found to superconduct at −297 °F (90 kelvins). If a superconductor is discovered that operates at room temperature, it could be used to save vast amounts of energy and to create a high-performance electrical power transmission system. Superconductors also expel all applied magnetic fields, which allows engineers to build magnetically levitated trains. Superconductivity is also used to create powerful electromagnets in MRI (magnetic-resonance imaging) scanners in hospitals.

SEE ALSO Discovery of Helium (1868), Third Law of Thermodynamics (1905), Superfluids (1937), Nuclear Magnetic Resonance (1938).

In 2008, physicists at the U.S. Department of Energy's Brookhaven National Laboratory discovered interface high-temperature superconductivity *in bilayer films of two cuprate materials—with potential for creating higher-efficiency electronic devices. In this artistic rendition, these thin films are built layer by layer.*

1911

Atomic Nucleus

Ernest Rutherford (1871–1937)

Today, we know that the atomic nucleus, which consists of protons and neutrons, is the very dense region at the center of an atom. However, in the first decade of the 1900s, scientists were unaware of the nucleus and thought of the atom as a diffuse web of positively charged material, in which negatively charged electrons were embedded like cherries in a cake. This model was utterly destroyed when Ernest Rutherford and his colleagues discovered the nucleus after they fired a beam of alpha particles at a thin sheet of gold foil. Most of the alpha particles (which we know today as helium nuclei) went through the foil, but a few bounced straight back. Rutherford said later that this was "quite the most incredible event that has ever happened to me. . . . It was almost as incredible as if you fired a 15-inch shell at a piece of tissue paper and it came back and hit you."

The cherry-cake model of the atom, which was a metaphor for a somewhat uniform spread of density across the gold foil, could never account for this behavior. Scientists would have observed the alpha particles slow down, perhaps, like a bullet being fired through water. They did not expect the atom to have a "hard center" like a pit in a peach. In 1911, Rutherford announced a model with which we are familiar today: an atom consisting of a positively charged nucleus encircled by electrons. Given the frequency of collisions with the nucleus, Rutherford could approximate the size of the nucleus with respect to the size of the atom. Author John Gribbin writes that the nucleus has "one hundred-thousandth of the diameter of the whole atom, equivalent to the size of a pinhead compared with the dome of St. Paul's cathedral in London. . . . And since everything on Earth is made of atoms, that means that your own body, and the chair you sit on, are each made up of a million billion times more empty space than 'solid matter.'"

SEE ALSO Atomic Theory (1808), Electron (1897), $E = mc^2$ (1905), Bohr Atom (1913), Cyclotron (1929), Neutron (1932), Nuclear Magnetic Resonance (1938), Energy from the Nucleus (1942), Stellar Nucleosynthesis (1946).

Artistic rendition of a classical model of the atom with its central nucleus. Only some of the nucleons (protons and neutrons) and electrons are seen in this view. In an actual atom, the diameter of the nucleus is very much smaller than the diameter of the entire atom. Modern depictions of the surrounding electrons often depict them as clouds that represent probability densities.

1911

Kármán Vortex Street

Henri Bénard (1874–1939), **Theodore von Kármán** (1881–1963)

One of the most visually beautiful phenomena in physics can also be among the most dangerous. Kármán vortex streets are a repeating set of swirling vortices caused by the unsteady separation of fluid flow over "bluff bodies"—that is, blunt bodies such as cylindrical shapes, airfoils at large angles of attack, or the shapes of some reentry vehicles. Author Iris Chang describes the phenomena researched by Hungarian physicist Theodore von Kármán, which included his "1911 discovery through mathematical analysis of the existence of a source of aerodynamical drag that occurs when the airstream breaks away from the airfoil and spirals off the sides in two parallel streams of vortices. A phenomena, now known as the Kármán Vortex Street, was used for decades to explain the oscillations of submarines, radio towers, power lines. . . ."

Various methods can be used to decrease unwanted vibrations from essentially cylindrical objects such as industrial chimneys, car antennas, and submarine periscopes. One such method employs helical (screw-like) projections that reduce the alternate shedding of vortices. The vortices can cause towers to collapse and are also undesirable because they are a significant source of wind resistance in cars and airplanes.

These vortices are sometimes observed in rivers downstream from columns supporting a bridge or in the small, circular motions of leaves on the ground when a car travels down the street. Some of their most beautiful manifestations occur in cloud configurations that form in the wake of giant obstacles on the Earth, such as tall volcanic islands. Similar vortices may help scientists study weather on other planets.

Von Kármán referred to his theory as "the theory whose name I am honored to bear," because, according to author István Hargittai, "he considered the discovery to be more important than the discoverer. When about twenty years later, a French scientist Henri Bénard claimed priority for the discovery of vortex streets, von Kármán did not protest. Rather, with characteristic humor he suggested that the term *Kármán Vortex Streets* be used in London and *Boulevard d'Henri Bénard* in Paris."

SEE ALSO Bernoulli's Law of Fluid Dynamics (1738), Baseball Curveball (1870), Golf Ball Dimples (1905), Fastest Tornado Speed (1999).

Kármán Vortex Street in a Landsat 7 image of clouds off the Chilean coast near the Juan Fernández Islands. A study of these patterns is useful for understanding flows involved in phenomena ranging from aircraft wing behavior to the Earth's weather.

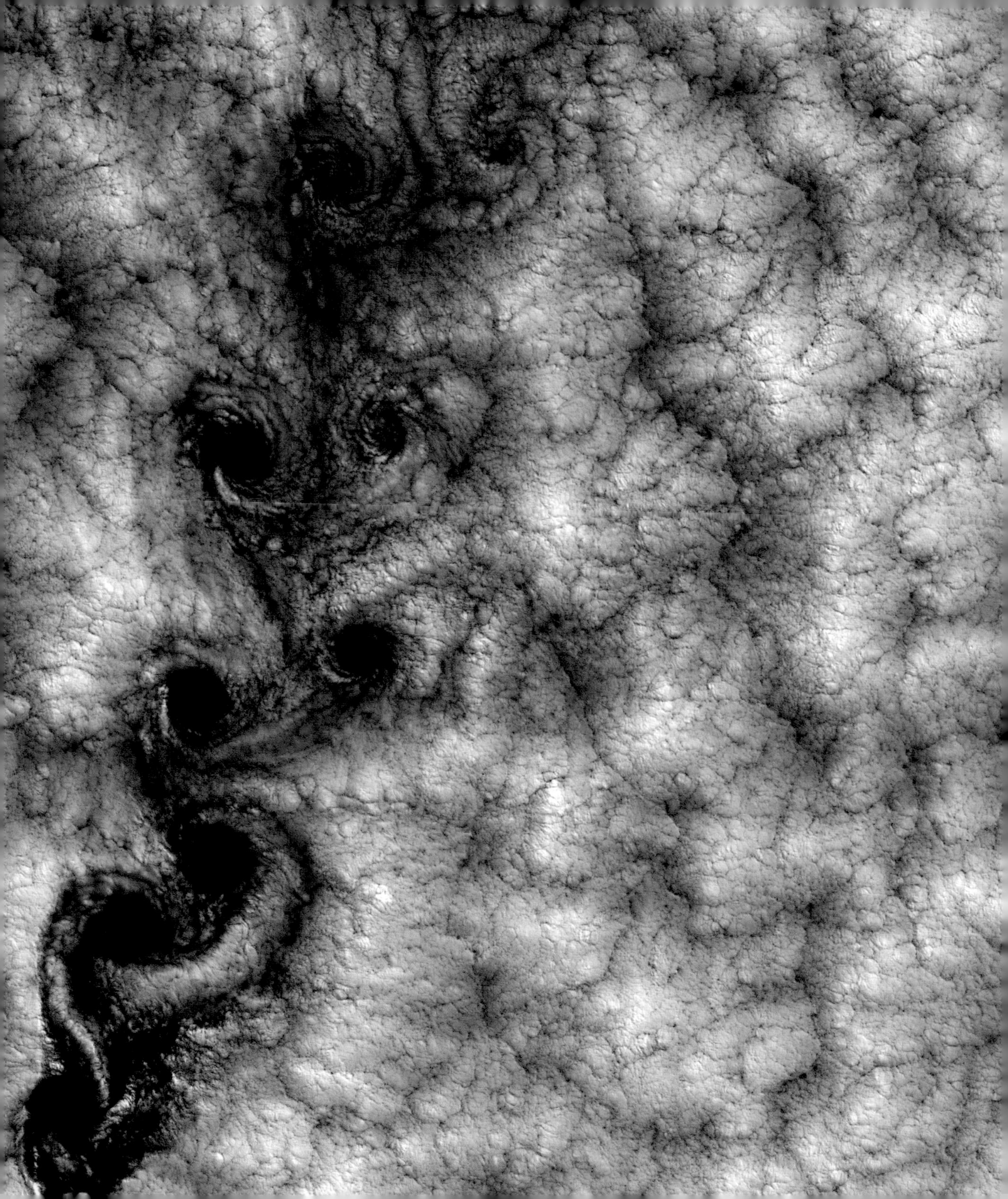

1911

Wilson Cloud Chamber

Charles Thomson Rees Wilson (1869–1959), **Alexander Langsdorf** (1912–1996), **Donald Arthur Glaser** (b. 1926)

At his Nobel Banquet in 1927, physicist Charles Wilson described his affection for cloudy Scottish hill-tops: "Morning after morning, I saw the sun rise above a sea of clouds and the shadow of the hills on the clouds below surrounded by gorgeous colored rings. The beauty of what I saw made me fall in love with clouds. . . ." Who would have guessed that Wilson's passion for mists would have been used to reveal the first form of **Antimatter** (the positron)—along with other particles—and change the world of particle physics forever?

Wilson perfected his first cloud chamber in 1911. First, the chamber was saturated with water vapor. Next, the pressure was lowered by the movement of a diaphragm, which expanded the air inside. This also cooled the air, creating conditions favorable to condensation. When an ionizing particle passed through the chamber, water vapor condensed on the resulting ion (charged particle), and a trail of the particle became visible in the vapor cloud. For example, when an alpha particle (a helium nucleus, which has a positive charge) moved through the cloud chamber, it tore off electrons from atoms in the gas, temporarily leaving behind charged atoms. Water vapor tended to accumulate on such ions, leaving behind a thin band of fog reminiscent of smoke-writing airplanes. If a uniform magnetic field was applied to the cloud chamber, positively and negatively charged particles curved in opposite directions. The radius of curvature could then be used to determine the momentum of the particle.

The cloud chamber was just the start. In 1936, physicist Alexander Langsdorf developed the *diffusion cloud chamber*, which employed colder temperatures and was sensitized for radiation detection for longer periods of time than traditional cloud chambers. In 1952, physicist Donald Glaser invented the *bubble chamber*. This chamber employed liquids, which are better able to reveal the tracks of more energetic particles than the gasses in traditional cloud chambers. More recent spark chambers employ a grid of electric wires to monitor charged particles by the detection of sparks.

SEE ALSO Geiger Counter (1908), Antimatter (1932), Neutron (1932).

In 1963, this Brookhaven National Laboratory bubble chamber was the largest particle detector of its type in the world. The most famous discovery made at this detector was of the omega-minus particle.

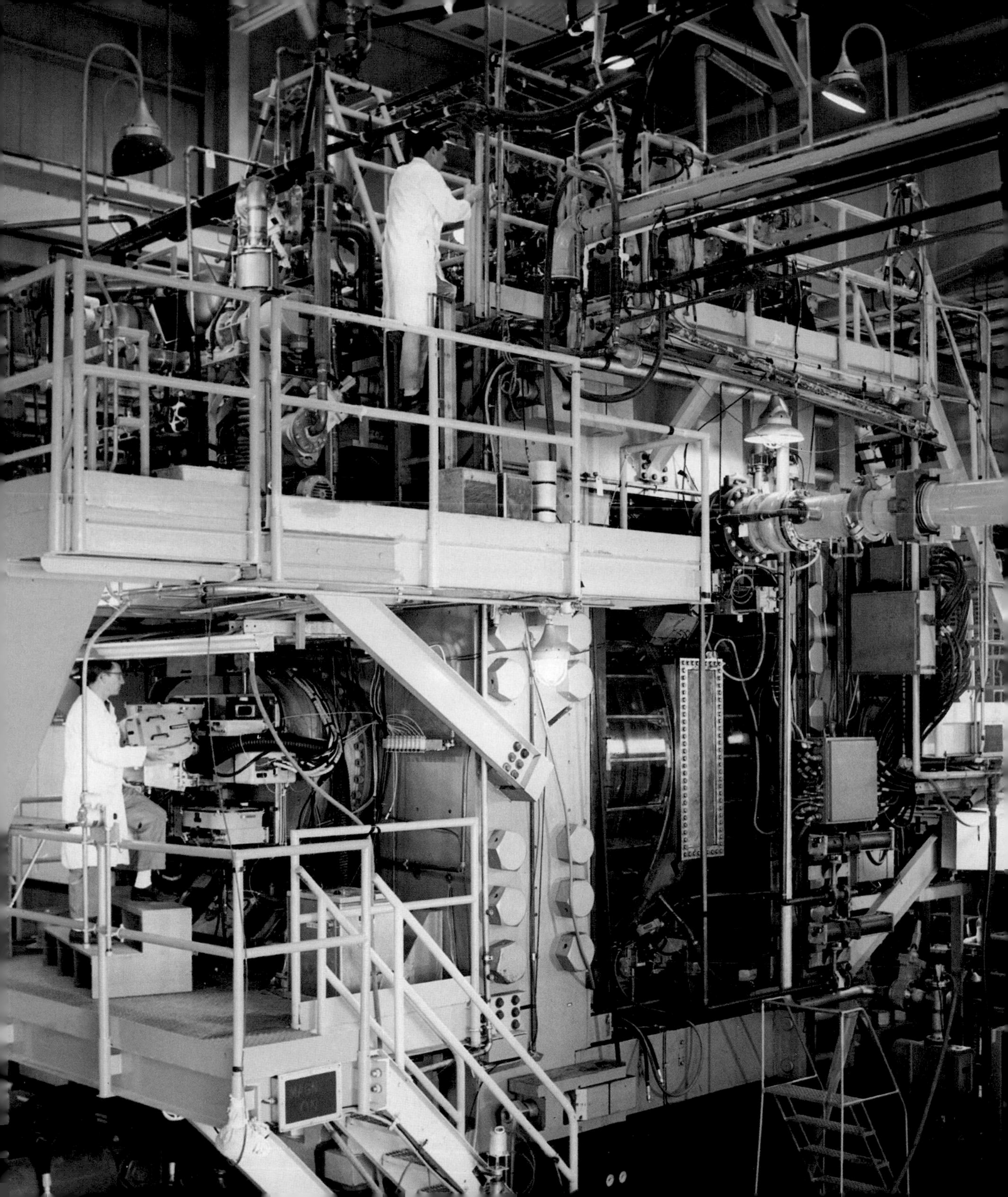

1912

Cepheid Variables Measure the Universe

Henrietta Swan Leavitt (1868–1921)

The poet John Keats once wrote "Bright star, were I as steadfast as thou art"—not realizing that some stars brighten and dim on periods ranging from days to weeks. Cepheid variable stars have periods (the time for one cycle of dimness and brightness) that are proportional to the stars' luminosity. Using a simple distance formula, this luminosity can be used to estimate interstellar and intergalactic distances. The American astronomer Henrietta Leavitt discovered the relationship between period and luminosity in Cepheid variables, thus making her perhaps the first to discover how to calculate the distance from the Earth to galaxies beyond the Milky Way. In 1902, Leavitt became a permanent staff member of the Harvard College Observatory and spent her time studying photographic plates of variable stars in the Magellanic Clouds. In 1904, using a time-consuming process called superposition, she discovered hundreds of variables in the Magellanic Clouds. These discoveries led Professor Charles Young of Princeton University to write, "What a variable-star 'fiend' Miss Leavitt is; one can't keep up with the roll of the new discoveries."

Leavitt's greatest discovery occurred when she determined the actual periods of 25 Cepheid variables, and in 1912 she wrote of the famous period-luminosity relation: "A straight line can be readily drawn among each of the two series of points corresponding to maxima and minima, thus showing that there is a simple relation between the brightness of the variables and their periods." Leavitt also realized that "since the variables are probably nearly the same distance from the earth, their periods are apparently associated with their actual emission of light, as determined by their mass, density, and surface brightness." Sadly, she died young of cancer before her work was complete. In 1925, completely unaware of her death, Professor Mittag-Leffler of the Swedish Academy of Sciences sent her a letter, expressing his intent to nominate her for the Nobel Prize in Physics. However, since Nobel Prizes are never awarded posthumously, Leavitt was not eligible for the honor.

SEE ALSO Eratosthenes Measures the Earth (240 B.C.), Measuring the Solar System (1672), Stellar Parallax (1838), Hubble Telescope (1990).

LEFT: *Henrietta Leavitt.* RIGHT: *Hubble Space Telescope image of spiral galaxy NGC 1309. Scientists can accurately determine the distance between the Earth and the galaxy (100 million light-years, or 30 Megaparsecs) by examining the light output of the galaxy's Cepheid variables.*

1912

Bragg's Law of Crystal Diffraction

William Henry Bragg (1862–1942), **William Lawrence Bragg** (1890–1971)

"I was captured for life by chemistry and by crystals," wrote X-ray crystallographer Dorothy Crowfoot Hodgkin whose research depended on Bragg's Law. Discovered by the English physicists Sir W. H. Bragg and his son Sir W. L. Bragg in 1912, Bragg's Law explains the results of experiments involving the diffraction of electromagnetic waves from crystal surfaces. Bragg's Law provides a powerful tool for studying crystal structure. For example, when X-rays are aimed at a crystal surface, they interact with atoms in the crystal, causing the atoms to re-radiate waves that may interfere with one another. The interference is constructive (reinforcing) for integer values of n according to Bragg's Law: $n\lambda = 2d \sin(\theta)$. Here, λ is the wavelength of the incident electromagnetic waves (e.g. X-rays); d is the spacing between the planes in the atomic lattice of the crystal; and θ is the angle between the incident ray and the scattering planes.

For example, X-rays travel down through crystal layers, reflect, and travel back over the same distance before leaving the surface. The distance traveled depends on the separation of the layers and the angle at which the X-ray entered the material. For maximum intensity of reflected waves, the waves must stay in phase to produce the constructive interferences. Two waves stay in phase, after both are reflected, when n is a whole number. For example, when $n = 1$, we have a "first order" reflection. For $n = 2$, we have a "second order" reflection. If only two rows were involved in the diffraction, as the value of θ changes, the transition from constructive to destructive interference is gradual. However, if interference from many rows occurs, then the constructive interference peaks become sharp, with mostly destructive interference occurring in between the peaks.

Bragg's Law can be used for calculating the spacing between atomic planes of crystals and for measuring the radiation's wavelength. The observations of X-ray wave interference in crystals, commonly known as X-ray diffraction, provided direct evidence for the periodic atomic structure of crystals that was postulated for several centuries.

SEE ALSO Wave Nature of Light (1801), X-rays (1895), Hologram (1947), Seeing the Single Atom (1955), Quasicrystals (1982).

LEFT: *Copper sulfate. In 1912, physicist Max von Laue used X-rays to record a diffraction pattern from a copper sulfate crystal, which revealed many well-defined spots. Prior to X-ray experiments, the spacing between atomic lattice planes in crystals was not accurately known.* RIGHT: *Bragg's Law eventually led to studies involving the X-ray scattering from crystal structures of large molecules such as DNA.*

1913

Bohr Atom

Niels Henrik David Bohr (1885–1962)

"Somebody once said about the Greek language that Greek flies in the writings of Homer," writes physicist Amit Goswami. "The quantum idea started flying with the work of Danish physicist Niels Bohr published in the year 1913." Bohr knew that negatively charged electrons are easily removed from atoms and that the positively charged nucleus occupied the central portion of the atom. In the Bohr model of the atom, the nucleus was considered to be like our central Sun, with electrons orbiting like planets.

Such a simple model was bound to have problems. For example, an electron orbiting a nucleus would be expected to emit electromagnetic radiation. As the electron lost energy, it should decay and fall into the nucleus. In order to avoid atomic collapse as well as to explain various aspects of the emission spectra of the hydrogen atom, Bohr postulated that the electrons could not be in orbits with an arbitrary distance from the nucleus. Rather, they were restricted to particular allowed orbits or shells. Just like climbing or descending a ladder, the electron could jump to a higher rung, or shell, when the electron received an energy boost, or it could fall to a lower shell, *if* one existed. This hopping between shells takes place only when a photon of the particular energy is absorbed or emitted from the atom. Today we know that the model has many shortcomings and does not work for larger atoms, and it violates the **Heisenberg Uncertainty Principle** because the model employs electrons with a definite mass and velocity in orbits with definite radii.

Physicist James Trefil writes, "Today, instead of thinking of electrons as microscopic planets circling a nucleus, we now see them as probability waves sloshing around the orbits like water in some kind of doughnut-shaped tidal pool governed by **Schrödinger's Equation**. . . . Nevertheless, the basic picture of the modern quantum mechanical atoms was painted back in 1913, when Niels Bohr had his great insight." *Matrix mechanics*—the first complete definition of quantum mechanics—later replaced the Bohr Model and better described the observed transitions in energy states of atoms.

SEE ALSO Electron (1897), Atomic Nucleus (1911), Pauli Exclusion Principle (1925), Schrödinger's Wave Equation (1926), Heisenberg Uncertainty Principle (1927).

These amphitheater seats in Ohrid, Macedonia, are a metaphor for Bohr's electron orbits. According to Bohr, electrons could not *be in orbits with an* arbitrary *distance from the nucleus; rather, electrons were restricted to particular allowed shells associated with discrete energy levels.*

1913

Millikan Oil Drop Experiment

Robert A. Millikan (1868–1953)

In his 1923 Nobel Prize address, American physicist Robert Millikan assured the world that he had detected individual electrons. "He who has seen that experiment," Millikan said, referring to his oil-drop research, "has literally *seen* the electron." In the early 1900s, in order to measure the electric charge on a single electron, Millikan sprayed a fine mist of oil droplets into a chamber that had a metal plate at top and bottom with a voltage applied between them. Because some of the oil droplets acquired electrons through friction with the sprayer nozzle, they were attracted to the positive plate. The mass of a single charged droplet can be calculated by observing how fast it falls. In fact, if Millikan adjusted the voltage between the metal plates, the charged droplet could be held stationary between the plates. The amount of voltage needed to suspend a droplet, along with knowledge of the droplet's mass, was used to determine the overall electric charge on the droplet. By performing this experiment many times, Millikan determined that the charged droplets did not acquire a continuous range of possible values for the charge but rather exhibited charges that were whole-number multiples of a lowest value, which he declared was the charge on an individual electron (about 1.592×10^{-19} coulomb). Millikan's experiment required extreme effort, and his value was slightly less than the actual value 1.602×10^{-19} accepted today because he had used an incorrect value for the viscosity of air in his measurements.

Authors Paul Tipler and Ralph Llewellyn write, "Millikan's measurement of the charge on the electron is one of the few truly crucial experiments in physics and . . . one whose simple directness serves as a standard against which to compare others. . . . Note that, while we have been able to measure the value of the quantized electric charge, there is no hint in any of the above as to *why* it has this value, nor do we know the answer to that question now."

SEE ALSO Coulomb's Law of Electrostatics (1785), Electron (1897).

Figure from Millikan's 1913 paper on his experiment. The atomizer (A) introduced oil droplets into chamber D. A voltage was maintained between parallel plates M and N. Droplets between M and N were observed.

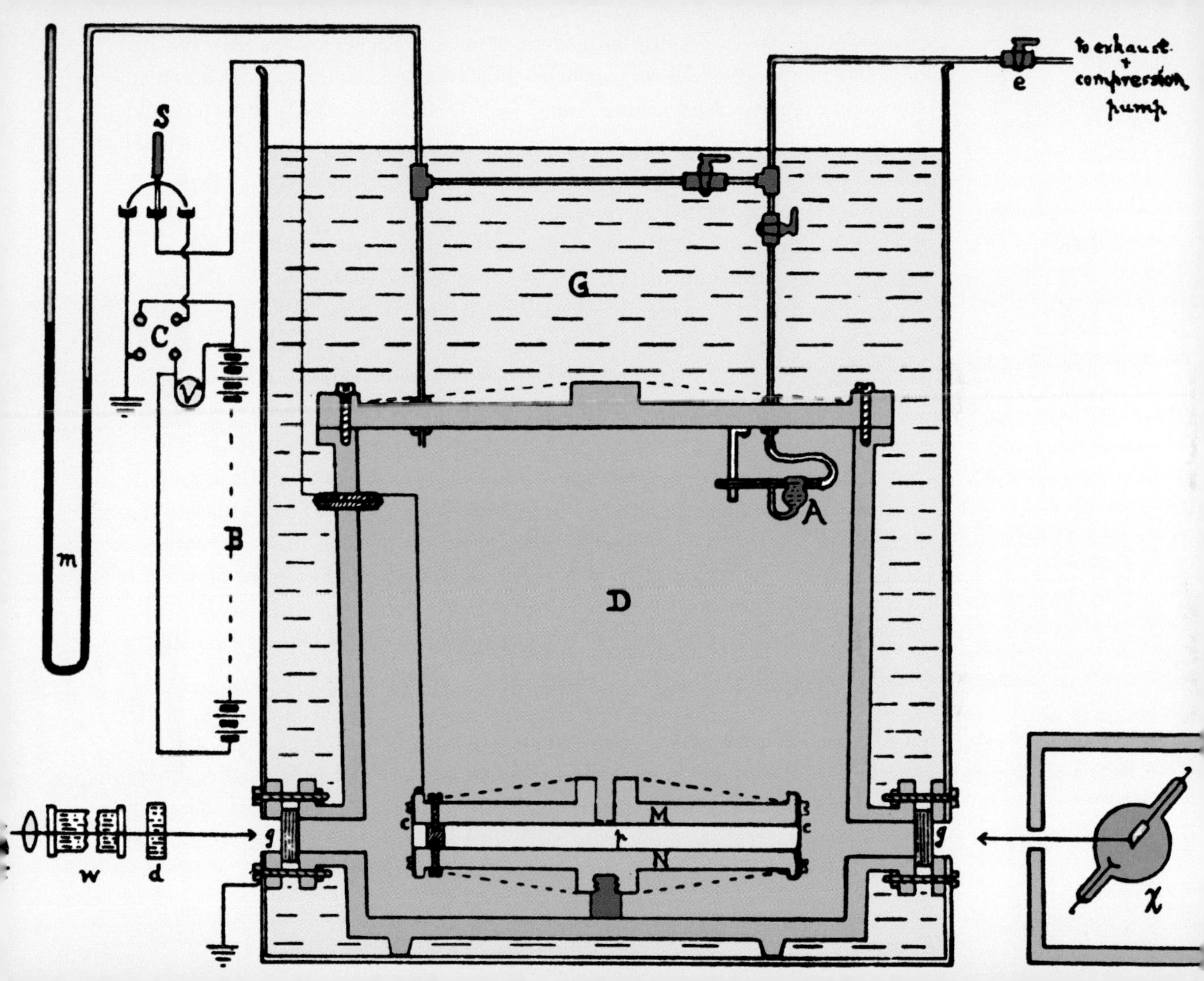

to exhaust + compression pump
e
S
C
V
B
m
G
A
D
M
N
r
c
c
g
g
w
d
X

1915

General Theory of Relativity

Albert Einstein (1879–1955)

Albert Einstein once wrote that "all attempts to obtain a deeper knowledge of the foundations of physics seem doomed to me unless the basic concepts are in accordance with general relativity from the beginning." In 1915, ten years after Einstein proclaimed his *Special* Theory of Relativity (which suggested that distance and time are not absolute and that one's measurements of the ticking rate of a clock depend on one's motion with respect to the clock), Einstein gave us an early form of his *General* Theory of Relativity (GTR), which explained gravity from a new perspective. In particular, Einstein suggested that gravity is not really a force like other forces, but results from the curvature of space-time caused by masses in space-time. Although we now know that GTR does a better job at describing motions in high gravitational fields than Newtonian mechanics (such as the orbit of Mercury about the Sun), Newtonian mechanics still is useful in describing the world of ordinary experience.

To better understand GTR, consider that wherever a mass exists in space, it warps space. Imagine a bowling ball sinking into a rubber sheet. This is a convenient way to visualize what stars do to the fabric of the universe. If you were to place a marble into the depression formed by the stretched rubber sheet, and give the marble a sideways push, it would orbit the bowling ball for a while, like a planet orbiting the Sun. The warping of the rubber sheet by the bowling ball is a metaphor for a star warping space.

GTR can be used to understand how gravity warps and slows time. In a number of circumstances, GTR also appears to permit **Time Travel.**

Einstein additionally suggested that gravitational effects move at the speed of light. Thus, if the Sun were suddenly plucked from the Solar System, the Earth would not leave its orbit about the Sun until about eight minutes later, the time required for light to travel from the Sun to the Earth. Many physicists today believe that gravitation must be quantized and take the form of particles called gravitons, just as light takes the form of photons, which are tiny quantum packets of electromagnetism.

SEE ALSO Newton's Laws of Motion and Gravitation (1687), Black Holes (1783), Time Travel (1949), Eötvös' Gravitational Gradiometry (1890), Special Theory of Relativity (1905), Randall-Sundrum Branes (1999).

Einstein suggested that gravity results from the curvature of space-time caused by masses in space-time. Gravity distorts both time and space.

1919

String Theory

Theodor Franz Eduard Kaluza (1885–1954), **John Henry Schwarz** (b. 1941), **Michael Boris Green** (b. 1946)

"The mathematics involved in string theory . . . ," writes mathematician Michael Atiyah, "in subtlety and sophistication . . . vastly exceeds previous uses of mathematics in physical theories. String theory has led to a whole host of amazing results in mathematics in areas that seem far removed from physics. To many this indicates that string theory must be on the right track. . . ." Physicist Edward Witten writes, "String theory is twenty-first century physics that fell accidentally into the twentieth century."

Various modern theories of "hyperspace" suggest that dimensions exist beyond the commonly accepted dimensions of space and time. For example, the Kaluza-Klein theory of 1919 made use of higher spatial dimensions in an attempt to explain electromagnetism and gravitation. Among the most recent formulations of these kinds of concepts is superstring theory, which predicts a universe of ten or eleven dimensions—three dimensions of space, one dimension of time, and six or seven more spatial dimensions. In many theories of hyperspace, the laws of nature become simpler and more elegant when expressed with these several extra spatial dimensions.

In string theory, some of the most basic particles, like quarks and fermions (which include electrons, protons, and neutrons) can be modeled by inconceivably tiny, essentially one-dimensional entities called strings. Although strings may seem to be mathematical abstractions, remember that atoms were once regarded as "unreal" mathematical abstractions that eventually became observables. However, strings are so tiny that there is no current way to directly observe them.

In some string theories, the loops of string move about in ordinary 3-space, but they also vibrate in higher spatial dimensions. As a simple metaphor, think of a vibrating guitar string whose "notes" correspond to different particles such as quarks and electrons or the hypothetical graviton, which may convey the force of gravity.

String theorists claim that a variety of higher spatial dimensions are "compactified"—tightly curled up (in structures known as Calabi-Yau spaces) so that the extra dimensions are essentially invisible. In1984, Michael Green and John H. Schwarz made additional breakthroughs in string theory.

SEE ALSO Standard Model (1961), Theory of Everything (1984), Randall-Sundrum Branes (1999), Large Hadron Collider (2009).

In string theory, the vibrational pattern of the string determines what kind of particle the string is. For a metaphor, consider a violin. Pluck the A string and an electron is formed. Pluck the E string and you create a quark.

1921

Einstein as Inspiration

Albert Einstein (1879–1955)

Nobel-prize winner Albert Einstein is recognized as one of the greatest physicists of all time and the most important scientist of the twentieth century. He proposed the **Special and General Theories of Relativity**, which revolutionized our understanding of space and time. He also made major contributions to the science of quantum mechanics, statistical mechanics, and cosmology.

"Physics has come to dwell at such a deep remove from everyday experiences," writes Thomas Levenson, author of *Einstein in Berlin*, "that it's hard to say whether most of us would be able to recognize an Einstein-like accomplishment should it occur [today]. When Einstein first came to New York in 1921, thousands lined the street for a motorcade. . . . Try to imagine any theoretician today getting such a response. It's impossible. The emotional connections between the physicist's conception of reality and the popular imagination has weakened greatly since Einstein."

According to many scholars I consulted, there will never be another individual on par with Einstein. Levenson suggested, "It seems unlikely that [science] will produce another Einstein in the sense of a broadly recognized emblem of genius. The sheer complexity of models being explored [today] confines almost all practitioners to *parts* of the problem." Unlike today's scientists, Einstein required little or no collaboration. Einstein's paper on special relativity contained no references to others or to prior work.

Bran Ferren, cochairman and chief creative officer of the Applied Minds technology company, affirms that "the *idea* of Einstein is perhaps more important than Einstein himself." Not only was Einstein the greatest physicist of the modern world, he was an "inspirational role model whose life and work ignited the lives of countless other great thinkers. The total of their contributions to society, and the contributions of the thinkers whom they will in turn inspire, will greatly exceed those of Einstein himself."

Einstein created an unstoppable "intellectual chain reaction," an avalanche of pulsing, chattering neurons and memes that will ring for an eternity.

SEE ALSO Newton as Inspiration (1687), Special Theory of Relativity (1905), Photoelectric Effect (1905), General Theory of Relativity (1915), Brownian Motion (1827), Stephen Hawking on *Star Trek* (1993).

Photo of Albert Einstein, while attending a lecture in Vienna in 1921 at the age of 42.

1922

Stern-Gerlach Experiment

Otto Stern (1888–1969), **Walter Gerlach** (1889–1979)

"What Stern and Gerlach found," writes author Louisa Gilder, "no one predicted. The Stern-Gerlach experiment was a sensation among physicists when it was published in 1922, and it was such an extreme result that many who had doubted the quantum ideas were converted."

Imagine carefully tossing little spinning red bar magnets between the north and south poles of a large magnet suspended in front of a wall. The fields of the big magnet deflect the little magnets, which collide with the wall and create a scattering of dents. In 1922, Otto Stern and Walter Gerlach conducted an experiment using neutral silver atoms, which contain a lone electron in the atom's outer orbital. Imagine that this electron is spinning and creating a small magnetic moment represented by the red bar magnets in our thought experiment. In fact, the spin of an unpaired electron causes the atom to have north and south poles, like a tiny compass needle. The silver atoms pass through a non-uniform magnetic field as they travel to a detector. If the magnetic moment could assume *all* orientations, we would expect a smear of impact locations at the detector. After all, the external magnetic field would produce a force at one end of the tiny "magnets" that might be slightly greater than at the other end, and over time, randomly oriented electrons would experience a range of forces. However, Stern and Gerlach found just two impact regions (above and below the original beam of silver atoms), suggesting that the electron spin was quantized and could assume just two orientations.

Note that the "spin" of a particle may have little to do with the classical idea of a rotating sphere having an angular momentum; rather, particle spin is a somewhat mysterious quantum mechanical phenomenon. For electrons as well as for protons and neutrons, there are two possible values for spin. However, the associated magnetic moment is much smaller for the heavier protons and neutrons, because the strength of a magnetic dipole is inversely proportional to the mass.

SEE ALSO *De Magnete* (1600), Gauss and the Magnetic Monopole (1835), Electron (1897), Pauli Exclusion Principle (1925), EPR Paradox (1935).

A memorial plaque honoring Stern and Gerlach, mounted near the entrance to the building in Frankfurt, Germany, where their experiment took place.

SCHNEIDE
$\frac{\partial B}{\partial z}$
N
S
RINNE
ATOMSTRAHL
Z
Z

IM FEBRUAR 1922 WURDE IN DIESEM GEBÄUDE DES
PHYSIKALISCHEN VEREINS, FRANKFURT AM MAIN,
VON OTTO STERN UND WALTHER GERLACH DIE
FUNDAMENTALE ENTDECKUNG DER RAUMQUANTISIERUNG
DER MAGNETISCHEN MOMENTE IN ATOMEN GEMACHT.
AUF DEM STERN-GERLACH-EXPERIMENT BERUHEN WICHTIGE
PHYSIKALISCH-TECHNISCHE ENTWICKLUNGEN DES 20. JHDTS.,
WIE KERNSPINRESONANZMETHODE, ATOMUHR ODER LASER.
OTTO STERN WURDE 1943 FÜR DIESE ENTDECKUNG
DER NOBELPREIS VERLIEHEN.

1923

Neon Signs

Georges Claude (1870–1960)

Discussions of neon signs are not complete without reminiscing on what I call the "physics of nostalgia." French chemist and engineer George Claude developed neon tubing and patented its commercial application in outdoor advertising signs. In 1923, Claude introduced the signs to the United States. "Soon, neon signs were proliferating along America's highways in the 1920s and '30s . . . piercing the darkness," writes author William Kaszynski. "The neon sign was a godsend for any traveler with a near-empty gas tank or anxiously searching for a place to sleep."

Neon lights are glass tubes that contain neon or other gases at low pressure. A power supply is connected to the sign through wires that pass through the walls of the tube. When the voltage supply is turned on, electrons are attracted to the positive electrode, occasionally colliding with the neon atoms. Sometimes, the electrons will knock an electron off of a neon atom, creating another free electron and a positive neon ion, or Ne^+. The mixture of free electrons, Ne^+, and neutral neon atoms form a conducting **Plasma** in which free electrons are attracted to the Ne^+ ions. Sometimes, the Ne^+ captures the electron in a high energy level, and as the electron drops from a high energy level to a lower one, light is emitted at specific wavelengths, or colors, for example, red-orange for neon gas. Other gases within the tube can yield other colors.

Author Holly Hughes writes in *500 Places to See Before They Disappear*, "The beauty of the neon sign was that you could twist those glass tubes into any shape you wanted. And as Americans took to the highways in the 1950s and 1960s, advertisers took advantage of that, spangling the nighttime streetscape with colorful whimsies touting everything from bowling alleys to ice-cream stands to Tiki bars. Preservationists are working to save neon signs for future generations, either on-site or in museums. After all, what would America be without a few giant neon donuts around?"

SEE ALSO Triboluminescence (1620), Stokes' Fluorescence (1852), Plasma (1879), Black Light (1903).

Retro fifties-style American car wash neon sign.

Valet
CAR
W

1923

Compton Effect

Arthur Holly Compton (1892–1962)

Imagine screaming at a distant wall, and your voice echoing back. You wouldn't expect your voice to sound an octave lower upon its return. The sound waves bounce back at the same frequency. However, as documented in 1923, physicist Arthur Compton showed that when X-rays were scattered off electrons, the scattered X-rays had a lower frequency and lower energy. This is not predicted using traditional wave models of electromagnetic radiation. In fact, the scattered X-rays were behaving as if the X-rays were billiard balls, with part of the energy transferred to the electron (modeled as a billiard ball). In other words, some of the initial momentum of an X-ray particle was gained by the electron. With billiard balls, the energy of the scattered balls after collision depends on the angle at which they leave each other, an angular dependence that Compton found when X-rays collided with electrons. The Compton Effect provided additional evidence for quantum theory, which implies that light has both wave and particle properties. Albert Einstein had provided earlier evidence for quantum theory when he showed that packets of light (now called photons) could explain the **Photoelectric Effect** in which certain frequencies of light shined on a copper plate causes the plate to eject electrons.

In a purely wave model of the X-rays and electrons, one might expect that the electrons would be triggered to oscillate with the frequency of the incident wave and hence reradiate the same frequency. Compton, who modeled the X-rays as photon particles, had assigned the photon a momentum of hf/c from two well-known physics relationships: $E = hf$ and $E = mc^2$. The energy of the scattered X-rays was consistent with these assumptions. Here, E is energy, f is frequency, c the speed of light, m the mass, and h is Planck's constant. In Compton's particular experiments, the forces with which electrons were bound to atoms could be neglected, and the electrons could be regarded as being essentially unbound and quite free to scatter in another direction.

SEE ALSO Photoelectric Effect (1905), Bremsstrahlung (1909), X-rays (1895), Electromagnetic Pulse (1962).

Arthur Compton (left) with graduate student Luis Alvarez at the University of Chicago in 1933. Both men won Nobel Prizes in physics.

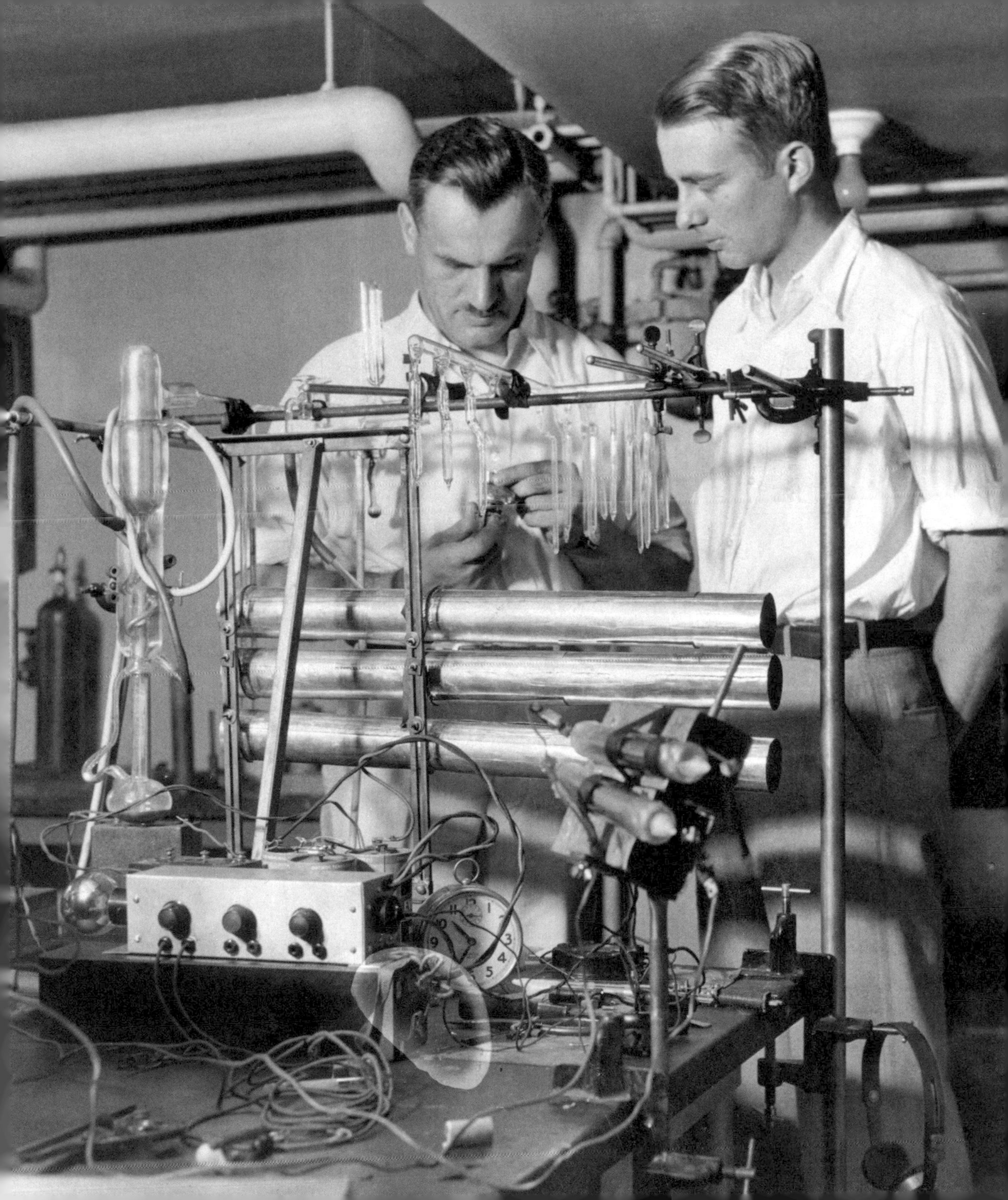

1924

De Broglie Relation

Louis-Victor-Pierre-Raymond, 7th duc de Broglie (1892–1987), **Clinton Joseph Davisson** (1881–1958), **Lester Halbert Germer** (1896–1971)

Numerous studies of the subatomic world have demonstrated that particles like electrons or photons (packets of light) are not like objects with which we interact in our everyday lives. These entities appear to possess characteristics of both waves and particles, depending on the experiment or phenomena being observed. Welcome to the strange realm of quantum mechanics.

In 1924, French physicist Louis-Victor de Broglie suggested that particles of matter could also be considered as waves and would posses properties commonly associated with waves, including a wavelength (the distance between successive crests of wave). In fact, all bodies have a wavelength. In 1927, American physicists Clinton Davisson and Lester Germer demonstrated the wave nature of electrons by showing that they could be made to diffract and interfere as if they were light.

De Broglie's famous relationship showed that the wavelength of a matter wave is inversely proportional to the particle's *momentum* (generally speaking, mass times velocity), and, in particular, $\lambda = h/p$. Here, λ is the wavelength, p is the momentum, and h is Planck's constant. According to author Joanne Baker, using this equation, it is possible to show that "Bigger objects, like ball bearings and badgers, have minuscule wavelengths, too small to see, so we cannot spot them behaving like waves. A tennis ball flying across a court has a wavelength of 10^{-34} meters, much smaller than a proton's width (10^{-15} m)." The wavelength of an ant is larger than for a human.

Since the original Davisson-Germer experiment for electrons, the de Broglie hypothesis has been confirmed for other particles like neutrons and protons and, in 1999, even for entire molecules such a buckyballs, soccer-ball-shaped molecules made of carbon atoms.

De Broglie had advanced his idea in his PhD thesis, but the idea was so radical that his thesis examiners were, at first, not sure if they should approve the thesis. He later won the Nobel Prize for this work.

SEE ALSO Wave Nature of Light (1801), Electron (1897), Schrödinger's Wave Equation (1926), Quantum Tunneling (1928), Buckyballs (1985).

In 1999, University of Vienna researchers demonstrated the wavelike behavior of buckminsterfullerene molecules formed of 60 carbon atoms (shown here). A beam of molecules (with velocities of around 200 m/sec, or 656 ft/sec) were sent through a grating, yielding an interference pattern characteristic of waves.

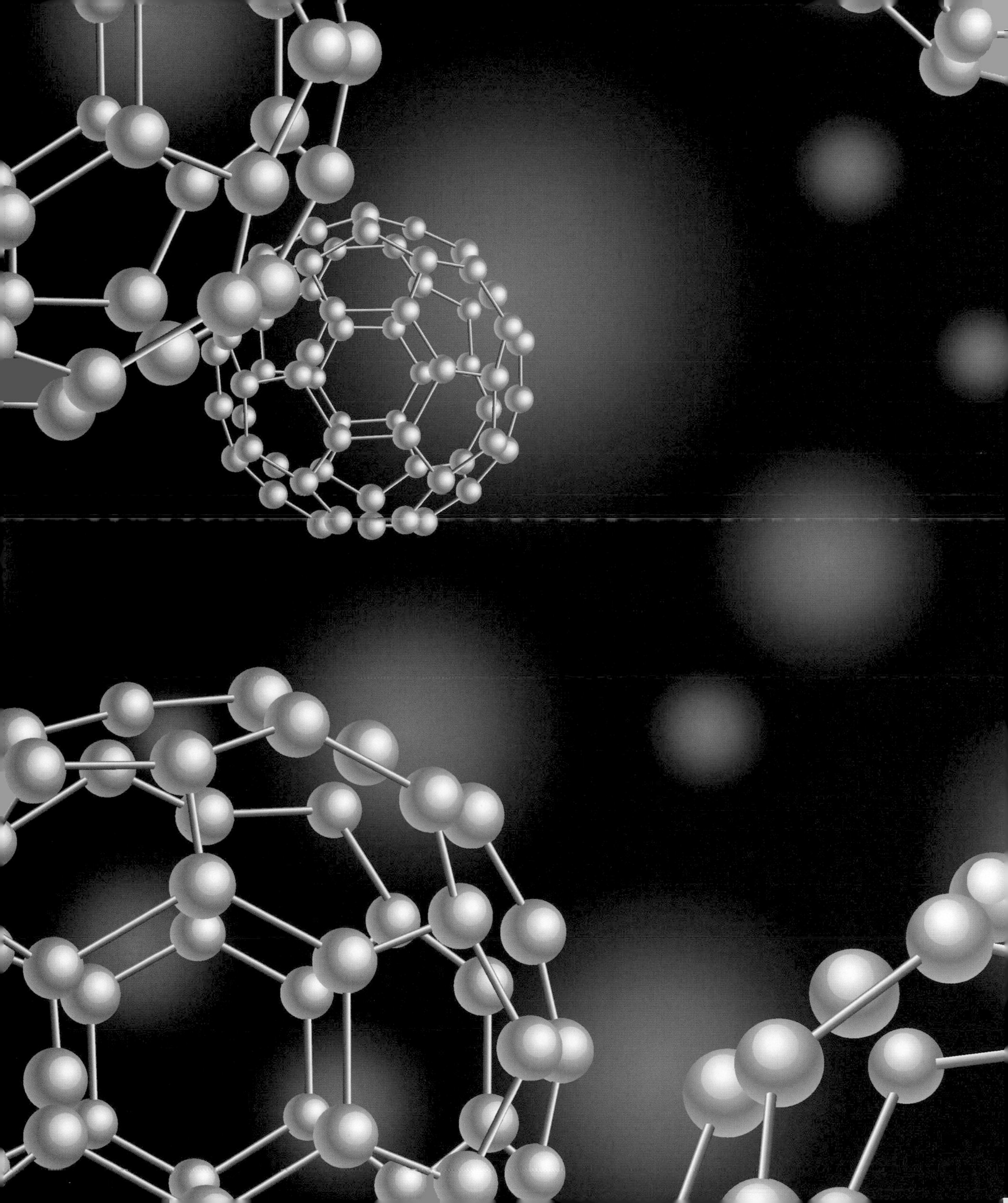

1925

Pauli Exclusion Principle

Wolfgang Ernst Pauli (1900–1958)

Imagine people who are beginning to fill the seats in a baseball stadium, starting at the rows nearest the playing field. This is a metaphor for electrons filling the orbitals of an atom—and in both baseball and atomic physics, there are rules that govern how many entities such as electrons or people can fill the allotted areas. After all, it would be quite uncomfortable if multiple people attempted to squeeze into a small seat.

Pauli's Exclusion Principle (PEP) explains why matter is rigid and why two objects cannot occupy the same space. It's why we don't fall through the floor and why **Neutron Stars** resist collapsing under their own incredible mass.

More specifically, PEP states that no pair of identical fermions (such as electrons, protons, or neutrons) can simultaneously occupy the same quantum state, which includes the spin of a fermion. For example, electrons occupying the same atomic orbital must have opposite spins. Once an orbital is occupied by a pair of electrons of opposite spin, no more electrons may enter the orbital until one leaves the orbital.

PEP is well-tested, and one of the most important principles in physics. According to author Michela Massimi, "From spectroscopy to atomic physics, from quantum field theory to high-energy physics, there is hardly another scientific principle that has more far-reaching implications than Pauli's exclusion principle." As a result of PEP, one can determine or understand electronic configurations underlying the classification of chemical elements in the Periodic Table as well as atomic spectra. Science-journalist Andrew Watson writes, "Pauli introduced this principle early in 1925, before the advent of modern quantum theory or the introduction of the idea of electron spin. His motivation was simple: there had to be something to prevent all the electrons in an atom collapsing down to a single lowest state. . . . So, Pauli's exclusion principle keeps electrons—and other fermions—from invading each other's space."

SEE ALSO Coulomb's Law of Electrostatics (1785), Electron (1897), Bohr Atom (1913), Stern-Gerlach Experiment (1922), White Dwarfs and Chandrasekhar Limit (1931), Neutron Stars (1933).

Artwork titled "Pauli's Exclusion Principle, or Why Dogs Don't Suddenly Fall Through Solids." PEP helps explain why matter is rigid, why we do not fall through solid floors, and why neutron stars resist collapsing under their humongous masses.

1926

Schrödinger's Wave Equation

Erwin Rudolf Josef Alexander Schrödinger (1887–1961)

"Schrödinger's Wave Equation enabled scientists to make detailed predictions about how matter behaves, while being able to visualize the atomic systems under study," writes physicist Arthur I. Miller. Schrödinger apparently developed his formulation while vacationing at a Swiss ski resort with his mistress who seemed to catalyze his intellectual and "erotic outburst," as he called it. The Schrödinger Wave Equation describes ultimate reality in terms of wave functions and probabilities. Given the equation, we can calculate the wave function of a particle:

$$i\hbar \frac{\partial}{\partial t}\psi(\mathbf{r},t) = -\frac{\hbar^2}{2m}\nabla^2\psi(\mathbf{r},t) + V(\mathbf{r})\psi(\mathbf{r},t)$$

Here, we need not worry about the details of this formula, except perhaps to note that $\psi(\mathbf{r}, t)$ is the wave function, which is the probability amplitude for a particle to have a given position $\mathbf{r}$ at any given time t. ∇^2 is used to describe how $\psi(\mathbf{r}, t)$ changes in space. $V(\mathbf{r})$ is the potential energy of the particle at each position $\mathbf{r}$. Just as an ordinary wave equation describes the progression of a ripple across a pond, Schrödinger's Wave Equation describes how a probability wave associated with a particle (e.g. an electron) moves through space. The peak of the wave corresponds to where the particle is most likely to be. The equation was also useful in understanding energy levels of electrons in atoms and became one of the foundations of quantum mechanics, the physics of the atomic world. Although it may seem odd to describe a particle as a wave, in the quantum realm such strange dualities are necessary. For example, light can act as either a wave or a particle (a photon), and particles such as electrons and protons can act as waves. As another analogy, think of electrons in an atom as waves on a drumhead, with the vibration modes of the wave equation associated with different energy levels of atoms.

Note that the *matrix mechanics* developed by Werner Heisenberg, Max Born, and Pascual Jordan in 1925 interpreted certain properties of particles in terms of matrices. This formulation is equivalent to the Schrödinger wave formulation.

SEE ALSO Wave Nature of Light (1801), Electron (1897), De Broglie Relation (1924), Heisenberg Uncertainty Principle (1927), Quantum Tunneling (1928), Dirac Equation (1928), Schrödinger's Cat (1935).

Erwin Schrödinger on a 1000 Austrian schilling banknote (1983).

TAUSEND
SCHILLING
OESTERREICHISCHE
NATIONALBANK
WIEN, AM 3. JÄNNER 1983
PRÄSIDENT
GENERALRAT
GENERALDIREKTOR
1000
ERWIN SCHRÖDINGER
1000
G 992654 H
992654 H
UNIVERSITÄT WIEN
TAUSEND SCHILLING
1000

1927

Heisenberg Uncertainty Principle

Werner Heisenberg (1901–1976)

"Uncertainty is the only certainty there is," wrote mathematician John Allen Paulos, "and knowing how to live with insecurity is the only security." The Heisenberg Uncertainty Principle states that the position and the velocity of a particle cannot both be known with high precision, at the same time. Specifically, the more precise the measurement of position, the more imprecise the measurement of momentum, and vice versa. The uncertainty principle becomes significant at the small size scales of atoms and subatomic particles.

Until this law was discovered, most scientists believed that the precision of any measurement was limited only by the accuracy of the instruments being used. German physicist Werner Heisenberg hypothetically suggested that even if we could construct an infinitely precise measuring instrument, we still could not accurately determine both the position and momentum (mass × velocity) of a particle. The principle is not concerned with the degree to which the measurement of the position of a particle may *disturb* the momentum of a particle. We could measure a particle's position to a high precision, but as a consequence, we could know little about the momentum.

For those scientists who accept the Copenhagen interpretation of quantum mechanics, the Heisenberg Uncertainty Principle means that the physical Universe literally does not exist in a deterministic form but is rather a collection of probabilities. Similarly, the path of an elementary particle such as a photon cannot be predicted, even in theory, by an infinitely precise measurement.

In 1935, Heisenberg was a logical choice to replace his former mentor Arnold Sommerfeld at the University of Munich. Alas, the Nazis required that "German physics" must replace "Jewish physics," which included quantum theory and relativity. As a result, Heisenberg's appointment to Munich was blocked even though he was not Jewish.

During World War II, Heisenberg led the unsuccessful German nuclear weapons program. Today, historians of science still debate as to whether the program failed because of lack of resources, lack of the right scientists on his team, Heisenberg's lack of a desire to give such a powerful weapon to the Nazis, or other factors.

SEE ALSO Laplace's Demon (1814), Third Law of Thermodynamics (1905), Bohr Atom (1913), Schrödinger's Wave Equation (1926), Complementarity Principle (1927), Quantum Tunneling (1928), Bose-Einstein Condensate (1995).

LEFT: *According to Heisenberg's Uncertainty Principle, particles likely exist only as a collection of probabilities, and their paths cannot be predicted even by an infinitely precise measurement.* RIGHT: *German postage stamp, 2001, featuring Werner Heisenberg.*

Begründer der
Quantenmechanik
300
Δp·Δq~h
Heisenbergsche
Unschärferelation
Deutschland
1,53 €
Werner Heisenberg
2001
Physiker
1901 – 1976

1927

Complementarity Principle

Niels Henrik David Bohr (1885–1962)

Danish physicist Niels Bohr developed a concept that he referred to as *complementarity* in the late 1920s, while trying to make sense of the mysteries of quantum mechanics, which suggested, for example, that light sometimes behaved like a wave and at other times like a particle. For Bohr, writes author Louisa Gilder, "complementarity was an almost religious belief that the paradox of the quantum world must be accepted as fundamental, not to be 'solved' or trivialized by attempts to find out 'what's really going on down there.' Bohr used the word in an unusual way: 'the complementarity' of waves and particles, for example (or of position and momentum), meant that when one existed fully, its complement did not exist at all." Bohr himself in a 1927 lecture in Como, Italy, said that waves and particles are "abstractions, their properties being definable and observable only through their interactions with other systems."

Sometimes, the physics and philosophy of complementarity seemed to overlap with theories in art. According to science-writer K. C. Kole, Bohr "was known for his fascination with cubism—especially 'that an object could be several things, could change, could be seen as a face, a limb, a fruit bowl,' as a friend of his later explained. Bohr went on to develop his philosophy of complementarity, which showed how an electron could change, could be seen as a wave [or] a particle. Like cubism, complementarity allowed contradictory views to coexist in the same natural frame."

Bohr thought that it was inappropriate to view the subatomic world from our everyday perspective. "In our description of nature," Bohr wrote, "the purpose is not to disclose the real essence of phenomena but only to track down, as far as it is possible, relations between the manifold aspects of experience."

In 1963, physicist John Wheeler expressed the importance of this principle: "Bohr's principle of complementarity is the most revolutionary scientific concept of this century and the heart of his fifty-year search for the full significance of the quantum idea."

SEE ALSO Wave Nature of Light (1801), Heisenberg Uncertainty Principle (1927), EPR Paradox (1935), Schrödinger's Cat (1935), Bell's Theorem (1964).

The physics and philosophy of complementarity often seemed to overlap with theories in art. Bohr was fascinated with Cubism, which sometimes allowed "contradictory" views to coexist, as in this artwork by Czech painter Eugene Ivanov.

1927

Hypersonic Whipcracking

A number of physics papers have been devoted to the supersonic booms created by whipcracking, and these papers have been accompanied by various recent, fascinating debates as to the precise mechanism involved. Physicists have known since the early 1900s that when a bullwhip handle is rapidly and properly moved, the tip of the whip can exceed the speed of sound. In 1927, physicist Z. Carrière used high-speed shadow photography to demonstrate that a sonic boom is associated with the crack of the whip. Ignoring frictional effects, the traditional explanation is that as a moving section of the whip travels along the whip and becomes localized in increasingly smaller portions, the moving section must travel increasingly fast due to the conservation of energy. The kinetic energy of motion of a point with mass m is given by $E = \frac{1}{2}mv^2$, and if E stays roughly constant, and m shrinks, velocity v must increase. Eventually, a portion of the whip near the end travels faster than the speed of sound (approximately 768 mph [1,236 kilometers per hour] at 68 °F [20 °C] in dry air) and creates a boom (see **Sonic Boom**), just like a jet plane exceeding the speed of sound in air. Whips were likely the first man-made devices to break the sound barrier.

In 2003, applied mathematicians Alain Goriely and Tyler McMillen modeled the impulse that creates a whip crack as a loop traveling along a tapered elastic rod. They write of the complexity of mechanisms, "The crack itself is a sonic boom created when a section of the whip at its tip travels faster than the speed of sound. The rapid acceleration of the tip of the whip is created when a wave travels to the end of the rod, and the energy consisting of the kinetic energy of the moving loop, the elastic energy stored in the loop, and the angular momentum of the rod is concentrated into a small section of the rod, which is then transferred into acceleration of the end of the rod." The tapering of the whip also increases the maximal speed.

SEE ALSO Atlatl (30,000 B.C.), Cherenkov Radiation (1934), Sonic Booms (1947).

A sonic boom is associated with the crack of a whip. Whips were likely the first man-made devices to break the sound barrier.

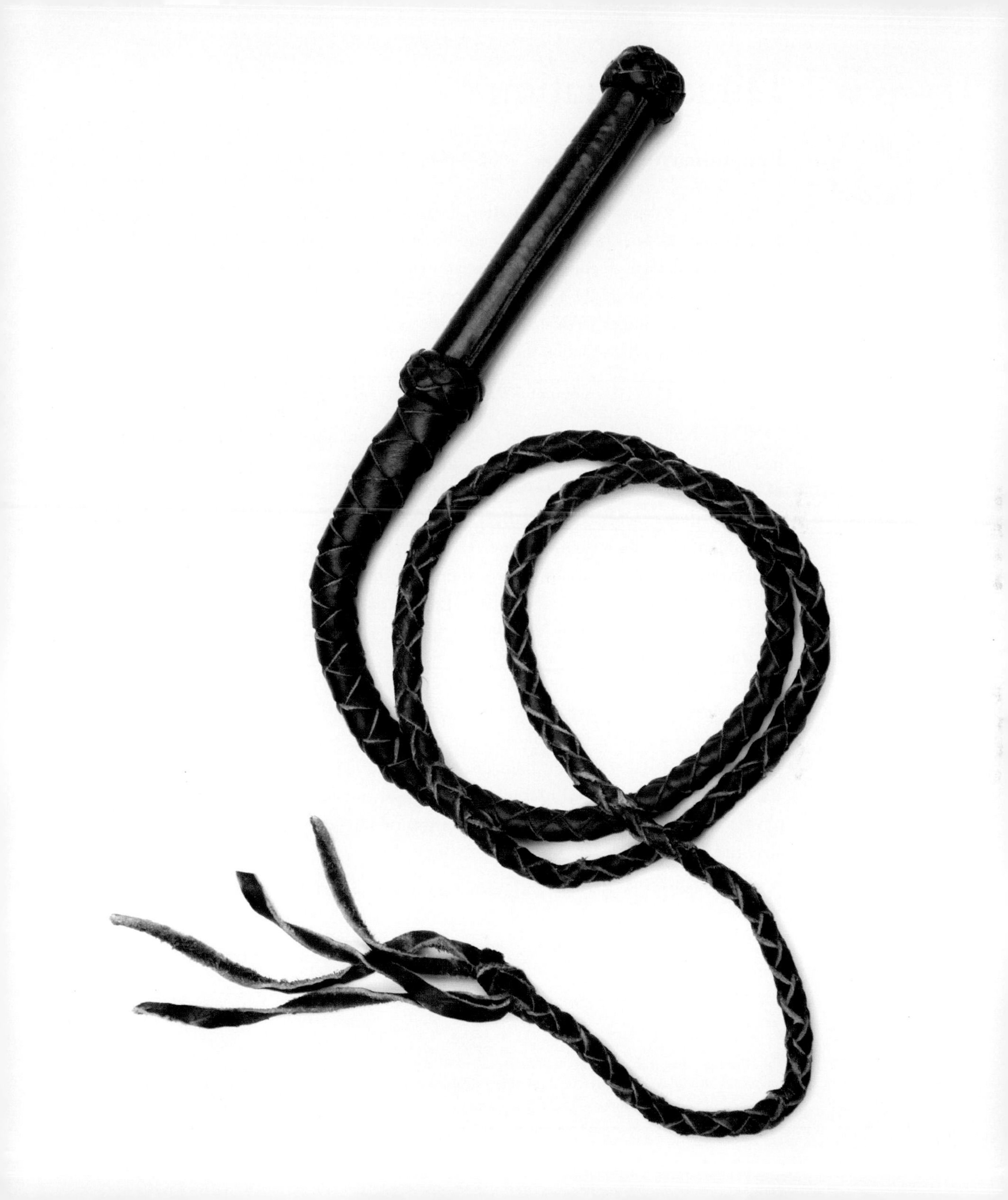

1928

Dirac Equation

Paul Adrien Maurice Dirac (1902–1984)

As discussed in the entry on **Antimatter**, the equations of physics can sometimes give birth to ideas or consequences that the discoverer of the equation did not expect. The power of these kinds of equations can seem magical, according to physicist Frank Wilczek in his essay on the Dirac Equation. In 1927, Paul Dirac attempted to find a version of **Schrödinger's Wave Equation** that would be consistent with the principles of special relativity. One way that the Dirac Equation can be written is

$$\left(\alpha_0 mc^2 + \sum_{j=1}^{3} \alpha_j p_j c\right)\Psi(\mathbf{x},t) = i\hbar \frac{\partial\Psi}{\partial t}(\mathbf{x},t)$$

Published in 1928, the equation describes electrons and other elementary particles in a way that is consistent with both quantum mechanics and the **Special Theory of Relativity**. The equation predicts the existence of antiparticles and in some sense "foretold" their experimental discovery. This feature made the discovery of the positron, the antiparticle of the electron, a fine example of the usefulness of mathematics in modern theoretical physics. In this equation, m is the rest mass of the electron, $\hbar$ is the reduced Planck's constant (1.054×10^{-34} J·s), c is the speed of light, p is the momentum operator, $\mathbf{x}$ and t are the space and time coordinates, and $\Psi(\mathbf{x},t)$ is a wave function. α is a linear operator that acts on the wave function.

Physicist Freeman Dyson has lauded this formula that represents a significant stage in humanity's grasp of reality. He writes, "Sometimes the understanding of a whole field of science is suddenly advanced by the discovery of a single basic equation. Thus it happened that the Schrödinger equation in 1926 and the Dirac equation in 1927 brought a miraculous order into the previously mysterious processes of atomic physics. Bewildering complexities of chemistry and physics were reduced to two lines of algebraic symbols."

SEE ALSO Electron (1897), Schrödinger's Wave Equation (1926), Special Theory of Relativity (1905), Antimatter (1932).

The Dirac equation is the only equation to appear in Westminster Abbey, London, where it is engraved on Dirac's commemorative plaque. Shown here is an artist's representation of the Westminster plaque, which depicts a simplified version of the formula.

1902
P·A·M
DIRAC O·M
PHYSICIST
iγ·∂ψ=mψ
1984

1928

Quantum Tunneling

George Gamow (1904–1968), **Ronald W. Gurney** (1898–1953), **Edward Uhler Condon** (1902–1974)

Imagine tossing a coin at a wall between two bedrooms. The coin bounces back because it does not have sufficient energy to penetrate the wall. However, according to quantum mechanics, the coin is actually represented by a fuzzy probability wave function that spills through the wall. This means that the coin has a very small chance of actually *tunneling* through the wall and ending up in the other bedroom. Particles can tunnel through such barriers due to **Heisenberg's Uncertainty Principle** applied to energies. According to the principle, it is not possible to say that a particle has precisely one amount of energy at precisely one instant in time. Rather, the energy of a particle can exhibit extreme fluctuations on short time scales and can have sufficient energy to traverse a barrier.

Some **Transistors** use tunneling to move electrons from one part of the device to another. The decay of some nuclei via particle emission employs tunneling. Alpha particles (helium nuclei) eventually tunnel out of uranium nuclei. According to the work published independently by George Gamow and the team of Ronald Gurney and Edward Condon in 1928, the alpha particle would never escape without the tunneling process.

Tunneling is also important in sustaining fusion reactions in the Sun. Without tunneling, stars wouldn't shine. *Scanning tunneling microscopes* employ tunneling phenomena to help scientists visualize microscopic surfaces, using a sharp microscope tip and a tunneling current between the tip and the specimen. Finally, the theory of tunneling has been applied to early models of the universe and to understand enzyme mechanisms that enhance reaction rates.

Although tunneling happens all the time at subatomic size scales, it would be quite unlikely (though possible) for you to be able to leak out through your bedroom into the adjacent kitchen. However, if you were to throw yourself into the wall every second, you would have to wait longer then the age of the universe to have a good chance of tunneling through.

SEE ALSO Radioactivity (1896), Schrödinger's Wave Equation (1926), Heisenberg Uncertainty Principle (1927), Transistor (1947), Seeing the Single Atom (1955).

Scanning tunneling microscope used at Sandia National Laboratory.

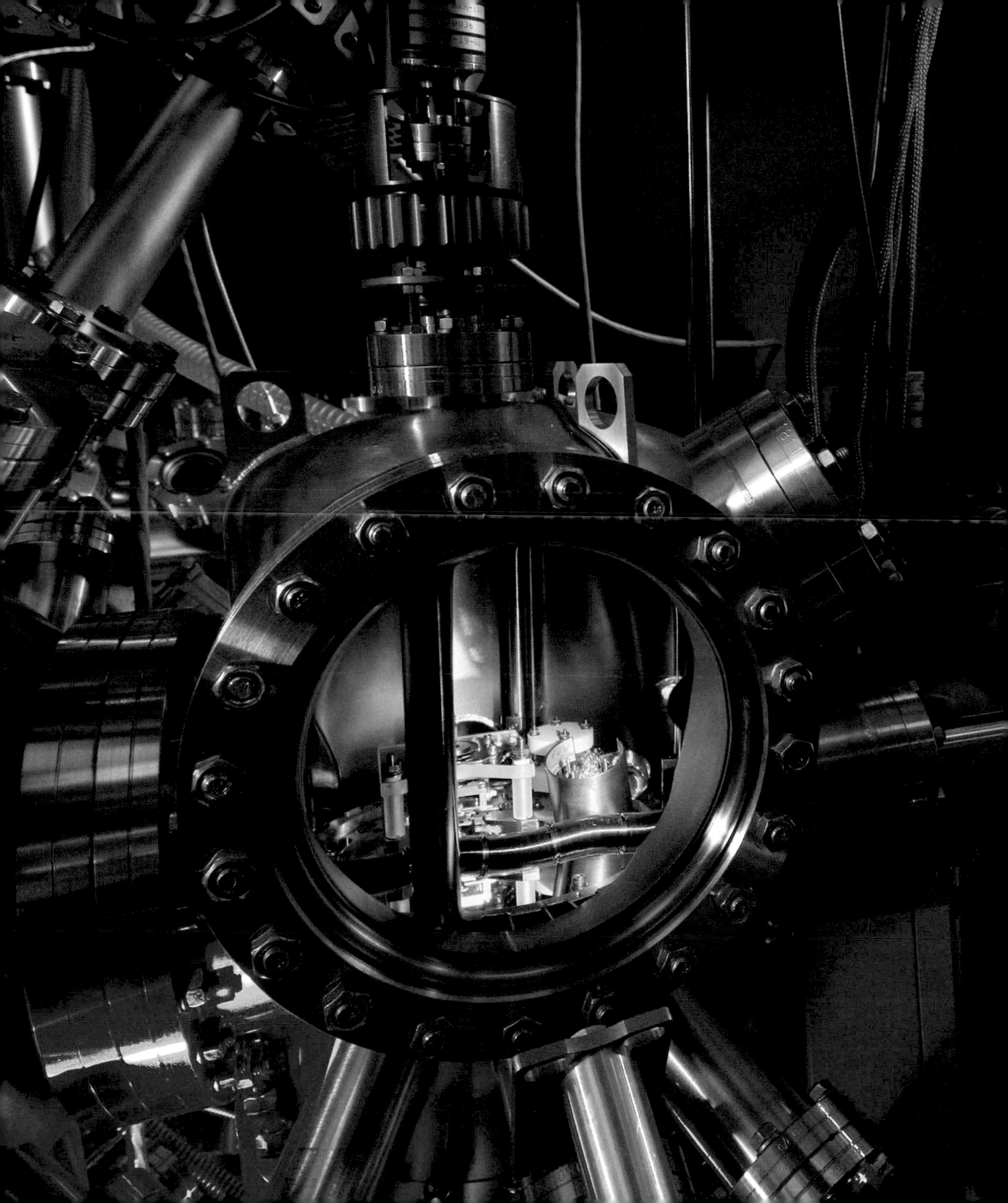

1929

Hubble's Law of Cosmic Expansion

Edwin Powell Hubble (1889–1953)

"Arguably the most important cosmological discovery ever made," writes cosmologist John P. Huchra, "is that our Universe is expanding. It stands, along with the Copernican Principle—that there is no preferred place in the Universe, and **Olbers' Paradox**—that the sky is dark at night, as one of the cornerstones of modern cosmology. It forced cosmologists to [consider] dynamic models of the Universe, and also implies the existence of a timescale or age for the Universe. It was made possible . . . primarily by Edwin Hubble's estimates of distances to nearby galaxies."

In 1929, American astronomer Edwin Hubble discovered that the greater the distance a galaxy is from an observer on the Earth, the faster it recedes. The distances between galaxies, or galactic clusters, are continuously increasing and, therefore, the Universe is expanding.

For many galaxies, the velocity (e.g. the movement of a galaxy away from an observer on the Earth) can be estimated from the red shift of a galaxy, which is an observed increase in the wavelength of electromagnetic radiation received by a detector on the Earth compared to that emitted by the source. Such red shifts occur because galaxies are moving away from our own galaxy at high speeds due to the expansion of space itself. The change in the wavelength of light that results from the relative motion of the light source and the receiver is an example of the **Doppler Effect**. Other methods also exist for determining the velocity for faraway galaxies. (Objects that are dominated by local gravitational interactions, like stars within a single galaxy, do not exhibit this apparent movement away from one another.)

Although an observer on the Earth finds that all distant galactic clusters are flying away from the Earth, our location in space is not special. An observer in another galaxy would also see the galactic clusters flying away from the observer's position because all of space is expanding. This is one of the main lines of evidence for the **Big Bang** from which the early Universe evolved and the subsequent expansion of space.

SEE ALSO Big Bang (13.7 Billion B.C.), Doppler Effect (1842), Olbers' Paradox (1823), Cosmic Microwave Background (1965), Cosmic Inflation (1980), Dark Energy (1998), Cosmological Big Rip (36 Billion).

For millennia, humans have looked to the skies and wondered about their place in the cosmos. Pictured here are Polish astronomer Johannes Hevelius and his wife Elisabeth making observations (1673). Elisabeth is considered to be among the first female astronomers.

Fig. M.

1929

Cyclotron

Ernest Orlando Lawrence (1901–1958)

"The first particle accelerators were linear," writes author Nathan Dennison, "but Ernest Lawrence bucked this trend and used many small electrical pulses to accelerate particles in circles. Starting off as a sketch on a scrap of paper, his first design cost just US $25 to make. Lawrence continued to develop his [cyclotron] using parts including a kitchen chair . . . until finally he was awarded the Nobel Prize in 1939."

Lawrence's cyclotron accelerated charged atomic or subatomic particles by using a constant magnetic field and changing electrical field to create a spiraling path of particles that started at the center of the spiral. After spiraling many times in an evacuated chamber, the high-energy particles could finally smash into atoms and the results could be studied using a detector. One advantage of the cyclotron over prior methods was its relatively compact size that could be used to achieve high energies.

Lawrence's first cyclotron was only a few inches in diameter, but by 1939, at the University of California, Berkeley, he was planning the largest and most expensive instrument thus far created to study the atom. The nature of the atomic nucleus was still mysterious, and this cyclotron—"requiring enough steel to build a good-sized freighter and electric current sufficient to light the city of Berkeley"—would allow the probing of this inner realm after the high-energy particles collided with nuclei and caused nuclear reactions. Cyclotrons were used to produce radioactive materials and for the production of tracers used in medicine. The Berkeley cyclotron created technetium, the first known artificially generated element.

The cyclotron was also important in that it launched the modern era of high-energy physics and the use of huge, costly tools that required large staffs to operate. Note that a cyclotron uses a constant magnetic field and a constant-frequency applied electric field; however, both of these fields are varied in the *synchrotron*, a circular accelerator developed later. In particular, synchrotron radiation was seen for the first time at the General Electric Company in 1947.

SEE ALSO Radioactivity (1896), Atomic Nucleus (1911), Neutrinos (1956), Large Hadron Collider (2009).

Physicists Milton Stanley Livingston (left) and Ernest O. Lawrence (right) in front of a 27-inch cyclotron at the old Radiation Laboratory at the University of California, Berkeley (1934).

1931

White Dwarfs and Chandrasekhar Limit

Subrahmanyan Chandrasekhar (1910–1995)

Singer Johnny Cash explained in his song "Farmer's Almanac" that God gave us the darkness so we could see the stars. However, among the most difficult luminous stars to find are the strange graveyard states of stars called *white dwarfs*. A majority of stars, like our Sun, end their lives as dense white dwarf stars. A teaspoon of white dwarf material would weigh several tons on the Earth.

Scientists first suspected the existence of white dwarf stars in 1844, when Sirius, the brightest star in the north sky, was discovered to wobble from side to side, as if being pulled by a neighbor in the sky that was too dim to be seen. The companion star was finally observed in 1862, and surprisingly seemed to be smaller than the Earth but almost as massive as our Sun. These white dwarf stars, still quite hot, result from the collapse of dying stars when they have exhausted all their nuclear fuel.

Non-rotating white dwarfs have a maximum mass of 1.4 solar masses—a value documented in 1931 by the young Subrahmanyan Chandrasekhar on a ship traveling from India to England, where he was about to begin his graduate studies at Cambridge University. As small or intermediate stars collapse, the electrons in the stars are crushed together and reach a state in which they can be no denser due to the **Pauli Exclusion Principle** that creates an outward *electron degeneracy* pressure. However, above 1.4 solar masses, electron degeneracy could no longer counteract that crushing gravitational force, and the star would continue to collapse (e.g. into a **Neutron Star**) or blow away excess mass from it surface in a supernova explosion. In 1983, Chandrasekhar won the Nobel Prize for his studies on stellar evolution.

After billions of years, a white dwarf will cool and no longer be visible, becoming a *black dwarf*. Although white dwarfs start out as a **Plasma**, at later stages of cooling, it is predicted that many become structurally like giant crystals.

SEE ALSO Black Holes (1783), Plasma (1879), Neutron Stars (1933), Pauli Exclusion Principle (1925).

This Hubble Space Telescope view of the Hourglass Nebula (MyCn 18) shows the glowing remains of a dying Sun-like star. The bright white spot slightly to the left of the center is the white-dwarf remnant of the original star that ejected this gaseous nebula.

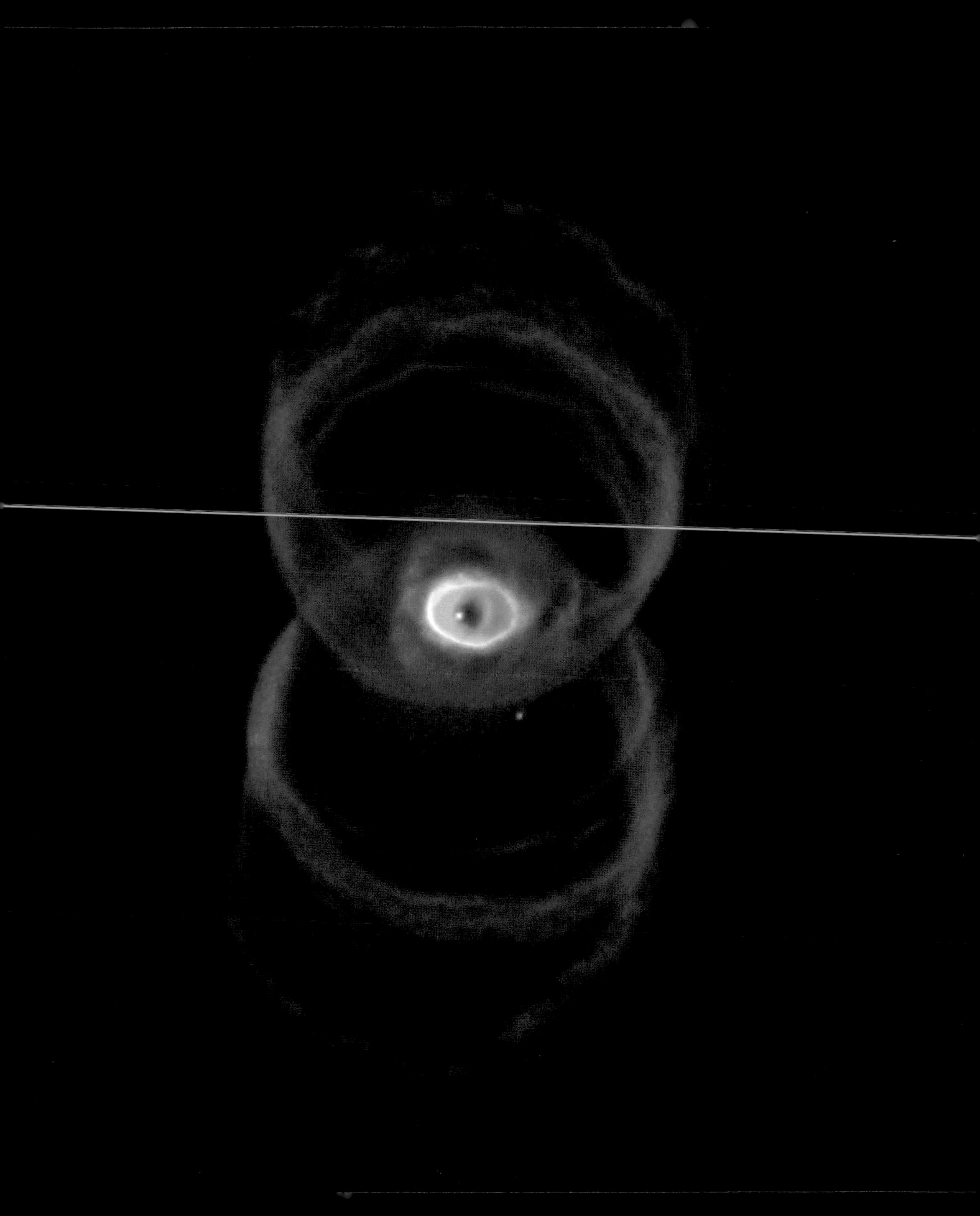

1931

Jacob's Ladder

Kenneth Strickfadden (1896–1984)

"It's alive!" cried Dr. Frankenstein when he first saw movement in the creature he had stitched together. The scene from the 1931 horror movie *Frankenstein* was replete with high-voltage special effects, the brainchild of electronics-expert Kenneth Strickfadden, who had probably gotten some of his ideas from inventor Nikola Tesla's earlier electrical demonstrations that included **Tesla Coils**. The Frankenstein device that struck the imagination of moviegoers for decades to come—as the icon for the mad scientists—was the V-shaped *Jacob's ladder*.

The Jacob's ladder is included in this book, partly because of its fame as a symbol for physics gone wild, but also because it continues to offer educators opportunities for illustrating concepts of electrical discharge, the variation of air density with temperature, **Plasmas** (ionized gases), and much more.

The idea of spark gaps consisting of two conducting electrodes separated by an air gap obviously predates its notoriety in *Frankenstein*. Generally, when sufficient voltage is supplied, a spark forms, which ionizes the gas, and reduces its electrical resistance. The atoms in the air become excited and fluoresce to produce the visible spark.

The Jacob's ladder produces a moving train of large sparks that rise upward. A spark first forms at the bottom of the ladder, where the two electrodes are quite close. The heated, ionized air rises, because it is less dense then the surrounding air, and carries the current path upward. Toward the top of the ladder, the current path becomes long and unstable. The arc's dynamic resistance increases, thus increasing power consumption and heat. When the arc finally breaks toward the top of the ladder, the voltage supply output momentarily exists in an open-circuit state, until the breakdown of the air dielectric at bottom causes another spark to form, and the cycle begins again.

In the early 1900s, the electrical arcs of similar ladders were used in chemical applications because they ionize nitrogen, yielding nitric oxide for producing nitrogen-based fertilizers.

SEE ALSO Von Guericke's Electrostatic Generator (1660), Leyden Jar (1744), Stokes' Fluorescence (1852), Plasma (1879), Tesla Coil (1891).

LEFT: *Jacob's ladder is named after the ladder in the Bible that Jacob sees reaching to heaven (painting by William Blake [1757–1827]).* RIGHT: *Photo of Jacob's ladder, showing the moving train of large sparks that rise upward. During operation, a spark first forms at the bottom of the ladder, where the two electrodes are quite close.*

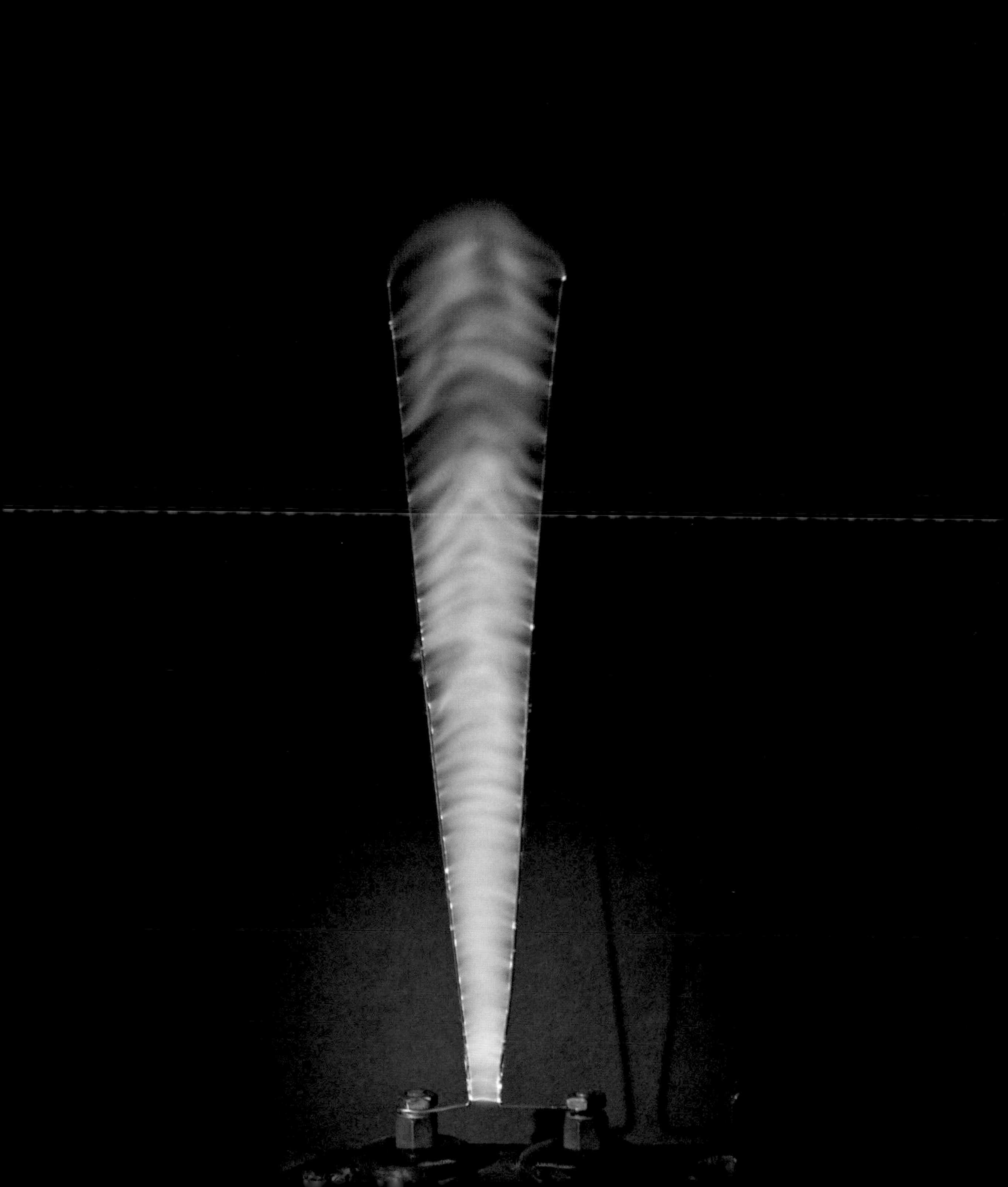

1932

Neutron

Sir James Chadwick (1891–1974), **Irène Joliot-Curie** (1897–1956), **Jean Frédéric Joliot-Curie** (1900–1958)

"James Chadwick's road to the discovery of the neutron was long and tortuous," writes chemist William H. Cropper. "Because they carried no electrical charge, neutrons did not leave observable trails of ions as they passed through matter, and left no tracks in Wilson's **Cloud Chamber**; to the experimenter, they were invisible." Physicist Mark Oliphant writes, "The neutron was discovered as a result of a persistent search by Chadwick, and not by accident as were **Radioactivity** and **X-rays**. Chadwick felt intuitively that it must exist and never gave up the chase."

The neutron is a subatomic particle that is part of every atomic nucleus except for ordinary hydrogen. It has no net electric charge and a mass slightly greater than that of a proton. Like the proton, it is composed of three quarks. When the neutron is inside the nucleus, the neutron is stable; however, free neutrons undergo beta decay, a type of radioactive decay, and have a mean lifetime of approximately 15 minutes. Free neutrons are produced during nuclear fission and fusion reactions.

In 1931, Irène Joliot-Curie (daughter of Marie Curie, the first person honored with two Nobel Prizes) and her husband Frédéric Joliot described a mysterious radiation produced by bombarding beryllium atoms with alpha particles (helium nuclei), and this radiation caused protons to be knocked loose from hydrogen-containing paraffin. In 1932, James Chadwick conducted additional experiments and suggested that this new kind of radiation was composed of uncharged particles of approximately the mass of the proton, namely, *neutrons*. Because the free neutrons are uncharged, they are not hindered by electrical fields and penetrate deeply into matter.

Later, researchers discovered that various elements, when bombarded by neutrons, undergo fission—a nuclear reaction that occurs when the nucleus of a heavy element splits into two nearly equal smaller pieces. In 1942, researchers in the U.S. showed that these free neutrons produced during fission can create a chain reaction and enormous amounts of energy, and could be used to create an atomic bomb—and nuclear power plants.

SEE ALSO Prehistoric Nuclear Reactor (2 Billion B.C.), Radioactivity (1896), Atomic Nucleus (1911), Wilson Cloud Chamber (1911), Neutron Stars (1933), Energy from the Nucleus (1942), Standard Model (1961), Quarks (1964).

The Brookhaven Graphite Research Reactor—the first peacetime reactor to be constructed in the United States following World War II. One purpose of the reactor was to produce neutrons via uranium fission for scientific experimentation.

1932

Antimatter

Paul Dirac (1902–1984), **Carl David Anderson** (1905–1991)

"Fictional spaceships are often powered by 'antimatter drives'," writes author Joanne Baker, "yet antimatter itself is real and has even been made artificially on the Earth. A 'mirror image' form of matter . . . , antimatter cannot coexist with matter for long—both annihilate in a flash of energy if they come into contact. The very existence of antimatter points at deep symmetries in particle physics."

The British physicist Paul Dirac once remarked that the abstract mathematics we study now gives us a glimpse of physics in the future. In fact, his equations from 1928 that dealt with electron motion predicted the existence of antimatter, which was subsequently discovered. According to the formulas, an electron must have an antiparticle with the same mass but a positive electrical charge. In 1932, U.S. physicist Carl Anderson observed this new particle experimentally and named it the *positron*. In 1955, the antiproton was produced at the Berkeley Bevatron (a particle accelerator). In 1995, physicists created the first anti-hydrogen atom at the CERN research facility in Europe. CERN (*Organisation Européenne pour la Recherche Nucléaire*), or the European Organization for Nuclear Research, is the largest particle physics laboratory in the world.

Antimatter-matter reactions have practical applications today in the form of positron emission tomography (PET). This medical imaging technique involves the detection of gamma rays (high-energy radiation) emitted by a positron-emitting tracer radionuclide, an atom with an unstable nucleus.

Modern physicists continue to offer hypotheses to explain why the observable universe appears to be nearly entirely composed of matter and not antimatter. Could regions of the universe exist in which antimatter predominates?

Upon casual inspection, antimatter would be almost indistinguishable from ordinary matter. Physicist Michio Kaku writes, "You can form antiatoms out of antielectrons and antiprotons. Even antipeople and antiplanets are theoretically possible. [However], antimatter will annihilate into a burst of energy upon contact with ordinary matter. Anyone holding a piece of antimatter in their hands would immediately explode with the force of thousands of hydrogen bombs."

SEE ALSO Wilson Cloud Chamber (1911), Dirac Equation (1928), CP Violation (1964).

In the 1960s, researchers at Brookhaven National Laboratory used detectors, such as this, for studying small brain tumors that absorbed injected radioactive material. Breakthroughs led to more practical devices for imaging areas of the brain, such as today's PET machines.

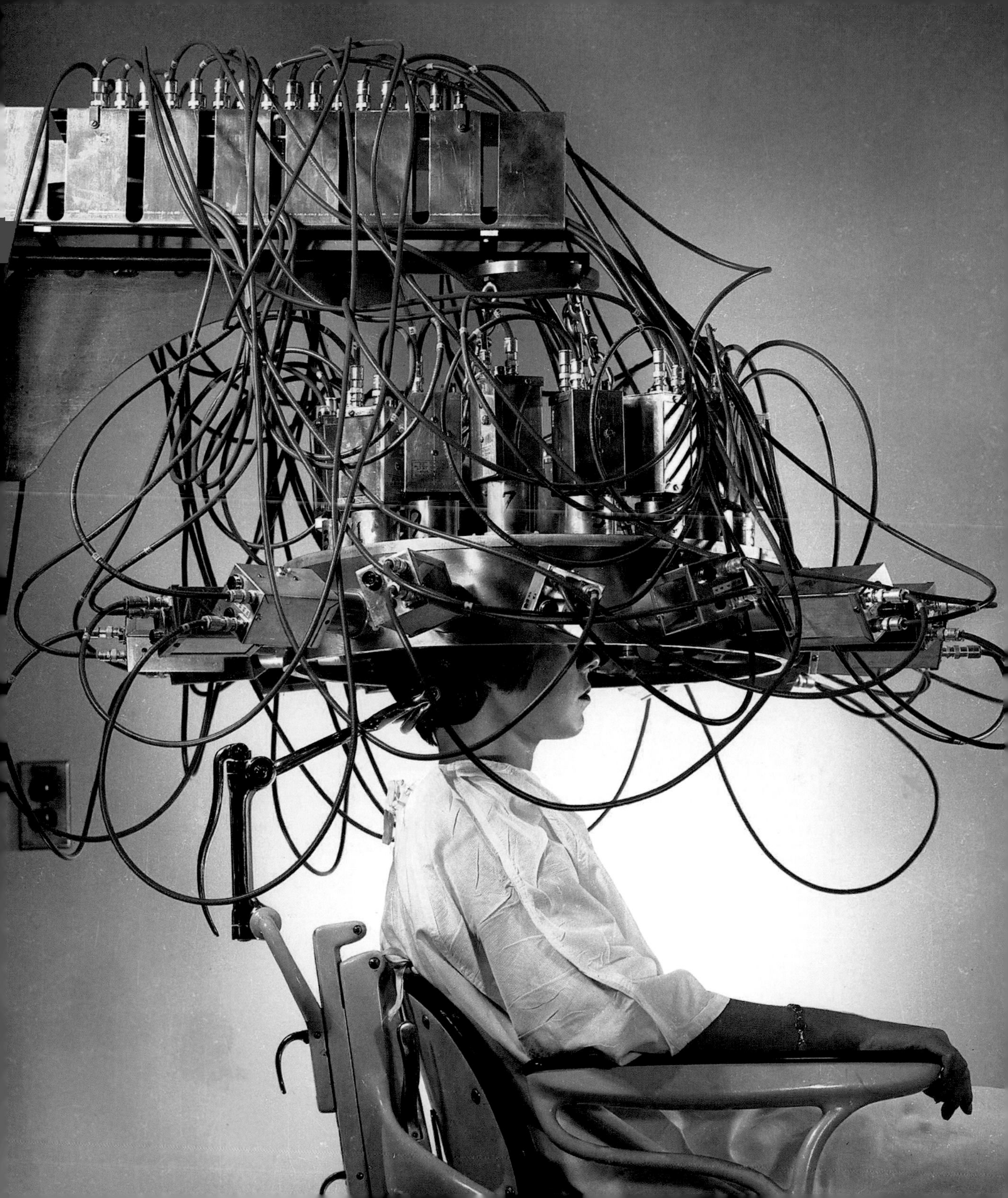

1933

Dark Matter

Fritz Zwicky (1898–1974), **Vera Cooper Rubin** (b. 1928)

Astronomer Ken Freeman and science-educator Geoff McNamara write, "Although science teachers often tell their students that the periodic table of the elements shows what the Universe is made of, this is not true. We now know that most of the Universe—about 96% of it—is made of dark material [dark matter and **Dark Energy**] that defies brief description. . . ." Whatever the composition of dark matter is, it does not emit or reflect sufficient light or other forms of electromagnetic radiation to be observed directly. Scientists infer its existence from its gravitational effects on visible matter such as the rotational speeds of galaxies.

Most of the dark matter probably does not consist of the standard elementary particles—such as protons, neutrons, electrons and known **Neutrinos**—but rather hypothetical constituents with exotic-sounding names such as sterile neutrinos, axions, and WIMPs (Weakly Interacting Massive Particles, including neutralinos), which do not interact with electromagnetism and thus cannot be easily detected. The hypothetical neutralinos are similar to neutrinos, but heavier and slower. Theorists also consider the wild possibility that dark matter includes gravitons, hypothetical particles that transmit gravity, leaking into our universe from neighboring universes. If our universe is on a membrane "floating" within a higher dimensional space, dark matter may be explained by ordinary stars and galaxies on nearby membrane sheets.

In 1933, the astronomer Fritz Zwicky provided evidence for the existence of dark matter through his studies of the motions of the edges of galaxies, which suggested a significant amount of galactic mass was undetectable. In the late 1960s, astronomer Vera Rubin showed that most stars in spiral galaxies orbit at approximately the same speed, which implied the existence of dark matter beyond the locations of stars in the galaxies. In 2005, astronomers from Cardiff University believed they had discovered a galaxy in the Virgo Cluster made almost entirely of dark matter.

Freeman and McNamara write, "Dark matter provides a further reminder that we humans are not essential to the Universe. . . . We are not even made of the same stuff as most of the universe. . . . Our Universe is made of darkness."

SEE ALSO Black Holes (1783), Neutrinos (1956), Supersymmetry (1971), Dark Energy (1998), Randall-Sundrum Branes (1999).

One early piece of evidence for the existence of dark matter is the 1959 observation of astronomer Louise Volders who demonstrated that spiral galaxy M33 (pictured here in a NASA Swift Satellite ultraviolet image) does not spin as expected according to standard Newtonian dynamics.

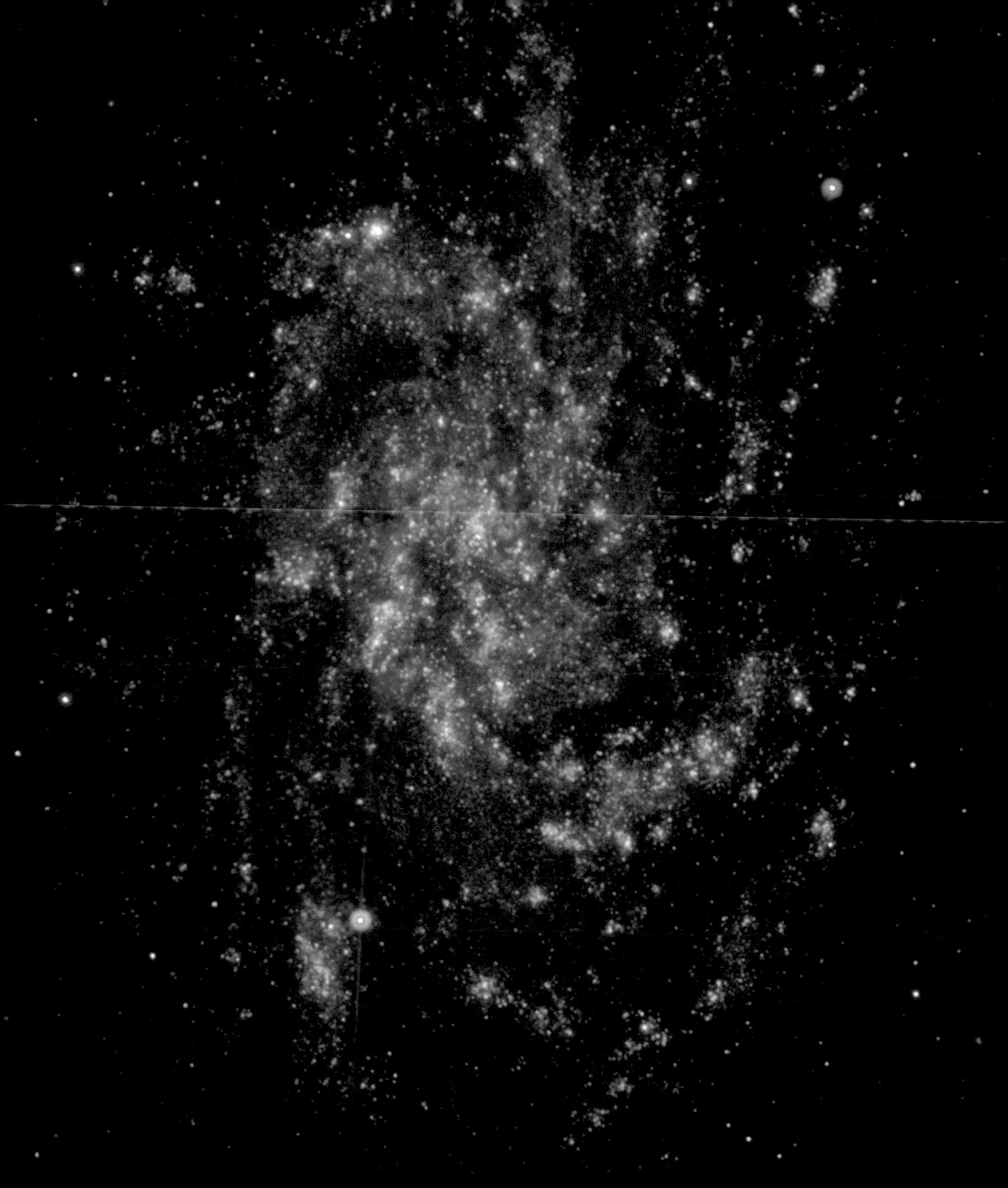

1933

Neutron Stars

Fritz Zwicky (1898–1974), **Jocelyn Bell Burnell** (b. 1943), **Wilhelm Heinrich Walter Baade** (1893–1960)

Stars are born when a large amount of hydrogen gas starts to collapse in on itself due to gravitational attraction. As the star coalesces, it heats up, produces light, and helium is formed. Eventually the star runs out of hydrogen fuel, starts to cool, and enters one of several possible "graveyard states" such as a **Black Hole** or one of its crushed cousins such as a **White Dwarf** (for relatively small stars) or a neutron star.

More particularly, after a massive star has finished burning its nuclear fuel, the central region collapses due to gravity, and the star undergoes a supernova explosion, blowing off its outer layers. A neutron star, made almost entirely of uncharged subatomic particles called **Neutrons**, may be created by this gravitational collapse. A neutron star is prevented from achieving the complete gravitational collapse of a black hole due to the **Pauli Exclusion Principle** repulsion between neutrons. A typical neutron star has a mass between about 1.4 and 2 times the mass of our Sun, but with a radius of only around 7.5 miles (12 kilometers). Interestingly, neutron stars are formed of an extraordinary material, known as *neutronium*, which is so dense that a sugar cube could contain the crushed mass of the entire human population.

Pulsars are rapidly rotating, highly magnetic neutron stars that send out steady electromagnetic radiation that arrives at the Earth as pulses due to the star's rotation. The pulses are at intervals ranging from milliseconds to several seconds. The fastest millisecond pulsars spin over 700 times per second! Pulsars were discovered in 1967 by graduate student Jocelyn Bell Burnell in the form of radio sources that seemed to blink at a constant frequency. In 1933, astrophysicists Fritz Zwicky and Walter Baade proposed the existence of a neutron star, only a year after the discovery of the neutron.

In the novel *Dragon's Egg*, creatures live on a neutron star, where the gravity is so strong that mountain ranges are about a centimeter high.

SEE ALSO Black Holes (1783), Pauli Exclusion Principle (1925), Neutron (1932), White Dwarfs and Chandrasekhar Limit (1931).

In 2004, a neutron star underwent a "star quake," causing it to flare brightly, temporarily blinding all X-ray satellites. The blast was created by the star's twisting magnetic fields that can buckle the surface of the neutron star. (Artist's concept from NASA.)

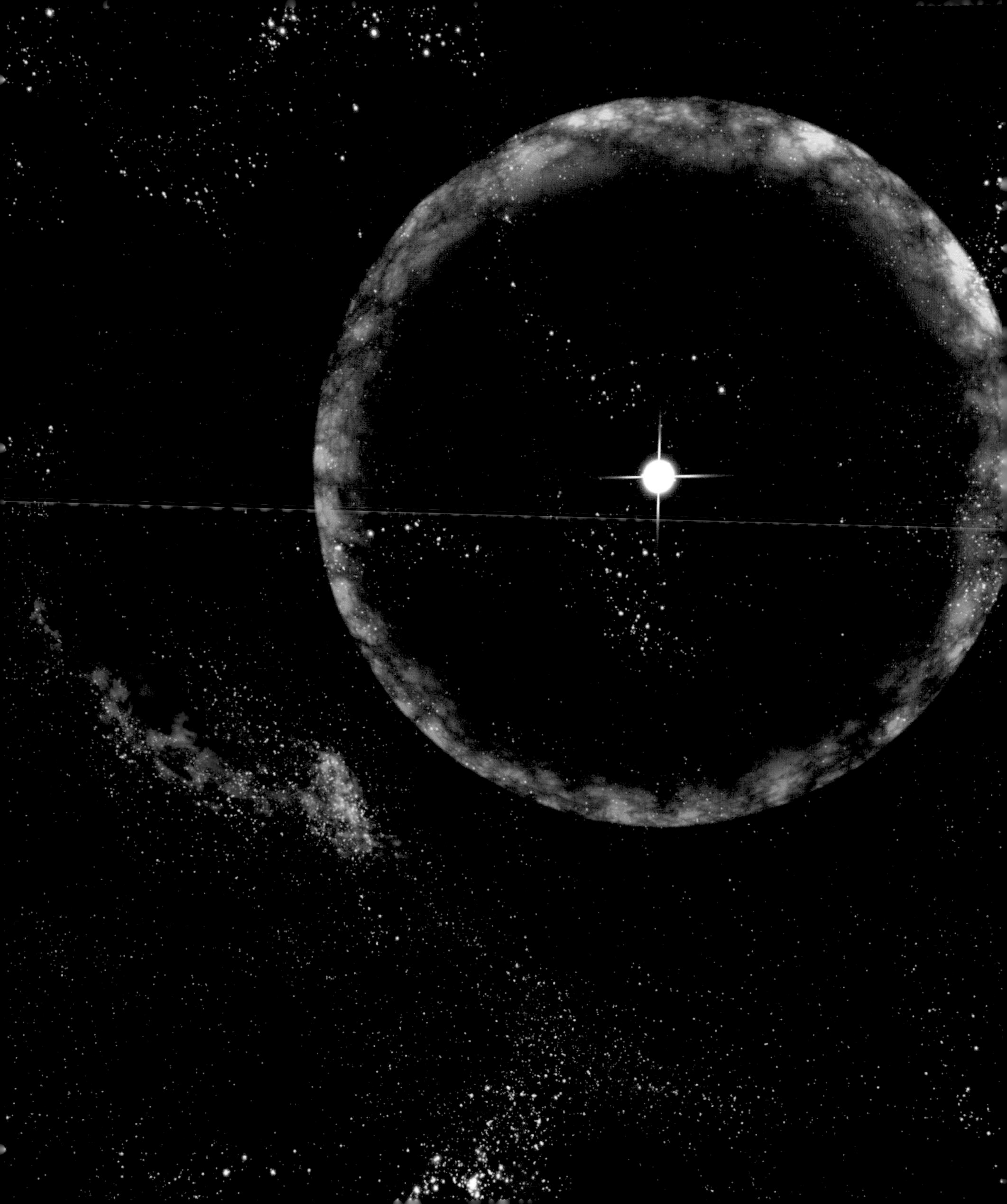

1934

Cherenkov Radiation

Igor Yevgenyevich Tamm (1895–1971), **Pavel Alekseyevich Cherenkov** (1904–1990), **Ilya Mikhailovich Frank** (1908–1990)

Cherenkov radiation is emitted when a charged particle, such as an electron, passes through a transparent medium, such as glass or water, at a speed greater than the speed of light in that medium. One of the most common examples of such radiation involves nuclear reactors, which are often placed within a pool of water for shielding. The reactor core may be bathed in an eerie blue glow caused by the Cherenkov radiation given off by particles produced by nuclear reactions. The radiation is named after Russian scientist Pavel Cherenkov, who studied this phenomenon in 1934.

When light moves through transparent materials, it moves more slowly overall than it does in a vacuum due to the interactions of the photons with the atoms in the medium. Consider the metaphor of a sports car traveling down a highway being stopped periodically by a police officer. The car cannot reach the end of the highway as fast as it could if no police officers were present. In glass or water, light typically travels around 70% of its speed in a vacuum, and it is possible for charged particles to travel faster than light in this medium.

In particular, the charged particle traveling through the medium displaces electrons in some of the atoms along its path. The displaced atomic electrons emit radiation to form a strong electromagnetic wave reminiscent of the bow wave caused by a speed boat in water or to a sonic boom produced by an airplane traveling faster than the speed of sound in air.

Because the emitted radiation is conical in shape, with a cone angle that depends on particle velocity and speed of light in the medium, Cherenkov radiation can provide particle physicists useful information on the speed of a particle under study. For pioneering work on this radiation, Cherenkov shared the Nobel Prize in 1958 with the physicists Igor Tamm and Ilya Frank.

SEE ALSO Snell's Law of Refraction (1621), Bremsstrahlung (1909), Hypersonic Whipcracking (1927), Sonic Booms (1947), Neutrinos (1956).

The blue glow of Cherenkov radiation in the core of the Advanced Test Reactor at the Idaho National Laboratory. The core is submerged in water.

1934

Sonoluminescence

Sonoluminescence, or SL, reminds me of those "light organs" of the 1970s, popular at dance parties to convert music to multicolored lighting that pulsates with the musical beat. However, the lights of SL are certainly much hotter and shorter than their psychedelic counterparts!

SL refers to the emission of short bursts of light caused by the implosion of bubbles in a liquid when stimulated by sound waves. German researchers H. Frenzel and H. Schultes discovered the effect in 1934 while experimenting with ultrasound in a tank of photographic developer. Their film showed tiny dots produced by bubbles in the fluid that had emitted light when the sound source was turned on. In 1989, physicist Lawrence Crum and his graduate student D. Felipe Gaitan produced stable SL in which a single bubble trapped in an acoustic standing wave emitted a pulse of light with each compression of the bubble within the standing wave.

In general, SL can occur when a sound wave stimulates a cavity of gas to form in the liquid (a process called *cavitation*). When the gas bubble collapses, a supersonic shockwave is produced, and the bubble temperature soars to higher than the temperature at the surface of the Sun, creating a **Plasma**. The collapse may be faster than 50 picoseconds (or 50 trillionths of a second) and produces blue and ultraviolet light and X-rays as a result of the collision of particles in the plasma. The collapsed bubble at the time it emits light is about a micrometer in diameter, about the size of a bacterium.

Temperature of at least 35,500 °F (20,000 kelvins) can be produced, hot enough to boil a diamond. Some researchers have speculated that if the temperature could be increased even further, SL might be used to cause thermonuclear fusion.

Pistol shrimp produce an SL-like phenomena. When they snap their claws, they create a collapsing bubble that generates a shockwave. This shockwave produces a loud sound to stun prey, as well as a dim light that is detectable using photomultiplier tubes.

SEE ALSO Triboluminesence (1620), Henry's Gas Law (1803), Plasma (1879).

Gas bubbles form and collapse when a liquid is energized by ultrasound. "Cavitation, which drives the implosive collapse of these bubbles, creates temperatures resembling those at the surface of the sun . . ." writes chemist Kenneth Suslick.

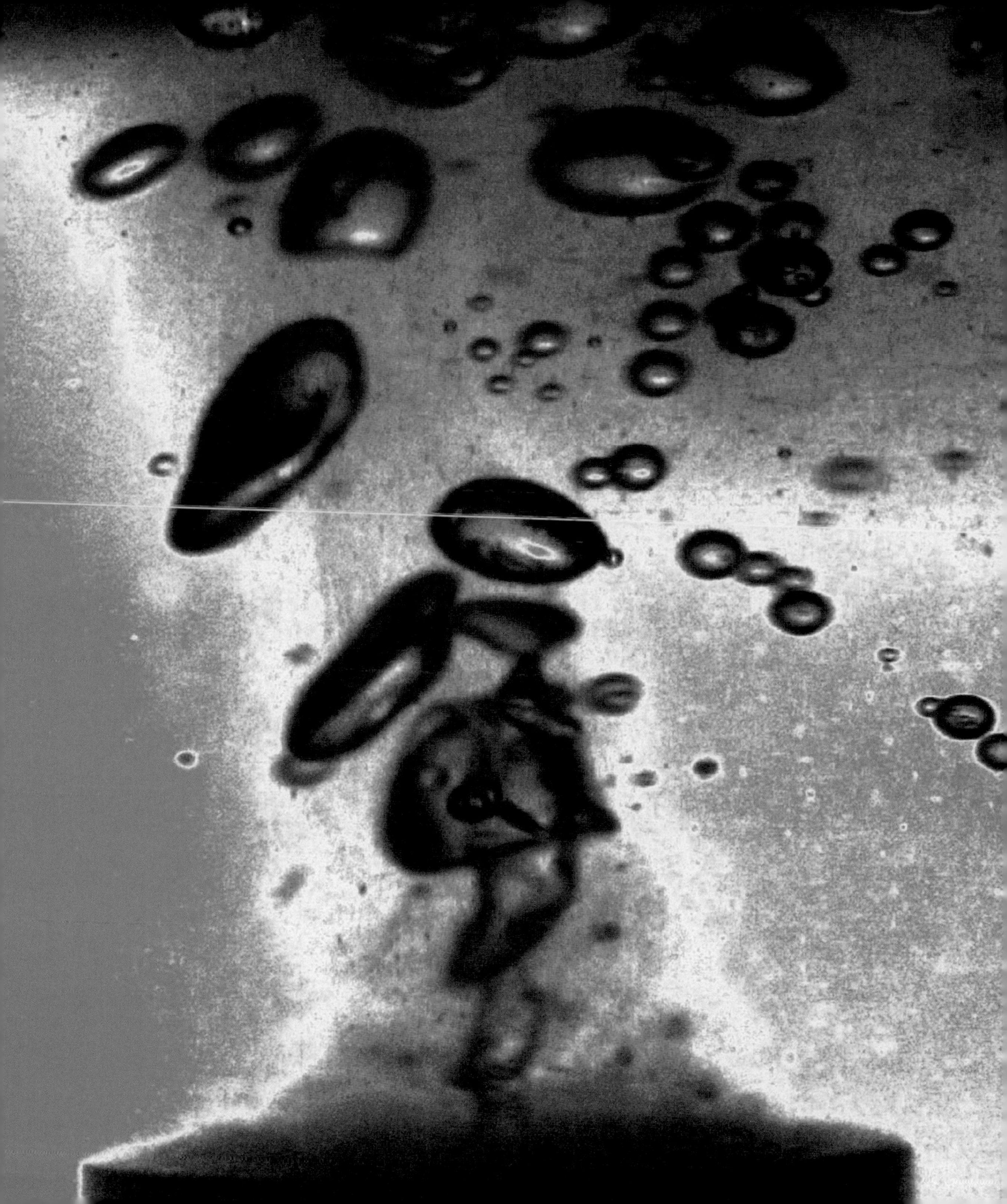

1935

EPR Paradox

Albert Einstein (1879–1955), **Boris Podolsky** (1896–1966), **Nathan Rosen** (1909–1995), **Alain Aspect** (b. 1947)

Quantum entanglement (QE) refers to an intimate connection between quantum particles, such as between two electrons or two photons. Once the pair of particles is *entangled*, a particular kind of change to one of them is reflected instantly in the other, and it doesn't matter if the pair is separated by inches or by interplanetary distances. This entanglement is so counterintuitive that Albert Einstein referred to it as being "spooky" and thought that it demonstrated a flaw in quantum theory, particularly the Copenhagen interpretation that suggested quantum systems, in a number of contexts, existed in a probabilistic limbo until observed and then reached a definite state.

In 1935, Albert Einstein, Boris Podolsky, and Nathan Rosen published a paper on their famous EPR paradox. Imagine two particles that are emitted by a source so that their spins are in a quantum superposition of opposite states, labeled + and −. Neither particle has a definite spin before measurement. The particles fly apart, one going to Florida and the other to California. According to QE, if scientists in Florida measure the spin and finds a +, the California particle *instantly* assumes the − state, even though the speed of light prohibits faster-than-light (FTL) communication of information. Note, however, that no FTL communication of information has actually occurred. Florida cannot use entanglement to send messages to California because Florida does not manipulate the spin of its particle, which has a 50-50 chance of being in a + or − state.

In 1982, physicist Alain Aspect performed experiments on oppositely directed photons that were emitted in a single event from the same atom, thus ensuring the photon pairs were correlated. He showed that the instantaneous connection in the EPR paradox actually did take place, even when the particle pair was separated by an arbitrarily large distance.

Today, quantum entanglement is being studied in the field of quantum cryptography to send messages that cannot be spied upon without leaving some kind of trace. Simple **Quantum Computers** are now being developed that perform calculations in parallel and more quickly than by traditional computers.

SEE ALSO Stern-Gerlach Experiment (1922), Complementarity Principle (1927), Schrödinger's Cat (1935), Bell's Theorem (1964), Quantum Computers (1981), Quantum Teleportation (1993).

Artist's rendition of "spooky action at a distance." Once a pair of particles is entangled, *a particular kind of change to one of them is reflected instantly in the other, even if the pair is separated by interplanetary distances.*

1935

Schrödinger's Cat

Erwin Rudolf Josef Alexander Schrödinger (1887–1961)

Schrödinger's cat reminds me of a ghost, or maybe a creepy zombie—a creature that appears to be alive and dead at the same time. In 1935, Austrian physicist Erwin Schrödinger published an article on this extraordinary paradox with consequences that are so striking that it continues to mystify and concern scientists to this day.

Schrödinger had been upset about the recently proposed *Copenhagen interpretation of quantum mechanics* that stated, in essence, that a quantum system (e.g. an electron) exists as a cloud of probability until an observation is made. At a higher level, it seemed to suggest that it is meaningless to ask precisely what atoms and particles were doing when unobserved; in some sense, reality is created by the observer. Before being observed, the system takes on all possibilities. What could this mean for our everyday lives?

Imagine that a live cat is placed in a box with a **Radioactive** source, a **Geiger Counter**, and a sealed glass flask containing deadly poison. When a radioactive decay event occurs, the Geiger counter measures the event, triggering a mechanism that releases the hammer that shatters the flask and releases the poison that kills the cat. Imagine that quantum theory predicts a 50 percent probability that one decay particle is emitted each hour. After an hour, there is an equal probability that the cat is alive or dead. According to some flavors of the Copenhagen interpretation, the cat seemed to be both alive and dead—a mixture of two states that is called a superposition of states. Some theorists suggested that if you open the box, the very act of observation "collapses the superposition" making the cat either alive or dead.

Schrödinger said that his experiment demonstrated the invalidity of the Copenhagen interpretation, and Albert Einstein agreed. Many questions spun from this thought experiment: What is considered to be a valid observer? The Geiger counter? A fly? Could the cat observe itself and so collapse its own state? What does the experiment really say about the nature of reality?

SEE ALSO Radioactivity (1896), Geiger Counter (1908), Complementarity Principle (1927), Quantum Tunneling (1928), EPR Paradox (1935), Parallel Universes (1956), Bell's Theorem (1964), Quantum Immortality (1987).

When the box is opened, the very act of observation may collapse the superposition, making Schrödinger's cat either alive or dead. Here, Schrödinger's cat thankfully emerges alive.

1937

Superfluids

Pyotr Leonidovich Kapitsa (1894–1984), **Fritz Wolfgang London** (1900–1954), **John "Jack" Frank Allen** (1908–2001), **Donald Misener** (1911–1996)

Like some living, creeping liquid out of a science-fiction movie, the eerie behavior of superfluids has intrigued physicists for decades. When liquid helium in a superfluid state is placed in a container, it climbs up the walls and leaves the container. Additionally, the superfluid remains motionless when its container is spun. It seems to seek out and penetrate microscopic cracks and pores, making traditionally adequate containers leaky for superfluids. Place your cup of coffee—with the liquid spinning in the cup—on the table, and a few minutes later the coffee is still. If you did this with superfluid helium, and your ancestors came back in a thousand years to view the cup, the superfluid might still be spinning.

Superfluidity is seen in several substances, but often studied in helium-4—the common, naturally occurring isotope of helium containing two protons, two neutrons, and two electrons. Below an extremely cold critical temperature called the lambda temperature (−455.49 °F, 2.17 K), this liquid helium-4 suddenly attains the ability to flow without apparent friction and achieves a thermal conductivity that is millions of times the conductivity of normal liquid helium and much greater than the best metal conductors. The term *helium I* refers to the liquid above 2.17 K, and *helium II* refers to the liquid below this temperature.

Superfluidity was discovered by physicists Pyotr Kapitsa, John F. Allen, and Don Misener in 1937. In 1938, Fritz London suggested that liquid helium beneath the lambda point temperature is composed of two parts, a normal fluid with characteristics of helium 1 and a superfluid (with a viscosity value essentially equal to 0).The transition from ordinary fluid to superfluid occurs when the constituent atoms begin to occupy the same quantum state, and their quantum wave functions overlap. As in the **Bose-Einstein Condensate**, the atoms lose their individual identities and behave as one large smeared entity. Because the superfluid has no internal viscosity, a vortex formed within the fluid continues to spin essentially forever.

SEE ALSO Ice Slipperiness (1850), Stokes' Law of Viscosity (1851), Discovery of Helium (1868), Superconductivity (1911), Silly Putty (1943), Bose-Einstein Condensate (1995).

Frame from Alfred Leitner's 1963 movie Liquid Helium, Superfluid. *The liquid helium is in the superfluid phase as a thin film creeps up the inside wall of the suspended cup and down on the outside to form a droplet at the bottom.*

1938

Nuclear Magnetic Resonance

Isidor Isaac Rabi (1898–1988), **Felix Bloch** (1905–1983), **Edward Mills Purcell** (1912–1997), **Richard Robert Ernst** (b. 1933), **Raymond Vahan Damadian** (b. 1936)

"Scientific research requires powerful tools for elucidating the secrets of nature," writes Nobel Laureate Richard Ernst. "Nuclear Magnetic Resonance (NMR) has proved to be one of the most informative tools of science with applications to nearly all fields from solid state physics, to material science, . . . and even to psychology, trying to understand the functioning of the human brain."

If an atomic nucleus has at least one neutron or proton unpaired, the nucleus may act like a tiny magnet. When an external magnetic field is applied, it exerts a force that can be visualized as causing the nuclei to precess, or wobble, like a spinning top. The potential energy difference between nuclear spin states can be made larger by increasing the external magnetic field. After turning on this static external magnetic field, a radio-frequency (RF) signal of the proper frequency is introduced that can induce transitions between spin states, and thus some of the spins are placed in their higher-energy states. If the RF signal is turned off, the spins relax back to the lower state and produce an RF signal at the *resonant* frequency associated with the spin flip. These NMR signals yield information beyond the specific nuclei present in a sample, because the signals are modified by the immediate chemical environment. Thus, NMR studies can yield a wealth of molecular information.

NMR was first described in 1937 by physicist Isidor Rabi. In 1945, physicists Felix Bloch, Edward Purcell, and colleagues refined the technique. In 1966, Richard Ernst further developed Fourier transform (FT) spectroscopy and showed how RF pulses could be used to create a spectrum of NMR signals as a function of frequency. In 1971, physician Raymond Damadian showed that the hydrogen relaxation rates of water could be different in normal and malignant cells, opening the possibility of using NMR for medical diagnosis. In the early 1980s, NMR methods started to be used in magnetic resonance imaging (MRI) to characterize the nuclear magnet moments of ordinary hydrogen nuclei of soft tissues of the body.

SEE ALSO Discovery of Helium (1868), X-rays (1895), Atomic Nucleus (1911), Superconductivity (1911).

An actual MRI/MRA (magnetic resonance angiogram) of the brain vasculature (arteries). This kind of MRI study is often used to reveal brain aneurysms.

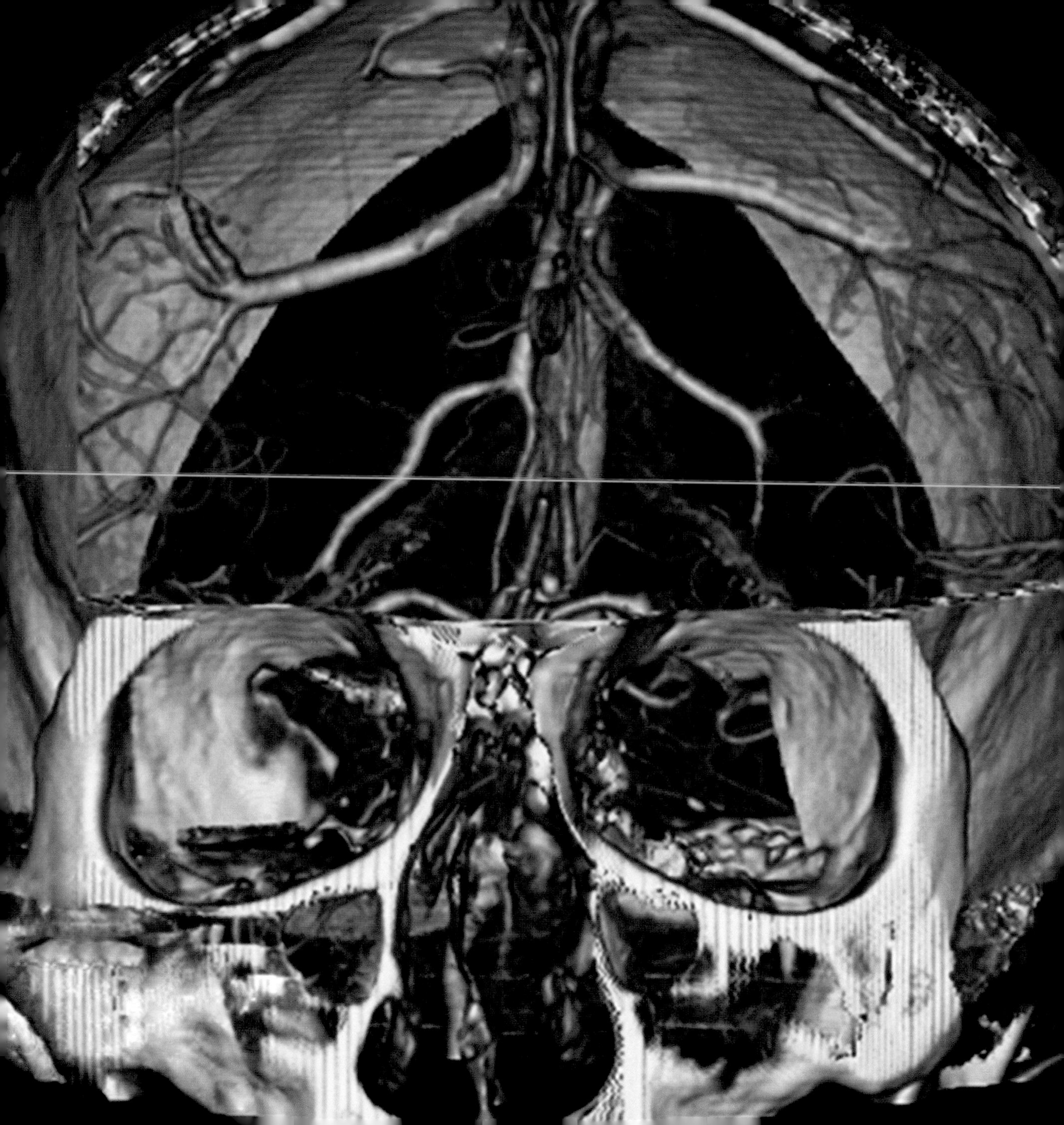

1942

Energy from the Nucleus

Lise Meitner (1878–1968), **Albert Einstein** (1879–1955), **Leó Szilárd** (1898–1964), **Enrico Fermi** (1901–1954), **Otto Robert Frisch** (1904–1979)

Nuclear fission is a process in which the nucleus of an atom, such as uranium, splits into smaller parts, often producing free neutrons, lighter nuclei, and a large release of energy. A chain reaction takes place when the neutrons fly off and split other uranium atoms, and the process continues. A nuclear reactor, used to produce energy, makes use of the process that is moderated so that it releases energy at a controlled rate. A nuclear weapon uses the process in a rapid, uncontrolled rate. The products of nuclear fission are themselves often radioactive and thus may lead to nuclear waste problems associated with nuclear reactors.

In 1942, in a squash court beneath the stadium at the University of Chicago, physicist Enrico Fermi and his colleagues produced a controlled nuclear chain reaction using uranium. Fermi had relied on the 1939 work of physicists Lise Meitner and Otto Frisch, who showed how the uranium nucleus breaks into two pieces, unleashing tremendous energy. In the 1942 experiment, metal rods absorbed the neutrons, allowing Fermi to control the rate of reaction. Author Alan Weismann explains, "Less than three years later, in a New Mexico desert, they did just the opposite. The nuclear reaction this time [which included plutonium] was intended to go completely out of control. Immense energy was released, and within a month, the act was repeated twice, over two Japanese cities. . . . Ever since, the human race has been simultaneously terrified and fascinated by the double deadliness of nuclear fission: fantastic destruction followed by slow torture."

The Manhattan Project, led by the U.S., was the codename for a project conducted during World War II to develop the first atomic bomb. Physicist Leó Szilárd had been so concerned about German scientists creating nuclear weapons that he approached Albert Einstein, who wrote a letter to President Roosevelt in 1939, alerting him to this danger. Note that a second kind of nuclear weapon (an "H-bomb") uses nuclear fusion reactions.

SEE ALSO Radioactivity (1896), $E = mc^2$ (1905), Atomic Nucleus (1911), Energy from the Nucleus (1942), Little Boy Atomic Bomb (1945), Tokamak (1956).

LEFT: *Lise Meitner was part of the team that discovered nuclear fission (1906 photo).* RIGHT: *Calutron (***Mass Spectrometer***) operators in the Y-12 plant at Oak Ridge, Tennessee, during World War II. The calutrons were used to refine uranium ore into fissile material. Workers toiled in secrecy during the Manhattan Project effort to construct an atomic explosive.*

853
855

1943

Silly Putty

"The Silly Putty collection at the [Smithsonian Institution] National Museum of American History tells many fascinating stories about how this unusual product became an American phenomenon," says chief archivist John Fleckner. "We were interested in this collection because Silly Putty is a case study of invention, business and entrepreneurship, and longevity."

Created by accident, that colorful goo that you probably played with as a child had its start in 1943 when engineer James Wright of the General Electric Corporation combined boric acid with silicone oil. To his surprise, the material had many surprising characteristics, and it could bounce like a rubber ball. Later, American marketing-guru Peter Hodgson saw its potential as a toy and launched the successful Silly Putty, sold in plastic eggs. Crayola owns the trademark.

Today, additional materials are added to the silicone polymer, and one recipe calls for approximately 65% dimethyl siloxane, 17% silica, 9% Thixatrol ST, 4% polydimethylsiloxane, 1% decamethyl cyclopentasiloxane, 1% glycerine, and 1% titanium dioxide. Not only does it bounce, but it tears when given a sharp ripping motion. Given sufficient time, it flows like a liquid and forms a puddle.

In 2009, students performed an experiment by dropping a 50-pound Silly Putty ball from the roof of an 11-story building at North Carolina State University. The ball struck the ground and shattered into many fragments. Silly Putty is an example of a non-Newtonian fluid with a viscosity that changes (e.g., it may depend on applied forces)—which is in contrast to a "Newtonian fluid" that flows like water. The viscosity of Newtonian fluids depends on temperature and pressure but not on the forces acting on the fluid.

Quicksand is another example of a non-Newtonian fluid. If you are ever stuck in quicksand and you move slowly, the quicksand will act like a liquid and you will be able to escape more easily than if you move around very quickly, because fast motions can cause the quicksand to act more like a solid from which it is harder to escape.

SEE ALSO Stokes' Law of Viscosity (1851), Superfluids (1937), Lava Lamp (1963).

Silly Putty and similar Plasticine-like materials are examples of non-Newtonian fluids with unusual flow characteristics and a non-constant value for viscosity. Silly Putty sometimes acts as a viscous liquid but can also act as an elastic solid.

1945

Drinking Bird

Miles V. Sullivan (b. 1917)

"If ever there was a perpetual motion machine, this must be it," Ed and Woody Sobey humorously write. "The bird moves apparently without a source of energy. But it turns out that it's a great example of a heat engine. You supply the heat from a lamp or the sun."

Invented in 1945 and patented in 1946 by Dr. Miles V. Sullivan, a scientist at Bell Labs in New Jersey, the drinking bird has fascinated physicists and educators ever since. Contributing to the bird's intrigue are the many physics principles at work as it seemingly bobs back and forth forever along a pivot point while lowering its head into a glass of water and then tilting back.

Here's how the bird works. The bird's head is covered in a felt-like material. Inside the bird's body is a colored solution of methylene chloride, a volatile liquid that vaporizes at a relatively low temperature. Air is removed from the bird's interior, and thus the bird's body is partially filled by methylene chloride vapor. The bird operates using a temperature differential between the bird's head and rear end. This creates a pressure differential inside the bird. When the head is wet, water evaporates from the felt and cools the head. As a result of this cooling, some of the vapor in the head condenses to form a liquid. The pressure drops in the head results from the cooling and the condensation. This lower pressure acts so as to draw liquid from the body. As the liquid rises into the head, the bird becomes top-heavy and tips into the glass of water. Once tipped, an internal tube allows a bubble of vapor from the body to rise and displace liquid in the head. Liquid flows back to the body, and the bird tilts back. The process repeats as long as enough water in the glass remains to wet the head with each tipping.

These bobbing birds can actually be harnessed to generate small amounts of energy.

SEE ALSO Siphon (250 B.C.), Perpetual Motion Machines (1150), Boyle's Gas Law (1662), Henry's Gas Law (1803), Carnot Engine (1824), Crookes Radiometer (1873).

Invented in 1945 and patented in 1946, the drinking bird has intrigued and amused physicists and educators ever since. Several physics principles help explain its incessant bobbing motion.

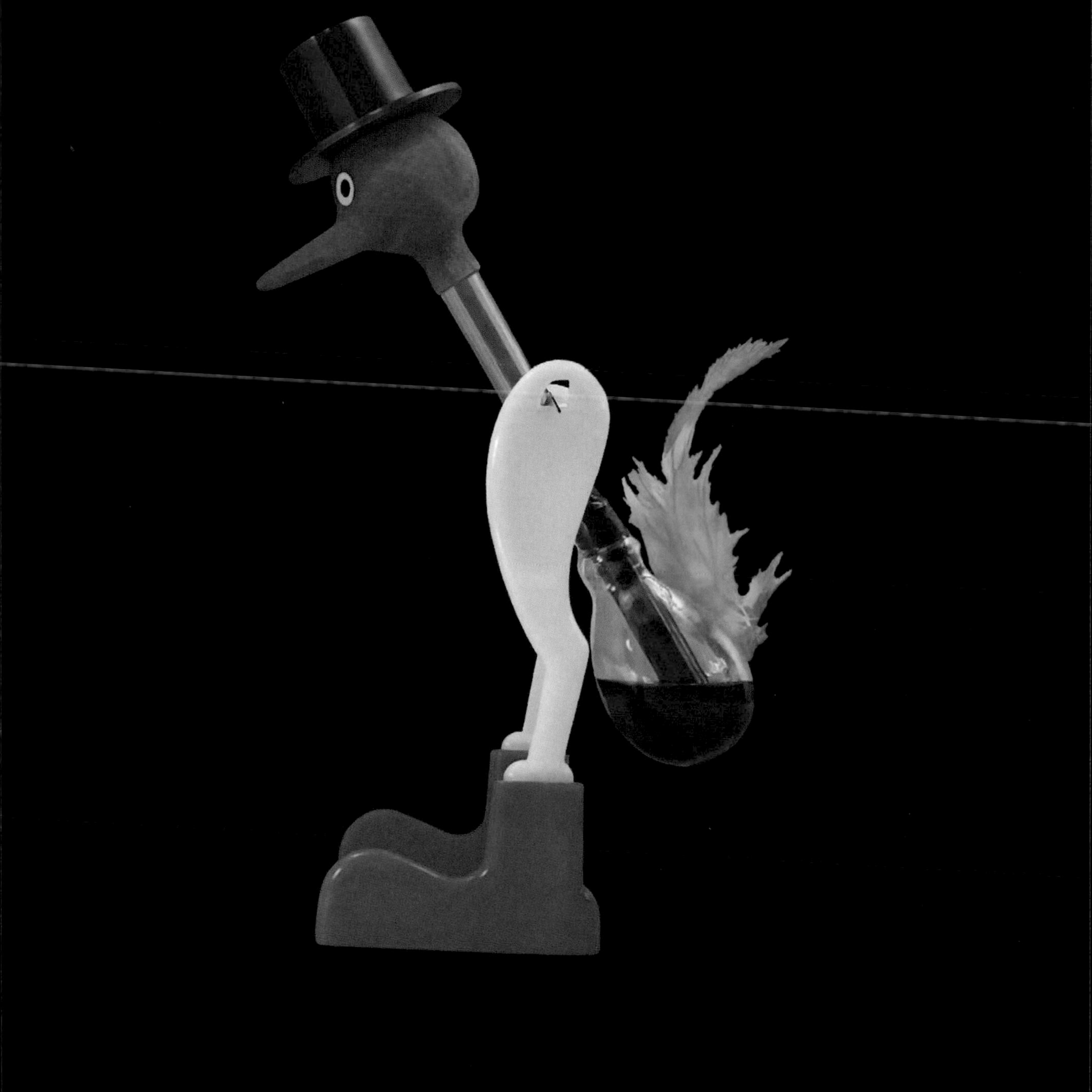

1945

Little Boy Atomic Bomb

J. Robert Oppenheimer (1904–1967), **Paul Warfield Tibbets, Jr.** (1915–2007)

On July 16, 1945, American physicist J. Robert Oppenheimer watched the first detonation of an atomic bomb in the New Mexico deserts, and he recalled a line from the *Bhagavad Gita*, "Now I become Death, the destroyer of worlds." Oppenheimer was the scientific director of the Manhattan Project, the World War II effort to develop the first nuclear weapon.

Nuclear weapons are exploded as a result of nuclear fission, nuclear fusion, or a combination of both processes. Atomic bombs generally rely on *nuclear fission* in which certain isotopes of uranium or plutonium split into lighter atoms, releasing neutrons and energy in a chain reaction. Thermonuclear bombs (or hydrogen bombs) rely on *fusion* for a portion of their destructive power. In particular, at very high temperatures, isotopes of hydrogen combine to form heavier elements and release energy. These high temperatures are achieved with a fission bomb to compress and heat fusion fuel.

Little Boy was the name of the atomic bomb dropped on Hiroshima, Japan on August 6, 1945, by the bomber plane *Enola Gay*, piloted by Colonel Paul Tibbets. Little Boy was about 9.8 feet long (3.0 meters) and contained 140 pounds (64 kilograms) of enriched uranium. After release from the plane, four radar altimeters were used to detect the altitude of the bomb. For greatest destructive power, the bomb was to explode at an altitude of 1,900 feet (580 meters). When any two of the four altimeters sensed the correct height, a cordite charge was to explode in the bomb, firing one mass of uranium-235 down a cylinder into another mass to create a self-sustaining nuclear reaction. After the explosion, Tibbets recalled the "awful cloud . . . boiling up, mushrooming terrible and incredibly tall." Over a period of time, as many as 140,000 people were killed—roughly half due the immediate blast and the other half due to gradual effects of the radiation. Oppenheimer later noted, "The deep things in science are not found because they are useful; they are found because it was possible to find them."

SEE ALSO Prehistoric Nuclear Reactor (2 Billion B.C.), Von Guericke's Electrostatic Generator (1660), Graham's Law of Effusion (1829), Dynamite (1866), Radioactivity (1896), Energy from the Nucleus (1942), Tokamak (1956), Electromagnetic Pulse (1962).

Little Boy on trailer cradle in pit, August 1945. Little Boy was about 9.8 feet (3.0 meters) long. Over a period of time, Little Boy may have killed as many as 140,000 people.

1946

Stellar Nucleosynthesis

Fred Hoyle (1915–2001)

"Be humble for you are made of dung. Be noble for you are made of stars." This old Serbian proverb serves to remind us today that all the elements heavier than hydrogen and helium would not exist in any substantial amounts in the universe were it not for their production in stars that eventually died and exploded and scattered the elements into the universe. Although light elements such as helium and hydrogen were created in the first few minutes of the **Big Bang**, the subsequent nucleosynthesis (atomic-nucleus creation) of the heavier elements required massive stars with their nuclear fusion reactions over long periods of time. Supernova explosions rapidly created even heavier elements due to an intense burst of nuclear reactions during the explosion of the core of the star. Very heavy elements, like gold and lead, are produced in the extremely high temperatures and neutron flux of a supernova explosion. The next time you look at the golden ring on a friend's finger, think of supernova explosions in massive stars.

Pioneering theoretical work into the mechanism with which heavy nuclei were created in stars was performed in 1946 by astronomer Fred Hoyle who showed how very hot nuclei could assemble into iron.

As I write this entry, I touch a saber-tooth tiger skull in my office. Without stars, there could be no skulls. As mentioned, most elements, like calcium in bones, were first cooked in stars and then blown into space when the stars died. Without stars, the tiger racing across the savanna fades away, ghostlike. There are no iron atoms for its blood, no oxygen for it to breathe, no carbon for its proteins and DNA. The atoms created in the dying ancient stars were blown across vast distances and eventually formed the elements in the planets that coalesced around our Sun. Without these supernova explosions, there are no mist-covered swamps, computer chips, trilobites, Mozarts, or the tears of a little girl. Without exploding stars, perhaps there could be a heaven, but there is certainly no Earth.

SEE ALSO Big Bang (13.7 Billion B.C.), Fraunhofer Lines (1814), $E = mc^2$ (1905), Atomic Nucleus (1911), Tokamak (1956).

The Moon passing in front of the Sun, captured by NASA's STEREO-B spacecraft on February 25, 2007, in four wavelengths of extreme ultraviolet light. Because the satellite is farther from the Sun than the Earth, the Moon appears smaller than usual.

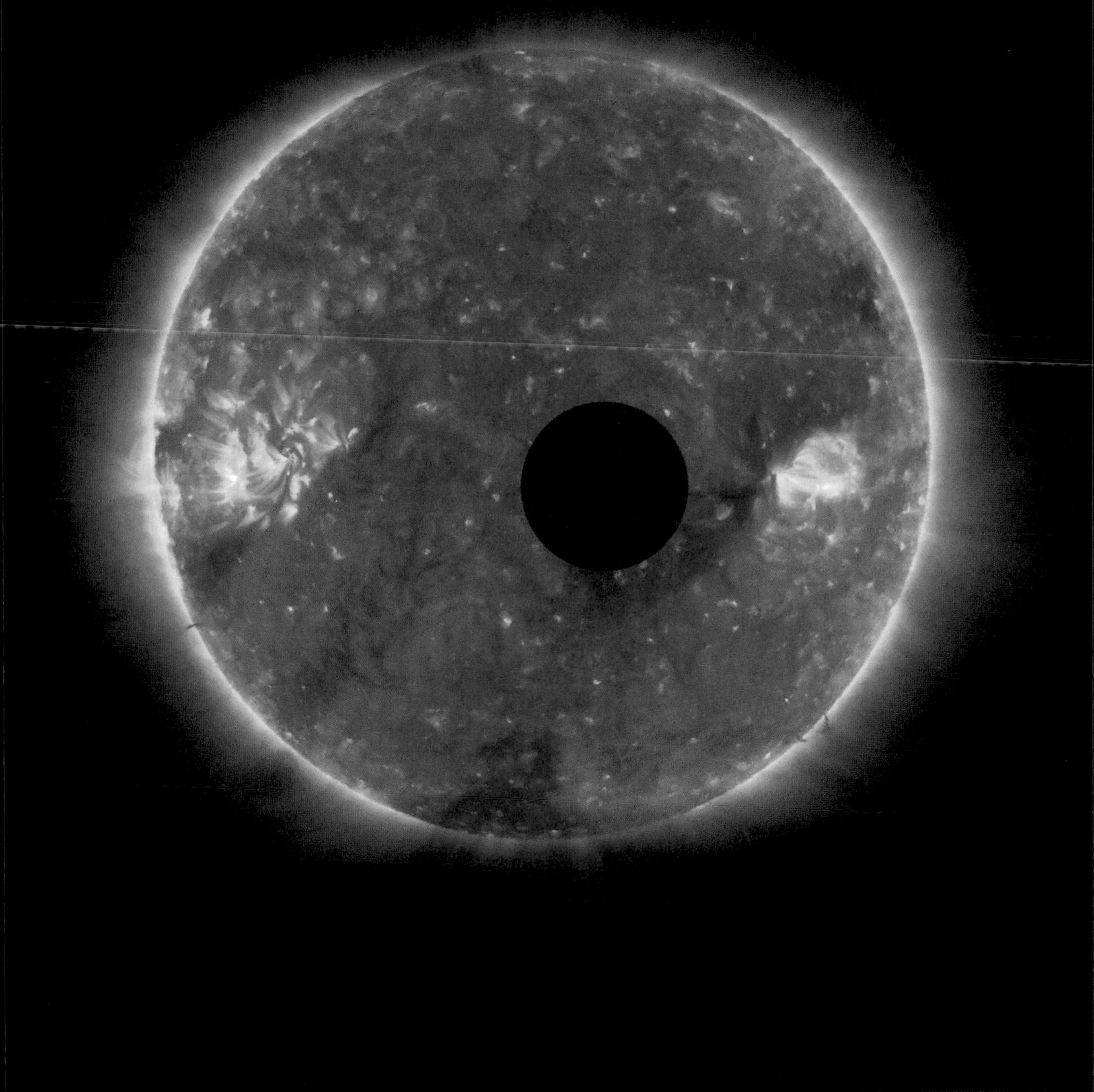

1947

Transistor

Julius Edgar Lilienfeld (1882–1963), **John Bardeen** (1908–1991), **Walter Houser Brattain** (1902–1987), **William Bradford Shockley** (1910–1989)

A thousand years from now, when our ancestors reflect upon history, they will mark December 16, 1947 as the start of humankind's Information Age—the day on which Bell Telephone Laboratories physicists John Bardeen and Walter Brattain connected two upper electrodes to a piece of specially treated germanium that sat on a third electrode (a metal plate attached to a voltage source). When a small current was introduced through one of the upper electrodes, another much stronger current flowed through the other two electrodes. The transistor was born.

Given the magnitude of the discovery, Bardeen's reaction was rather sedate. That evening, walking in through the kitchen door of his home, he mumbled to his wife, "We discovered something important today," and he said no more. Their fellow scientist William Shockley understood the device's great potential and also contributed to the knowledge of semiconductors. Later, when Shockley was angered by being left out of the Bell Lab's transistor patent that had only the names of Bardeen and Brattain, he created a better transistor design.

A transistor is a semiconductor device that may be used to amplify or switch electronic signals. The conductivity of a semiconductor material can be controlled by introduction of an electrical signal. Depending on the transistor design, a voltage or current applied to one pair of the transistor's terminals changes the current flowing through another terminal.

Physicists Michael Riordan and Lillian Hoddeson write, "It is hard to imagine any device more crucial to modern life than the microchip and the transistor from which it sprang. Every waking hour, people of the world take their vast benefits for granted—in cellular phones, ATMs, wrist watches, calculators, computers, automobiles, radios, televisions, fax machines, copiers, stoplights, and thousands of electronic devices. Without a doubt, the transistor is the most important artifact of the 20th century and the 'nerve cell' of our electronic age." In the future, fast transistors made from graphene (sheets of carbon atoms) and carbon nanotubes may become practical. Note that, in 1925, physicist Julius Lilienfeld was the first to actually file a patent for an early version of the transistor.

SEE ALSO Vacuum Tube (1906), Quantum Tunneling (1928), Integrated Circuit (1958), Quantum Computers (1981), Buckyballs (1985).

The Regency TR-1 radio, announced in October 1954, was the first practical transistor radio made in bulk quantities. Shown here is a figure from Richard Koch's transistor-radio patent. Koch was employed by the company that made the TR-1.

June 30, 1959 R. C. KOCH **2,892,931**

TRANSISTOR RADIO APPARATUS

Filed March 25, 1955 2 Sheets-Sheet 1

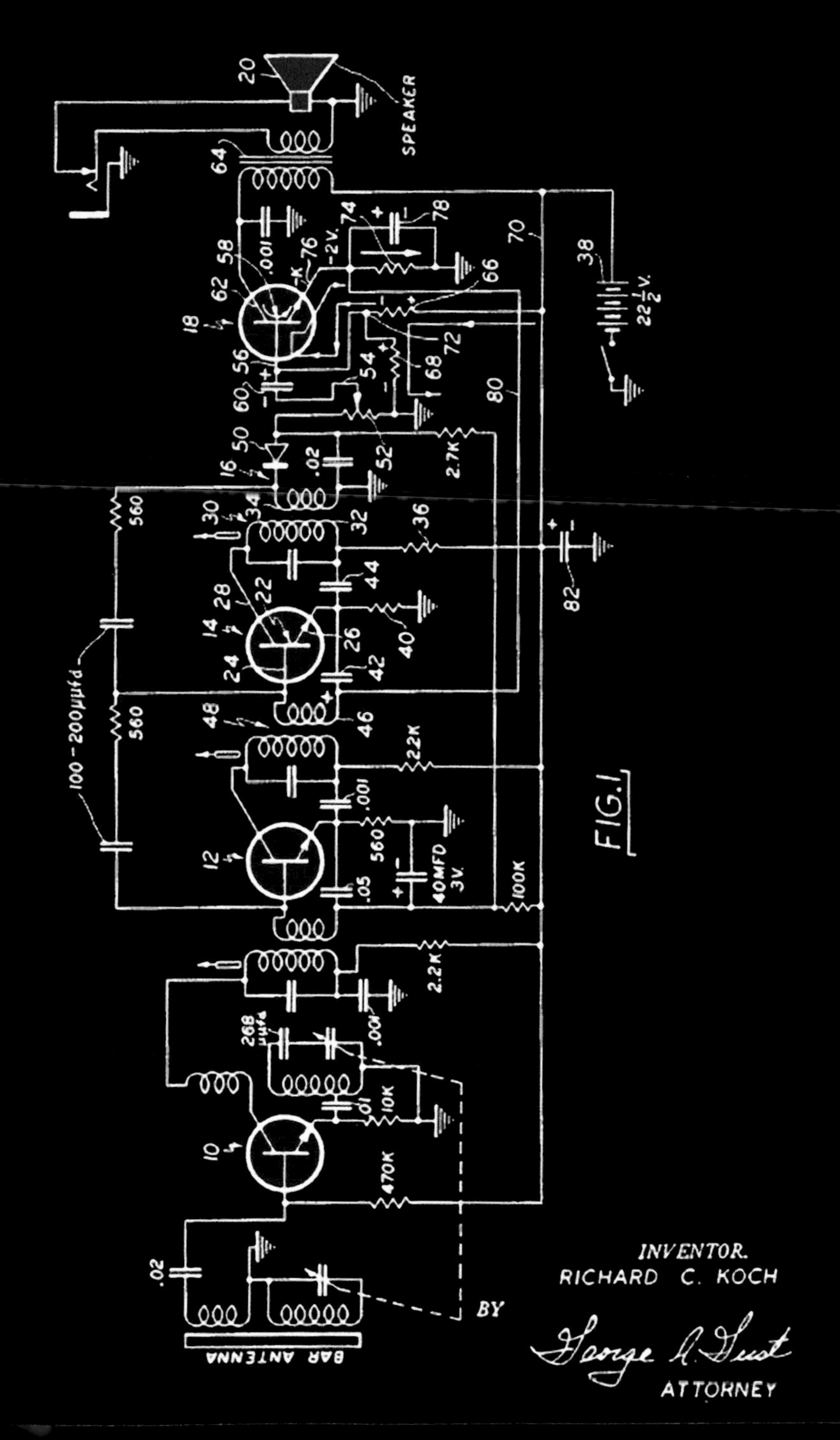

1947

Sonic Booms

Charles Elwood "Chuck" Yeager (b. 1923)

The term *sonic boom* usually refers to the sound caused by the supersonic flight of planes. The boom forms as a result of the displaced air becoming highly compressed, which creates a shock wave. Thunder is an example of a natural sonic boom and is produced when lightning ionizes air, causing it to expand supersonically. The cracking sound of a bullwhip is produced by a small sonic boom as the end of the whip moves faster than the speed of sound.

To visualize the wave phenomena of a sonic boom, imagine a speed boat leaving behind a V-shaped wake. Place your hand in the water. When the wake hits your hand, you feel a slap. When a plane in the air travels faster than the speed of sound (faster than Mach 1, or approximately 660 mph or 1,062 kilometers per hour, in the cold air in which jets usually fly), the shock waves in the air take the form of a cone that stretches behind the aircraft. When the edge of the cone finally reaches your ear, you hear the loud boom. Not only does the nose of a plane create shock waves, but so do other leading edges, such as at the tail and front of the wings. The boom is generated the entire time that the plane flies at Mach 1 or faster.

In the 1940s, there was speculation that it might be impossible to "break the sound barrier," given that pilots in Korea who attempted to force their aircraft to travel at speeds near Mach 1experienced a severe buffeting, and several pilots died when their planes broke apart. The first person to officially break the sound barrier was American pilot Chuck Yeager on October 14, 1947 in a Bell X-1 aircraft. The sonic barrier on land was not broken until 1997—using a British jet car. Interestingly, Yeager had encountered the same problems as the Korean pilots as he neared Mach 1 but then experienced a strange quiet as he raced ahead of the noise and shock wave his airplane was generating.

SEE ALSO Doppler Effect (1842), Tsiolkovsky Rocket Equation (1903), Hypersonic Whipcracking (1927), Cherenkov Radiation (1934), Fastest Tornado Speed (1999).

The Prandtl–Glauert singularity, manifest as a vapor cone, sometimes surrounds an aircraft traveling at transonic speeds. Shown here is an F/A-18 Hornet jet breaking the sound barrier in the skies over the Pacific Ocean.

1947

Hologram

Dennis Gabor (1900–1979)

Holography, the process by which a three-dimensional image can be recorded and later reproduced, was invented in 1947 by physicist Dennis Gabor. In his Nobel Prize acceptance speech for the invention, he said of holography, "I need not write down a single equation or show an abstract graph. One can, of course, introduce almost any amount of mathematics into holography, but the essentials can be explained and understood from physical arguments."

Consider an object such as a pretty peach. Holograms can be stored on photographic film as a record of the peach from many viewpoints. To produce a transmission hologram, a beam-splitter is used to divide the **Laser** light into a reference beam and an object beam. The reference beam does not interact with the peach and is directed toward the recording film with a mirror. The object beam is aimed at the peach. The light reflected off the peach meets the reference beam to create an interference pattern in the film. This pattern of stripes and whorls is totally unrecognizable. After the film is developed, a 3D image of the peach can be reconstructed in space by directing light toward the hologram at the same angle used for the reference beam. The finely spaced fringes on the hologram film act to diffract, or deflect, the light to form the 3D image.

"Upon seeing your first hologram," write physicists Joseph Kasper and Steven Feller, "you are certain to feel puzzlement and disbelief. You may place your hand where the scene apparently lies, only to find nothing tangible is there."

Transmission holograms employ light shining through the developed film from behind, and *reflection holograms* make use of light shining on the film with the light source in front of the film. Some holograms require laser light for viewing, while rainbow holograms (such as those with a reflective coating commonly seen on credit cards) can be viewed without the use of lasers. Holography can also be used to optically store large amounts of data.

SEE ALSO Snell's Law of Refraction (1621), Bragg's Law of Crystal Diffraction (1912), Laser (1960).

The hologram on the 50 euro banknote. Security holograms are very difficult to forge.

EURO
EYPΩ
EYPΩ 50 EURO
EYPΩ 50 EURO 50
EURO
EYPΩ
EYPΩ
EURO

1948

Quantum Electrodynamics

Paul Adrien Maurice Dirac (1902–1984), **Sin-Itiro Tomonaga** (1906–1979), **Richard Phillips Feynman** (1918–1988), **Julian Seymour Schwinger** (1918–1994)

"Quantum electrodynamics (QED) is arguably the most precise theory of natural phenomena ever advanced," writes physicist Brian Greene. "Through quantum electrodynamics, physicists have been able to solidify the role of photons as the 'smallest possible bundles of light' and to reveal their interactions with electrically charged particles such as electrons, in a mathematically complete, predictable and convincing framework." QED mathematically describes interactions of light with matter and also the interactions of charged particles with one another.

In 1928, the English physicist Paul Dirac established the foundations for QED, and the theory was refined and developed in the late 1940s by physicists Richard P. Feynman, Julian S. Schwinger, and Sin-Itiro Tomonaga. QED relies on the idea that charged particles (such as electrons) interact by emitting and absorbing photons, which are the particles that transmit electromagnetic forces. Interestingly, these photons are "virtual" and cannot be detected, yet they provide the "force" of the interaction as the interacting particles change their speed and direction of travel when absorbing or releasing the energy of a photon. The interactions can be graphically represented and understood through the use of squiggly Feynman diagrams. These drawings also help physicists to calculate the probability that particular interactions take place.

According to QED theory, the greater the number of virtual photons exchanged in an interaction (i.e., a more complex interaction), the less likely is the chance of occurrence of the process. The accuracy of predictions made by QED is astonishing. For example, the predicted strength of the magnetic field carried by an electron is so close to the experimental value, that if you could measure the distance from New York to Los Angeles with this accuracy, you would be accurate to within the thickness of a human hair.

QED has served as the launchpad for subsequent theories, such as *quantum chromodynamics*, which began in the early 1960s and involves the strong forces that hold **Quarks** together through the exchange of particles called gluons. Quarks are particles that combine to form other subatomic particles such as protons and neutrons.

SEE ALSO Electron (1897), Photoelectric Effect (1905), Standard Model (1961), Quarks (1964), Theory of Everything (1984).

Feynman diagram depicting the annihilation of an electron and a positron and creating a photon that decays into a new electron-positron pair. Feynman was so delighted with his diagrams that he painted them on the side of his van.

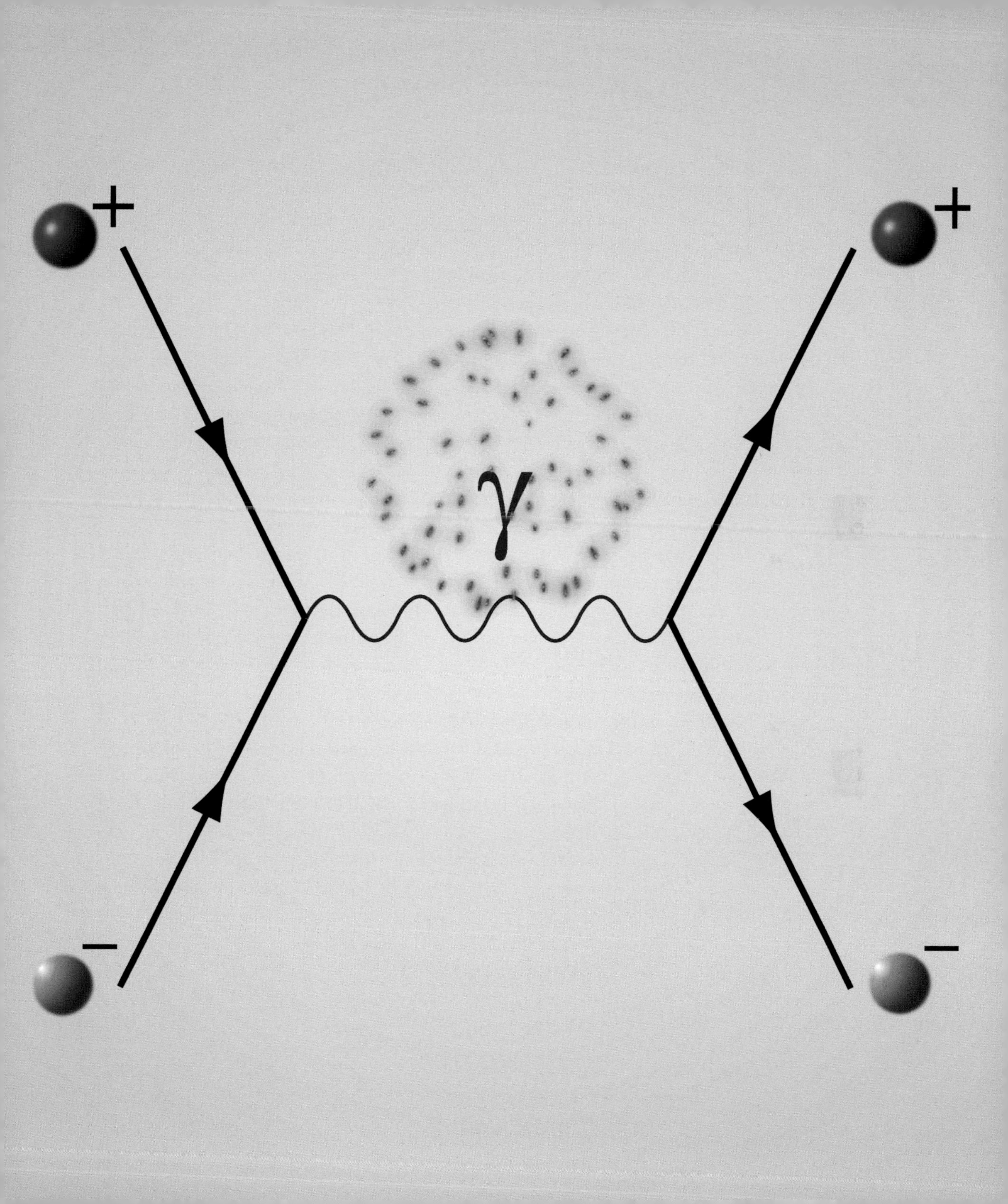
+
+
γ
−
−

1948

Tensegrity

Kenneth Snelson (b. 1927), **Richard Buckminster "Bucky" Fuller** (1895–1983)

The ancient Greek philosopher Heraclitus of Ephesus once wrote that the world is a "harmony of tensions." One of the most intriguing embodiments of this philosophy is the tensegrity system, which inventor Buckminster Fuller described as "islands of compressions in an ocean of tension."

Imagine a structure composed only of rods and of cables. The cables connect the ends of rods to other ends of rods. The rigid rods never touch one another. The structure is stable against the force of gravity. How can such a flimsy-looking assemblage persist?

The structural integrity of such structures is maintained by a balance of tension forces (e.g., pulling forces exerted by a wire) and compression forces (e.g., forces that tend to compress the rods). As an example of such forces, when we push the two ends of a dangling spring together, we compress it. When we pull apart the two ends, we create more tension in the spring.

In tensegrity systems, compression-bearing rigid struts tend to stretch (or tense) the tension-bearing cables, which in turn compress the struts. An increase in tension in one of the cables can result in increased tensions throughout the structure, balanced by an increase in compression in the struts. Overall, the forces acting in all directions in a tensegrity structure sum to a zero net force. If this were not the case, the structure might fly away (like an arrow shot from a bow) or collapse.

In 1948, artist Kenneth Snelson produced his kite-like "X-Piece" tensegrity structure. Later, Buckminster Fuller coined the term *tensegrity* for these kinds of structures. Fuller recognized that the strength and efficiency of his huge geodesic domes were based on a similar kind of structural stabilization that distributes and balances mechanical stresses in space.

To some extent, we are tensegrity systems in which bones are under compression and balanced by the tension-bearing tendons. The cytoskeleton of a microscopic animal cell also resembles a tensegrity system. Tensegrity structures actually mimic some behaviors observed in living cells.

SEE ALSO Truss (2500 B.C.), Arch (1850 B.C.), I-Beams (1844), Leaning Tower of Lire (1955).

A diagram from U.S. Patent 3,695,617, issued in 1972 to G. Mogilner and R. Johnson, for "Tensegrity Structure Puzzle." Rigid columns are shown in dark green. One objective is to try to remove the inner sphere by sliding the columns.

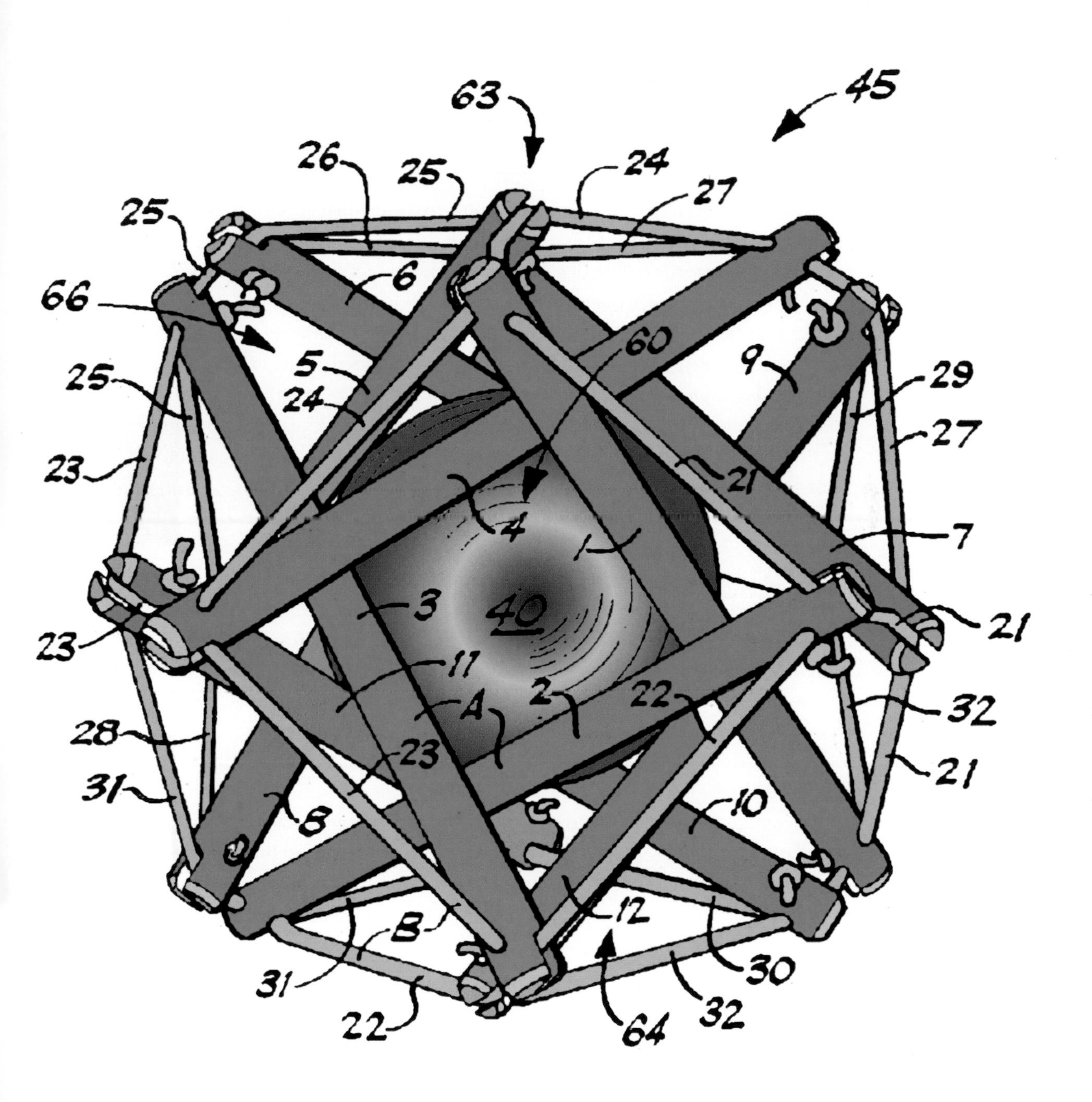

45
63
26
25
24
27
25
6
66
60
5
9
25
24
29
27
23
21
4
7
1
3
40
23
21
11
2
22
32
28
A
23
21
31
8
10
12
B
30
31
22
64
32

1948

Casimir Effect

Hendrik Brugt Gerhard Casimir (1909–2000), **Evgeny Mikhailovich Lifshitz** (1915–1985)

The Casimir effect often refers to a weird attractive force that appears between two uncharged parallel plates in a vacuum. One possible way to understand the Casimir effect is to imagine the nature of a vacuum in space according to quantum field theory. "Far from being empty," write physicists Stephen Reucroft and John Swain, "modern physics assumes that a vacuum is full of fluctuating electromagnetic waves that can never be completely eliminated, like an ocean with waves that are always present and can never be stopped. These waves come in all possible wavelengths, and their presence implies that empty space contains a certain amount of energy" called *zero-point energy*.

If the two parallel plates are brought very close together (e.g. a few nanometers apart), longer waves will not fit between them, and the total amount of vacuum energy between the plates will be smaller than outside of the plates, thus causing the plates to attract each other. One may imagine the plates as prohibiting all of the fluctuations that do not "fit" into the space between the plates. This attraction was first predicted in 1948 by physicist Hendrik Casimir.

Theoretical applications of the Casimir effect have been proposed, ranging from using its "negative energy density" for propping open traversable wormholes between different regions of space and time to its use for developing levitating devices—after physicist Evgeny Lifshitz theorized that the Casimir effect can give rise to repulsive forces. Researchers working on micromechanical or nanomechanical robotic devices may need to take the Casimir effect into account as they design tiny machines.

In quantum theory, the vacuum is actually a sea of ghostly *virtual particles* springing in and out of existence. From this viewpoint, one can understand the Casimir effect by realizing that fewer virtual photons exist between the plates because some wavelengths are forbidden. The excess pressure of photons outside the plates squeezes the plates together. Note that Casimir forces can also be interpreted using other approaches without reference to zero-point energy.

SEE ALSO Third Law of Thermodynamics (1905), Wormhole Time Machine (1988), Quantum Resurrection (100 Trillion).

The sphere shown in this scanning electron microscope image is slightly over one tenth of a millimeter in diameter and moves toward a smooth plate (not shown) due to the Casimir Effect. Research on the Casimir Effect helps scientists better predict the functioning of micro-mechanical machine parts. (Photo courtesy of Umar Mohideen.)

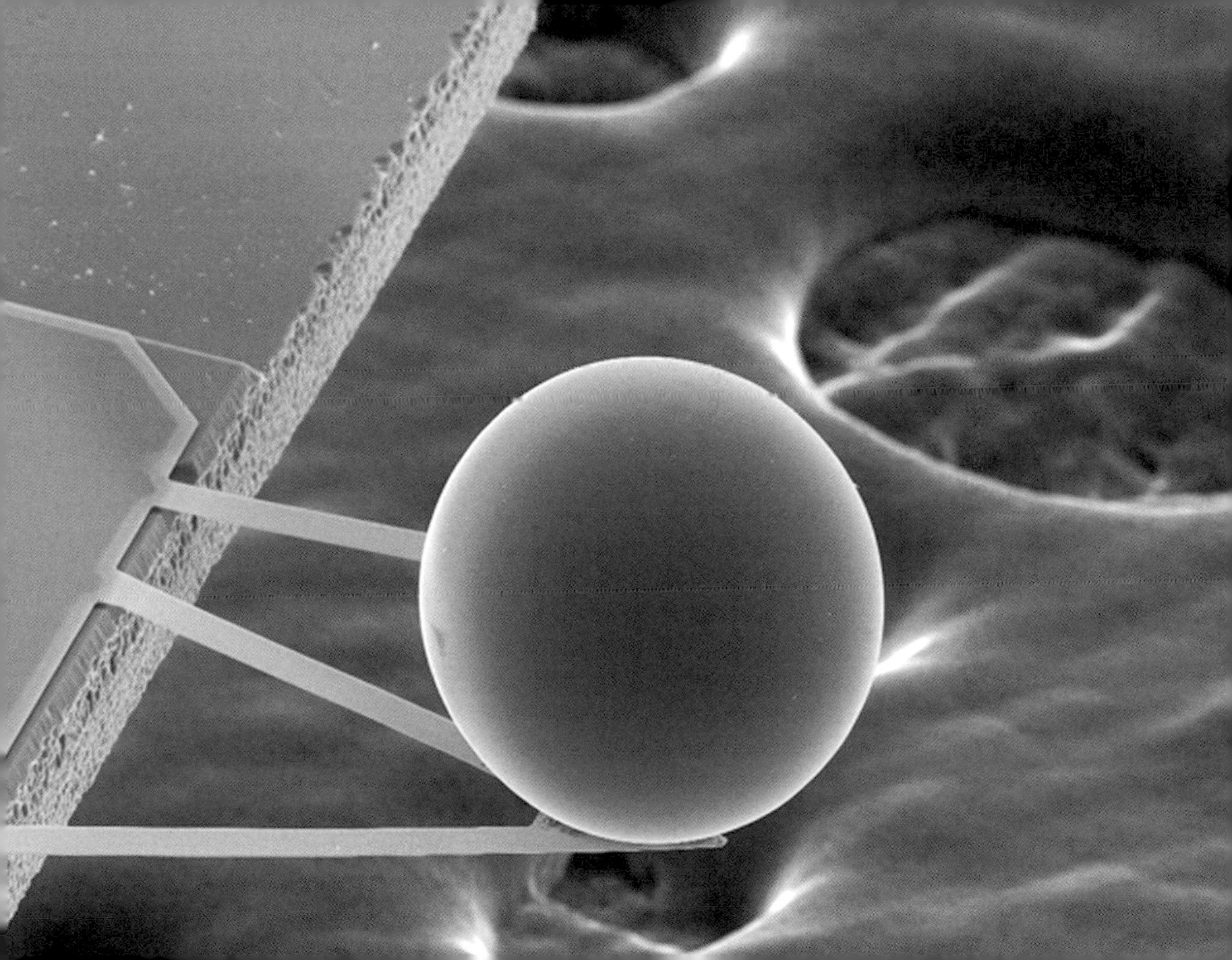

1949

Time Travel

Albert Einstein (1879–1955), **Kurt Gödel** (1906–1978), **Kip Stephen Thorne** (b. 1940)

What is time? Is time travel possible? For centuries, these questions have intrigued philosophers and scientists. Today, we know for certain that time travel is possible. For example, scientists have demonstrated that objects traveling at high speeds age more slowly than a stationary object sitting in a laboratory frame of reference. If you could travel on a near light-speed rocket into outer space and return, you could travel thousands of years into the Earth's future. Scientists have verified this time slowing or "dilation" effect in a number of ways. For example, in the 1970s, scientists used atomic clocks on airplanes to show that these clocks had a slight slowing of time with respect to clocks on the Earth. Time is also significantly slowed near regions of very large masses.

Although seemingly more difficult, numerous ways exist in which time machines for travel to the past can theoretically be built that do not seem to violate any known laws of physics. Most of these methods rely on high gravities or on wormholes (hypothetical "shortcuts" through space and time). To Isaac Newton, time was like a river flowing straight. Nothing could deflect the river. Einstein showed that the river could curve, although it could never circle back on itself, which would be a metaphor for backwards time travel. In 1949, mathematician Kurt Gödel went even further and showed that the river could circle back on itself. In particular, he found a disturbing solution to Einstein's equations that allows backward time travel in a universe that rotated. For the first time in history, backward time travel had been given a mathematical foundation!

Throughout history, physicists have found that if phenomena are not expressly forbidden, they are often eventually found to occur. Today, designs for time travel machines are proliferating in top science labs and include such wild concepts as Thorne **Wormhole Time Machines**, Gott loops that involve cosmic strings, Gott shells, Tipler and van Stockum cylinders, and Kerr Rings. In the next few hundred years, perhaps our heirs will explore space and time to degrees we cannot currently fathom.

SEE ALSO Tachyons (1967), Wormhole Time Machine (1988), Special Theory of Relativity (1905), General Theory of Relativity (1915), Chronology Protection Conjecture (1992).

If time is like space, might the past, in some sense, still exist "back there" as surely as your home still exists even after you have left it? If you could travel back in time, which genius of the past would you visit?

Yes, Archimedes' Principle can be used to measure the density of a solid object as well as the density of a fluid

Density of fluid =

Volume of fluid displac

the solid = $\rho_s V_s g - \rho_l V_s g$

$= (\rho_s - \rho_l) V_s g$

the air & then i

the loss in weight

n calculate

1949

Radiocarbon Dating

Willard Frank Libby (1908–1980)

"If you were interested in finding out the age of things, the University of Chicago in the 1940s was the place to be," writes author Bill Bryson. "Willard Libby was in the process of inventing radiocarbon dating, allowing scientists to get an accurate reading of the age of bones and other organic remains, something they had never been able to do before. . . ."

Radiocarbon dating involves the measuring of the abundance of the radioactive element carbon-14 (^{14}C) in a carbon-containing sample. The method relies on the fact that ^{14}C is created in the atmosphere when cosmic rays strike nitrogen atoms. The ^{14}C is then incorporated into plants, which animals subsequently eat. While an animal is alive, the abundance of ^{14}C in its body roughly matches the atmospheric abundance. ^{14}C continually decays at a known exponential rate, converting to nitrogen-14, and once the animal dies and no longer replenishes its ^{14}C supply from the environment, the animal's remains slowly lose ^{14}C. By detecting the amount of ^{14}C in a sample, scientists can estimate its age if the sample is not older than 60,000 years. Older samples generally contain too little of ^{14}C to measure accurately. ^{14}C has a half-life of about 5,730 years due to radioactive decay. This means that every 5,730 years, the amount of ^{14}C in a sample has dropped by half. Because the amount of atmospheric ^{14}C undergoes slight variations through time, small calibrations are made to improve the accuracy of the dating. Also, atmospheric ^{14}C increased during the 1950s due to atomic bomb tests. Accelerator **Mass Spectrometry** can be used to detect ^{14}C abundances in milligram samples.

Before radiocarbon dating, it was very difficult to obtain reliable dates before the First Dynasty in Egypt, around 3000 B.C. This was quite frustrating for archeologists who were feverish to know, for example, when Cro-Magnon people painted the caves of Lascaux in France or when the last Ice Age finally ended.

SEE ALSO Olmec Compass (1000 B.C.), Hourglass (1338), Radioactivity (1896), Mass Spectrometer (1898), Atomic Clocks (1955).

Because carbon is very common, numerous kinds of materials are potentially useable for radiocarbon investigations, including ancient skeletons found during archeological digs, charcoal, leather, wood, pollen, antlers, and much more.

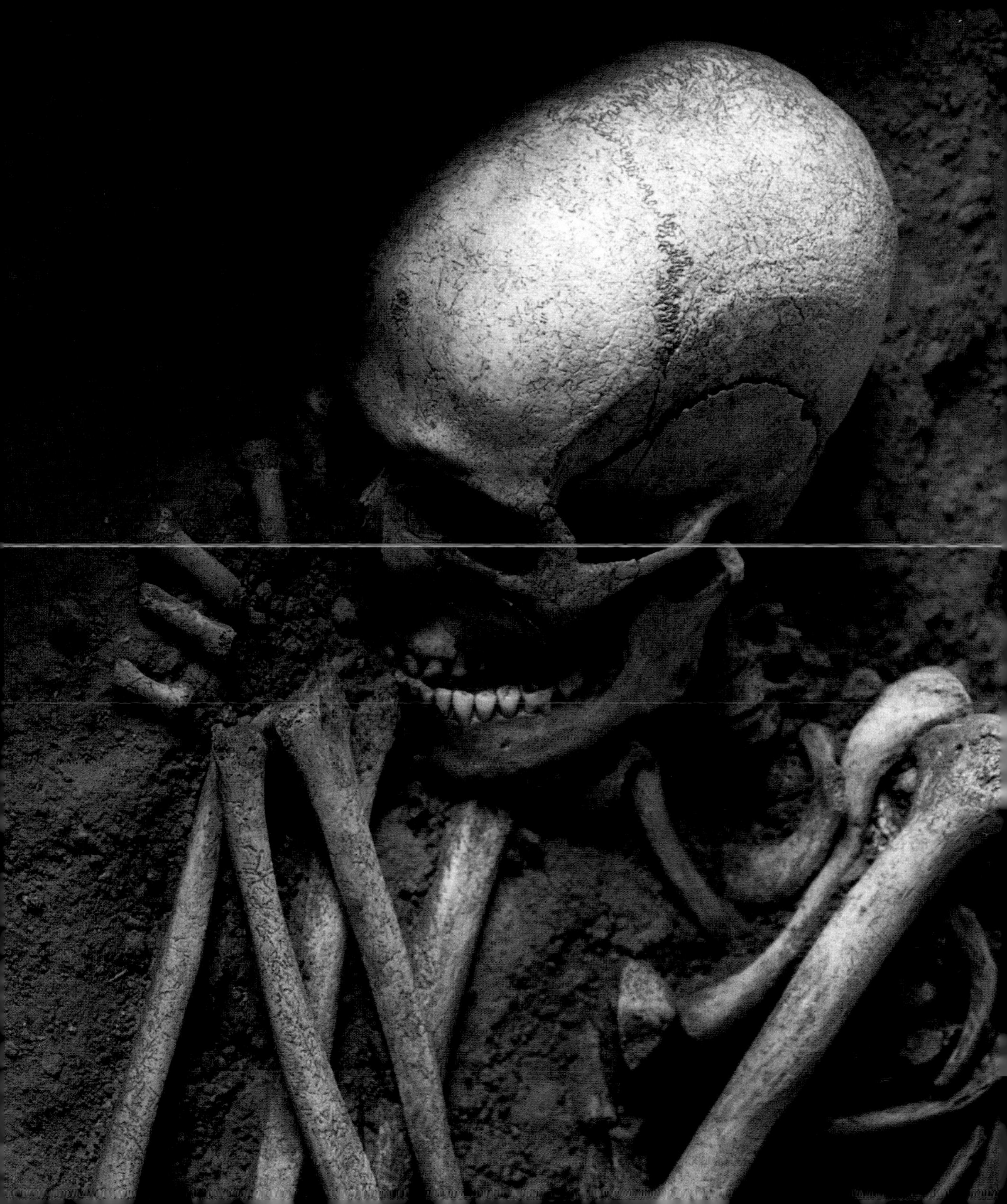

1950

Fermi Paradox

Enrico Fermi (1901–1954), **Frank Drake** (b. 1930)

During our Renaissance, rediscovered ancient texts and new knowledge flooded medieval Europe with the light of intellectual transformation, wonder, creativity, exploration, and experimentation. Imagine the consequences of making contact with an alien race. Another, far more profound Renaissance would be fueled by the wealth of alien scientific, technical, and sociological information. Given that our universe is both ancient and vast—there are an estimated 250 billion stars in our Milky Way galaxy alone—the physicist Enrico Fermi asked in 1950, "Why have we not yet been contacted by an extraterrestrial civilization?" Of course, many answers are possible. Advanced alien life could exist, but we are unaware of their presence. Alternatively, intelligent aliens may be so rare in the universe that we may never make contact with them. The Fermi Paradox, as it is known today, has given rise to scholarly works attempting to address the question in fields ranging from physics and astronomy to biology.

In 1960, astronomer Frank Drake suggested a formula to estimate the number of extraterrestrial civilizations in our galaxy with whom we might come into contact:

$$N = R^{*} \times f_p \times n_e \times f_l \times f_i \times f_c \times L$$

Here, N is the number of alien civilizations in the Milky Way with which communication might be possible; for example, alien technologies may produce detectable radio waves. R^{*} is the average rate of star formation per year in our galaxy. f_p is the fraction of those stars that have planets (hundreds of extra solar planets have been detected). n_e is the average number of "Earth-like" planets that can potentially support life per star that has planets. f_l is the fraction of these n_e planets that actually yield life forms. f_i is the fraction of f_l that actually produce intelligent life. The variable f_c represents the fraction of civilizations that develop a technology that releases detectable signs of their existence into outer space. L is the length of time such civilizations release signals into space that we can detect. Because many of the parameters are very difficult to determine, the equation serves more to focus attention on the intricacies of the paradox than to resolve it.

SEE ALSO Tsiolkovsky Rocket Equation (1903), Time Travel (1949), Dyson Sphere (1960), Anthropic Principle (1961), Living in a Simulation (1967), Chronology Protection Conjecture (1992), Cosmic Isolation (100 Billion).

Given that our universe is both ancient and vast, the physicist Enrico Fermi asked in 1950, "Why have we not yet been contacted by an extraterrestrial civilization?"

1954

Solar Cells

Alexandre-Edmond Becquerel (1820–1891), **Calvin Souther Fuller** (1902–1994)

In 1973, British chemist George Porter said, "I have no doubt that we will be successful in harnessing the sun's energy. . . . If sunbeams were weapons of war, we would have had solar energy centuries ago." Indeed, the quest to efficiently create energy from sunshine has had a long history. Back in 1839, nineteen-year-old French physicist Edmund Becquerel discovered the *photovoltaic effect* in which certain materials produce small amounts of electric current when exposed to light. However, the most important breakthrough in solar power technology did not take place until 1954 when three scientists from Bell Laboratories—Daryl Chapin, Calvin Fuller, and Gerald Pearson—invented the first practical silicon solar cell for converting sunlight into electrical power. Its efficiency was only around 6% in direct sunlight, but today efficiencies of advanced solar cells can exceed 40%.

You may have seen solar panels on the roofs of buildings or used to power warning signs on highways. Such panels contain solar cells, commonly composed of two layers of silicon. The cell also has an antireflective coating to increase the absorption of sunlight. To ensure that the solar cell creates a useful electrical current, small amounts of phosphorus are added to the top silicon layer, and boron is added to the bottom layer. These additions cause the top layer to contain more electrons and the bottom layer to have fewer electrons. When the two layers are joined, electrons in the top layer move into the bottom layer very close to the junction between layers, thus creating an electric field at the junction. When photons of sunlight hit the cell, they knock loose electrons in both layers. The electric field pushes electrons that have reached the junction toward the top layer. This "push" or "force" can be used to move electrons out of the cell into attached metal conductor strips in order to generate electricity. To power a home, this direct-current electricity is converted to alternating current by a device called an inverter.

SEE ALSO Archimedes' Burning Mirrors (212 B.C.), Battery (1800), Fuel Cell (1839), Photoelectric Effect (1905), Energy from the Nucleus (1942), Tokamak (1956), Dyson Sphere (1960).

LEFT: *Solar panels used to power the equipment of a vineyard.* RIGHT: *Solar panels on the roof of a home.*

1955

Leaning Tower of Lire

Martin Gardner (1914–2010)

One day while walking through a library, you notice a stack of books leaning over the edge of a table. You wonder if it would be possible to stagger a stack of many books so that the top book would be far out into the room—say five feet—with the bottom book still resting on the table? Or would such a stack fall under its own weight? For simplicity, the books are assumed to be identical, and you are only allowed to have one book at each level of the stack; in other words, each book rests on at most one other book.

The problem has puzzled physicists since at least the early 1800s, and in 1955 was referred to as the *Leaning Tower of Lire* in the *American Journal of Physics*. The problem received extra attention in 1964 when Martin Gardner discussed it in *Scientific American*.

The stack of n books will not fall if the stack has a center of mass that is still above the table. In other words, the center of mass of all books above any book B must lie on a vertical axis that "cuts" through B. Amazingly, there is no limit to how far you can make the stack jut out beyond the table's edge. Martin Gardner referred to this arbitrarily large overhang as the *infinite-offset paradox*. For an overhang of just 3 book lengths, you'd need a walloping 227 books! For 10 books, you'd need 272,400,600 books. And for 50 book lengths you'd need more than 1.5×10^{44} books. A formula for the amount of overhang attainable with n books, in book lengths, can be used: $0.5 \times (1 + 1/2 + 1/3 + \ldots + 1/n)$. This harmonic series diverges very slowly; thus, a modest increase in book overhang requires many more books. Additional fascinating work has since been conducted on this problem after removing the constraint of having only one book at each level of the stack.

SEE ALSO Arch (1850 B.C.), Truss (2500 B.C.), Tensegrity (1948).

Would it be possible to stagger a stack of many books so that the top book would be many feet into the room, with the bottom book still resting on the table? Or would such a stack fall under its own weight?

1955

Seeing the Single Atom

Max Knoll (1897–1969), **Ernst August Friedrich Ruska** (1906–1988), **Erwin Wilhelm Müller** (1911–1977), **Albert Victor Crewe** (1927–2009)

"Dr. Crewe's research opened a new window into the Lilliputian world of the fundamental building blocks of nature," writes journalist John Markoff, "giving [us] a powerful new tool to understand the architecture of everything from living tissue to metal alloys."

The world had never "seen" an atom using an electron microscope before University of Chicago Professor Albert Crewe used his first successful version of the *scanning transmission electron microscope*, or STEM. Although the concept of an elementary particle had been proposed in the fifth century B.C. by the Greek philosopher Democritus, atoms were far too small to be visualized using optical microscopes. In 1970, Crewe published his landmark paper titled "Visibility of Single Atoms" in the journal *Science*, which presented photographic evidence of atoms of uranium and thorium.

"After attending a conference in England and forgetting to buy a book at the airport for the flight home, he pulled out a pad of paper on the plane and sketched two ways to improve existing microscopes," writes Markoff. Later, Crewe designed an improved source of electrons (a field emission gun) for scanning the specimen.

Electron microscopes employ a beam of electrons to illuminate a specimen. Using the *transmission electron microscope*, invented around 1933 by Max Knoll and Ernst Ruska, electrons pass through a thin sample followed by a magnetic lens produced by a current-carrying coil. A *scanning electron microscope* employs electric and magnetic lenses before the sample, allowing electrons to be focused onto a small spot that is then scanned across the surface. The STEM is a hybrid of both approaches.

In 1955, physicist Erwin Müller used a *field ion microscope* to visualize atoms. The device used a large electric field, applied to a sharp metal tip in a gas. The gas atoms arriving at the tip are ionized and detected. Physicist Peter Nellist writes, "Because this process is more likely to occur at certain places on the surface of the tip, such as at steps in the atomic structure, the resulting image represents the underlying atomic structure of the sample."

SEE ALSO Von Guericke's Electrostatic Generator (1660), *Micrographia* (1665), Atomic Theory (1808), Bragg's Law of Crystal Diffraction (1912), Quantum Tunneling (1928), Nuclear Magnetic Resonance (1938).

A field ion microscope (FIM) image of a very sharp tungsten needle. The small roundish features are individual atoms. Some of the elongated features are caused by atoms moving during the imaging process (approximately 1 second).

1955

Atomic Clocks

Louis Essen (1908–1997)

Clocks have become more accurate through the centuries. Early mechanical clocks, such as the fourteenth-century Dover Castle clock, varied by several minutes each day. When pendulum clocks came into general use in the 1600s, clocks became accurate enough to record minutes as well as hours. In the 1900s, vibrating quartz crystals were accurate to fractions of a second per day. In the 1980s, cesium atom clocks lost less than a second in 3,000 years, and, in 2009, an atomic clock known as NIST-F1—a cesium fountain atomic clock—was accurate to a second in 60 million years!

Atomic clocks are accurate because they involve the counting of periodic events involving two different energy states of an atom. Identical atoms of the same isotope (atoms having the same number of nucleons) are the same everywhere; thus, clocks can be built and run independently to measure the same time intervals between events. One common type of atomic clock is the cesium clock, in which a microwave frequency is found that causes the atoms to make a transition from one energy state to another. The cesium atoms begin to fluoresce at a natural resonance frequency of the cesium atom (9,192,631,770 Hz, or cycles per second), which is the frequency used to define the second. Measurements from many cesium clocks throughout the world are combined and averaged to define an international time scale.

One important use of atomic clocks is exemplified by the GPS (global positioning system). This satellite-based system enables users to determine their positions on the ground. To ensure accuracy, the satellites must send out accurately timed radio pulses, which receiving devices need to determine their positions.

English physicist Louis Essen created the first accurate atomic clock in 1955, based on energy transitions of the cesium atom. Clocks based on other atoms and methods are continually being researched in labs worldwide in order to increase accuracy and decrease cost.

SEE ALSO Hourglass (1338), Anniversary Clock (1841), Stokes' Fluorescence (1852), Time Travel (1949), Radiocarbon Dating (1949).

In 2004, scientists at the National Institute of Standards and Technology (NIST) demonstrated a tiny atomic clock, the inner workings of which were about the size of a grain of rice. The clock included a laser and a cell containing a vapor of cesium atoms.

1956

Parallel Universes

Hugh Everett III (1930–1982), **Max Tegmark** (b. 1967)

A number of prominent physicists now suggest that universes exist that are parallel to ours and that might be visualized as layers in a cake, bubbles in a milkshake, or buds on an infinitely branching tree. In some theories of parallel universes, we might actually detect these universes by gravity leaks from one universe to an adjacent universe. For example, light from distant stars may be distorted by the gravity of invisible objects residing in parallel universes only millimeters away. The entire idea of multiple universes is not as far-fetched as it may sound. According to a poll of 72 leading physicists conducted by the American researcher David Raub and published in 1998, 58% of physicists (including Stephen Hawking) believe in some form of multiple universes theory.

Many flavors of parallel-universe theory exist. For example, Hugh Everett III's 1956 doctoral thesis "The Theory of the Universal Wavefunction" outlines a theory in which the universe continually "branches" into countless parallel worlds. This theory is called the many-worlds interpretation of quantum mechanics and posits that whenever the universe ("world") is confronted by a choice of paths at the quantum level, it actually follows the various possibilities. If the theory is true, then all kinds of strange worlds may "exist" in some sense. In a number of worlds, Hitler won World War II. Sometimes, the term "multiverse" is used to suggest the idea that the universe that we can readily observe is only part of the reality that comprises the multiverse, the set of possible universes.

If our universe is *infinite*, then identical copies of our visible universe may exist, with an exact copy of our Earth and of you. According to physicist Max Tegmark, on average, the nearest of these identical copies of our visible universe is about 10 to the 10^{100} meters away. Not only are there infinite copies of you, there are infinite copies of variants of you. Chaotic **Cosmic Inflation** theory also suggests the creation of different universes—with perhaps countless copies of you existing but altered in fantastically beautiful and ugly ways.

SEE ALSO Wave Nature of Light (1801), Schrödinger's Cat (1935), Anthropic Principle (1961), Living in a Simulation (1967), Cosmic Inflation (1980), Quantum Computers (1981), Quantum Immortality (1987), Chronology Protection Conjecture (1992).

Some interpretations of quantum mechanics posit that whenever the universe is confronted by a choice of paths at the quantum level, it actually follows the various possibilities. Multiverse *implies that our observable universe is part of a reality that includes other universes.*

1956

Neutrinos

Wolfgang Ernst Pauli (1900–1958), **Frederick Reines** (1918–1998), **Clyde Lorrain Cowan, Jr.** (1919–1974)

In 1993, physicist Leon Lederman wrote, "Neutrinos are my favorite particles. A neutrino has almost no properties: no mass (or very little), no electric charge . . . and, adding insult to injury, no strong force acts on it. The euphemism used to describe a neutrino is 'elusive.' It is barely a fact, and it can pass through millions of miles of solid lead with only a tiny chance of being involved in a measurable collision."

In 1930, physicist Wolfgang Pauli predicted the essential properties of the neutrino (no charge, very little mass)—to explain the loss of energy during certain forms of radioactive decay. He suggested that the missing energy might be carried away by ghostlike particles that escaped detection. Neutrinos were first detected in 1956 by physicists Frederick Reines and Clyde Cowan in their experiments at a nuclear reactor in South Carolina.

Each second, over 100 billion neutrinos from the Sun pass through every square inch (6.5 square centimeters) of our bodies, and virtually none of them interact with us. According to the **Standard Model** of particle physics, neutrinos do not have a mass; however, in 1998, the subterranean Super Kamiokande neutrino detector in Japan was used to determine that they actually have a minuscule mass. The detector used a large volume of water surrounded by detectors for **Cherenkov Radiation** emitted from neutrino collisions. Because neutrinos interact so weakly with matter, neutrino detectors must be huge to increase the chances of detection. The detectors also reside beneath the Earth's surface to shield them from other forms of background radiation such as **Cosmic Rays**.

Today, we know that there are three known types, or flavors, of neutrinos and that neutrinos are able to oscillate between the three flavors while they travel through space. For years, scientists wondered why they detected so few of the expected neutrinos from the Sun's energy-producing fusion reactions. However, the solar neutrino flux only *appears* to be low because the other neutrino flavors are not easily observed by some neutrino detectors.

SEE ALSO Radioactivity (1896), Cherenkov Radiation (1934), Standard Model (1961), Quarks (1964).

The Fermi National Accelerator Laboratory, near Chicago, uses protons from an accelerator to produce an intense beam of neutrinos that allow physicists to observe neutrino oscillations at a distant detector. Shown here are "horns" that help focus the particles that decay and produce neutrinos.

1956

Tokamak

Igor Yevgenyevich Tamm (1895–1971), **Lev Andreevich Artsimovich** (1909–1973), **Andrei Dmitrievich Sakharov** (1921–1989)

Fusion reactions in the Sun bathe the Earth in light and energy. Can we learn to safely generate energy by fusion here on Earth to provide more direct power for human needs? In the Sun, four hydrogen nuclei (four protons) are fused into a helium nucleus. The resulting helium nucleus is less massive than the hydrogen nuclei that combine to make it, and the missing mass is converted to energy according to Einstein's $E = mc^2$. The huge pressures and temperatures needed for the Sun's fusion are aided by its crushing gravitation.

Scientists wish to create nuclear fusion reactions on Earth by generating sufficiently high temperatures and densities so that gases consisting of hydrogen isotopes (deuterium and tritium) become a plasma of free-floating nuclei and electrons—and then the resultant nuclei may fuse to produce helium and neutrons with the release of energy. Unfortunately, no material container can withstand the extremely high temperatures needed for fusion. One possible solution is a device called a *tokamak*, which employs a complex system of magnetic fields to confine and squeeze the plasmas within a hollow, doughnut-shaped container. This hot plasma may be created by magnetic compression, microwaves, electricity, and neutral particle beams from accelerators. The plasma then circulates around the tokamak without touching its walls. Today, the world's largest tokamak is ITER, under construction in France.

Researchers continue to perfect their tokamaks with the goal of creating a system that generates more energy than is required for the system's operation. If such a tokamak can be built, it would have many benefits. First, small amounts of fuel required are easy to obtain. Second, fusion does not have the high-level radioactive waste problems of current fission reactors in which the nucleus of an atom, such as uranium, splits into smaller parts, with a large release of energy.

The tokamak was invented in the 1950s by Soviet physicist Igor Yevgenyevich Tamm and Andrei Sakharov, and perfected by Lev Artsimovich. Today, scientists also study the possible use of *inertial confinement* of the hot plasma using laser beams.

SEE ALSO Plasma (1879), $E = mc^2$ (1905), Energy from the Nucleus (1942), Stellar Nucleosynthesis (1946), Solar Cells (1954), Dyson Sphere (1960).

A photo of the National Spherical Torus Experiment (NSTX), an innovative magnetic fusion device, based on a spherical tokamak concept. NSTX was constructed by the Princeton Plasma Physics Laboratory (PPPL) in collaboration with the Oak Ridge National Laboratory, Columbia University, and the University of Washington at Seattle.

1958

Integrated Circuit

Jack St. Clair Kilby (1923–2005), **Robert Norton Noyce** (1927–1990)

"It seems that the integrated circuit was destined to be invented," writes technology-historian Mary Bellis. "Two separate inventors, unaware of each other's activities, invented almost identical integrated circuits, or ICs, at nearly the same time."

In electronics, an IC, or microchip, is a miniaturized electronic circuit that relies upon semiconductor devices and is used today in countless examples of electronic equipment, ranging from coffeemakers to fighter jets. The conductivity of a semiconductor material can be controlled by introduction of an electric field. With the invention of the monolithic IC (formed from a single crystal), the traditionally separate transistors, resistors, capacitors, and all wires could now be placed on a single crystal (or chip) made of semiconductor material. Compared with the manual assembly of discrete circuits of individual components, such as resistors and transistors, an IC can be made more efficiently using the process of photolithography, which involves selectively transferring geometric shapes on a mask to the surface of a material such as a silicon wafer. The speed of operations is also higher in ICs because the components are small and tightly packed.

Physicist Jack Kilby invented the IC in 1958. Working independently, physicist Robert Noyce invented the IC six months later. Noyce used silicon for the semiconductor material, and Kilby used germanium. Today, a postage-stamp-sized chip can contain over a billion transistors. The advances in capability and density—and decrease in price—led technologist Gordon Moore to say, "If the auto industry advanced as rapidly as the semiconductor industry, a Rolls Royce would get a half a million miles per gallon, and it would be cheaper to throw it away than to park it."

Kilby invented the IC as a new employee at Texas Instruments during the company's late-July vacation time when the halls of his employer were deserted. By September, Kilby had built a working model, and on February 6, Texas Instruments filed a patent.

SEE ALSO Kirchhoff's Circuit Laws (1845), Transistor (1947), Cosmic Rays (1910), Quantum Computers (1981).

The exterior packaging of microchips (e.g., large rectangular shape at left) house the integrated circuits inside that contain the tiny components such as transistor devices. The housing protects the much smaller integrated circuit and provides a means of connecting the chip to a circuit board.

R58
JP1

1959

Dark Side of the Moon

John Frederick William Herschel (1792–1871), **William Alison Anders** (b. 1933)

Due to the particular gravitational forces between the Moon and the Earth, the Moon takes just as long to rotate around its own axis as it does to revolve around the Earth; thus, the same side of the Moon always faces the Earth. The "dark side of the moon" is the phrase commonly used for the far side of the Moon that can never be seen from the Earth. In 1870, the famous astronomer Sir John Herschel wrote that the Moon's far side might contain an ocean of ordinary water. Later, flying saucer buffs speculated that the far side could be harboring a hidden base for extraterrestrials. What secrets did it actually hold?

Finally, in 1959, we had our first glimpse of the far side when it was first photographed by the Soviet Luna 3 probe. The first atlas of the far side of the Moon was published by the USSR Academy of Sciences in 1960. Physicists have suggested that the far side might be used for a large radio telescope that is shielded from terrestrial radio interference.

The far side is actually not always dark, and both the side that faces us and the far side receive similar amounts of sunlight. Curiously, the near and far sides have vastly different appearances. In particular, the side toward us contains many large "maria" (relatively smooth areas that looked like seas to ancient astronomers). In contrast, the far side has a blasted appearance with more craters. One reason for this disparity arises from the increased volcanic activity around three billion years ago on the near side, which created the relatively smooth basaltic lavas of the maria. The far side crust may be thicker and thus was able to contain the interior molten material. Scientists still debate the possible causes.

In 1968, humans finally gazed directly on the far side of the moon during America's Apollo 8 mission. Astronaut William Anders, who traveled to the moon, described the view: "The backside looks like a sand pile my kids have played in for some time. It's all beat up, no definition, just a lot of bumps and holes."

SEE ALSO Telescope (1608), Discovery of Saturn's Rings (1610).

The far side of the moon, with its strangely rough and battered surface, looks very different than the surface facing the Earth. This view was photographed in 1972 by Apollo 16 astronauts while in lunar orbit.

1960

Dyson Sphere

William Olaf Stapledon (1886–1950), **Freeman John Dyson** (b. 1923)

In 1937, British philosopher and author Olaf Stapledon described an immense artificial structure in his novel *Star Maker*: "As the eons advanced . . . many a star without natural planets came to be surrounded by concentric rings of artificial worlds. In some cases the inner rings contained scores, the outer rings thousands of globes adapted to life at some particular distance from the Sun."

Stimulated by *Star Maker*, in 1960, physicist Freeman Dyson published a technical paper in the prestigious journal *Science* on a hypothetical spherical shell that might encompass a star and capture a large percentage of its energy. As technological civilizations advanced, such structures would be desired in order to meet their vast energy needs. Dyson actually had in mind a swarm of artificial objects orbiting a star, but science-fiction authors, physicists, teachers, and students have ever since wondered about the possible properties of a rigid shell with a star at its center, with aliens potentially inhabiting the inner surface of the sphere.

In one embodiment, the Dyson Sphere would have a radius equal to the distance of the Earth to the Sun and would thus have a surface area 550 million times the Earth's surface area. Interestingly, the central star would have no net gravitational attraction on the shell, which may dangerously drift relative to the star unless adjustments could be made to the shell's position. Similarly, any creatures or objects on the inner surface of the sphere would not be gravitationally attracted to the sphere. In a related concept, creatures could still reside on a planet, and the shell could be used to capture energy from the star. Dyson had originally estimated that sufficient planetary and other material existed in the solar system to create such a sphere with a 3-meter thick shell. Dyson also speculated that Earthlings might be able to detect a far-away Dyson Sphere because it would absorb starlight and re-radiate energy in readily definable ways. Researchers have already attempted to find possible evidence of such constructs by searching for their infrared signals.

SEE ALSO Measuring the Solar System (1672), Fermi Paradox (1950), Solar Cells (1954), Tokamak (1956).

Artistic depiction of a Dyson sphere that might encompass a star and capture a large percentage of its energy. The lightning effects depicted here represent the capture of energy at the inner surface of the sphere.

1960

Laser

Charles Hard Townes (b. 1915), **Theodore Harold "Ted" Maiman** (1927–2007)

"Laser technology has become important in a wide range of practical applications," writes laser-expert Jeff Hecht, "ranging from medicine and consumer electronics to telecommunications and military technology. Lasers are also vital tools on the cutting edge of research—18 recipients of the Nobel Prize received the award for laser-related research, including the laser itself, holography, laser cooling, and **Bose-Einstein Condensates**."

The word *laser* stands for light amplification by stimulated emission of radiation, and lasers make use of a subatomic process known as *stimulated emission*, first considered by Albert Einstein in 1917. In stimulated emission, a photon (a particle of light) of the appropriate energy causes an electron to drop to a lower energy level, which results in the creation of another photon. This second photon is said to be coherent with the first and has the same phase, frequency, polarization, and direction of travel as the first photon. If the photons are reflected so that they repeatedly traverse the same atoms, an amplification can take place and an intense radiation beam is emitted. Lasers can be created so that they emit electromagnetic radiations of various kinds, and thus there are X-ray lasers, ultraviolet lasers, infrared lasers, etc. The resultant beam may be highly collimated—NASA scientists have bounced laser beams generated on the Earth from reflectors left on the moon by astronauts. At the Moon's surface, this beam is a mile or two wide (about 2.5 kilometers), which is actually a rather small spread when compared to ordinary light from a flashlight!

In 1953, physicist Charles Townes and students produced the first microwave laser (a *maser*), but it was not capable of continuous radiation emission. Theodore Maiman created the first practical working laser in 1960, using pulsed operation, and today the largest applications of lasers include DVD and CD players, fiber-optic communications, bar-code readers, and laser printers. Other uses include bloodless surgery and target-marking for weapons use. Research continues in the use of lasers capable of destroying tanks or airplanes.

SEE ALSO Brewster's Optics (1815), Hologram (1947), Bose-Einstein Condensate (1995).

An optical engineer studies the interaction of several lasers that will be used aboard a laser weapons system being developed to defend against ballistic missile attacks. The U.S. Directed Energy Directorate conducts research into beam-control technologies.

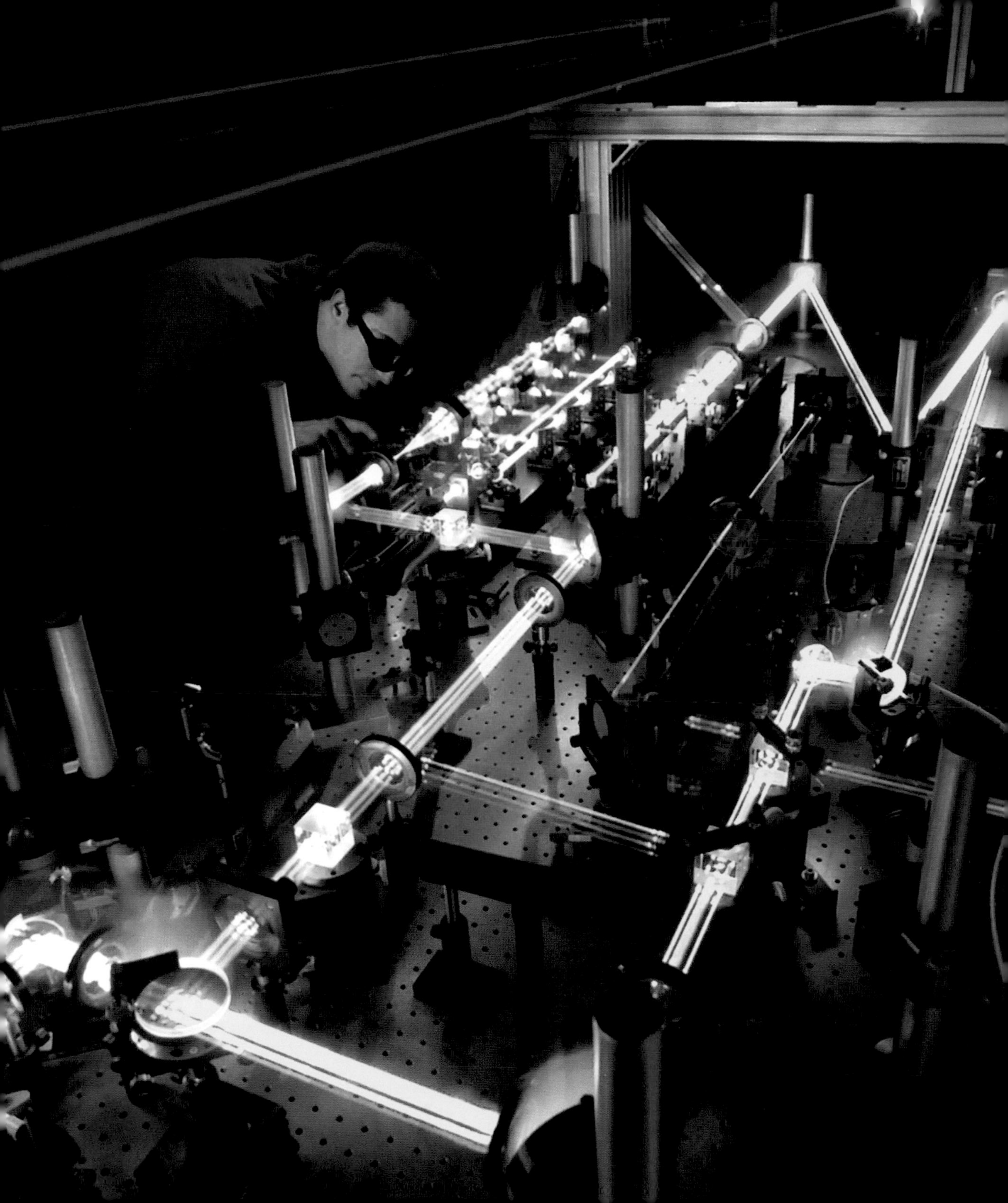

1960

Terminal Velocity

Joseph William Kittinger II (b. 1928)

Perhaps many of you have heard the gruesome legend of the killer penny. If you were to drop a penny from the Empire State Building in New York City, the penny would gain speed and could kill someone walking on the street below, penetrating the brain.

Fortunately for pedestrians below, the physics of terminal velocity saves them from this ghastly demise. The penny falls about 500 feet (152 meters) before reaching its maximum velocity: about 50 miles an hour (80 kilometers/hour). A bullet travels at ten times this speed. The penny is not likely to kill anyone, thus debunking the legend. Updrafts also tend to slow the penny. The penny is not shaped like a bullet, and the odds are that the tumbling coin would barely break the skin.

As an object moves through a medium (such as air or water), it encounters a resistive force that slows it down. For objects in free-fall through air, the force depends on the square of the speed, the area of the object, and the density of air. The faster the object falls, the greater the opposing force becomes. As the penny accelerates downward, eventually the resistive force grows to such an extent that the penny finally falls at a constant speed, known as the terminal velocity. This occurs when the viscous drag on the object is equal to the force of gravity.

Skydivers reach a terminal velocity of about 120 miles per hour (190 kilometers/hour) if they spread their feet and arms. If they assume a heads-down diving position, they reach a velocity of about 150 miles per hour (240 kilometers/hour).

The highest terminal velocity ever to have been reached by a human in free fall was achieved in 1960 by U.S. military officer Joseph Kittinger II, who is estimated to have reached 614 miles per hour (988 kilometers/hour) due to the high altitude (and hence lower air density) of his jump from a balloon. His jump started at 102,800 feet (31,300 meters), and he opened his parachute at 18,000 feet (5,500 meters).

SEE ALSO Acceleration of Falling Objects (1638), Escape Velocity (1728), Bernoulli's Law of Fluid Dynamics (1738), Baseball Curveball (1870), Super Ball (1965).

Skydivers reach a terminal velocity of about 120 miles per hour (190 kilometers/hour) if they spread their feet and arms.

1961

Anthropic Principle

Robert Henry Dicke (1916–1997), **Brandon Carter** (b. 1942)

"As our knowledge of the cosmos has increased," writes physicist James Trefil, ". . . it has become apparent that, had the universe been put together just a little differently, we could not be here to contemplate it. It is as though the universe has been made for us—a Garden of Eden of the grandest possible design."

While this statement is subject to continued debate, the anthropic principle fascinates both scientists and laypeople and was first elucidated in detail in a publication by astrophysicist Robert Dicke in 1961 and later developed by physicist Brandon Carter and others. The controversial principle revolves around the observation that at least some physical parameters appear to be tuned to permit life forms to evolve. For example, we owe our very lives to the element carbon, which was first manufactured in stars before the Earth formed. The nuclear reactions that facilitate the production of carbon have the appearance, at least to some researchers, of being "just right" to facilitate carbon production.

If all of the stars in the universe were heavier than three solar masses, they would live for only about 500 million years, and multicellular life would not have time to evolve. If the rate of the universe's expansion one second after the **Big Bang** had been smaller by even one part in a hundred thousand million million, the universe would have recollapsed before reaching its present size. On the other hand, the universe might have expanded so rapidly that protons and electrons never united to make hydrogen atoms. An extremely small change in the strength of gravity or of the nuclear weak force could prevent advanced life forms from evolving.

An infinite number of random (non-designed) universes could exist, ours being just one that permits carbon-based life. Some researchers have speculated that child universes are constantly budding off from parent universes and that the child universe inherits a set of physical laws similar to the parent, a process reminiscent of evolution of biological characteristics of life on Earth. Universes with many stars may be long-lived and have the opportunity to have many children with many stars; thus, perhaps our star-filled universe is not quite so unusual after all.

SEE ALSO Parallel Universes (1956), Fermi Paradox (1950), Living in a Simulation (1967).

If values for certain fundamental constants of physics were a little different, intelligent carbon-based life may have had great difficulty evolving. To some religious individuals, this gives the impression that the universe was fine-tuned to permit our existence.

1961

Standard Model

Murray Gell-Mann (b. 1929), **Sheldon Lee Glashow** (b. 1932), **George Zweig** (b. 1937)

"Physicists had learned, by the 1930s, to build all matter out of just three kinds of particle: electrons, neutrons, and protons," author Stephen Battersby writes. "But a procession of unwanted extras had begun to appear—neutrinos, the positron and antiproton, pions and muons, and kaons, lambdas and sigmas—so that by the middle of the 1960s, a hundred supposedly fundamental particles have been detected. It was a mess."

Through a combination of theory and experiment, a mathematical model called the Standard Model explains most of particle physics observed so far by physicists. According to the model, elementary particles are grouped into two classes: *bosons* (e.g., particles that often transmit forces) and *fermions*. Fermions include various kinds of **Quarks** (3 quarks make up both the proton and **Neutrons**) and leptons (such as the **Electron** and **Neutrino**, the latter of which was discovered in 1956). Neutrinos are very difficult to detect because they have a minute (but not zero) mass and pass through ordinary matter almost undisturbed. Today, we know about many of these subatomic particles by smashing apart atoms in particle accelerators and observing the resulting fragments.

As suggested, the Standard Model explains forces as resulting from matter particles exchanging boson force-mediating particles that include photons and gluons. The Higgs particle is the only fundamental particle predicted by the Standard Model that has yet to be observed—and it explains why other elementary particles have masses. The force of gravity is thought to be generated by the exchange of massless gravitons, but these have not yet been experimentally detected. In fact, the Standard Model is incomplete, because it does not include the force of gravity. Some physicists are trying to add gravity to the Standard Model to produce a grand unified theory, or GUT.

In 1964, physicists Murray Gell-Mann and George Zweig proposed the concept of quarks, just a few years after Gell-Mann's 1961 formulation of a particle classification system known as the Eightfold Way. In 1960, physicist Sheldon Glashow's unification theories provided an early step toward the Standard Model.

SEE ALSO String Theory (1919), Neutron (1932), Neutrinos (1956), Quarks (1964), God Particle (1964), Supersymmetry (1971), Theory of Everything (1984), Large Hadron Collider (2009).

The Cosmotron. This was the first accelerator in the world to send particles with energies in the billion electron volt, or GeV, region. The Cosmotron synchrotron reached its full design energy of 3.3 GeV in 1953 and was used for studying subatomic particles.

1962

Electromagnetic Pulse

In William R. Forstchen's best-selling novel *One Second After*, a high-altitude nuclear bomb explodes, unleashing a catastrophic electromagnetic pulse (EMP) that instantly disables electrical devices such as those in airplanes, heart pacemakers, modern cars, and cell phones, and the U.S. descends into "literal and metaphorical darkness." Food becomes scarce, society turns violent, and towns burn—all following a scenario that is completely plausible.

EMP usually refers to the burst of electromagnetic radiation that results from a nuclear explosion that disables many forms of electronic devices. In 1962, the U.S. conducted a nuclear test 250 miles (400 kilometers) above the Pacific Ocean. The test, called Starfish Prime, caused electrical damage in Hawaii, about 898 miles (1,445 kilometers) away. Streetlights went out. Burglar alarms went off. The microwave link of a telephone company was damaged. It is estimated today that if a single nuclear bomb was exploded 250 miles above Kansas, the entire continental U.S would be affected, due to the greater strength of the Earth's magnetic field over the U.S. Even the water supply would be affected, given that it often relies on electrical pumps.

After nuclear detonation, the EMP starts with a short, intense burst of gamma rays (high-energy electromagnetic radiation). The gamma rays interact with the atoms in air molecules, and electrons are released through a process called the **Compton Effect**. The electrons ionize the atmosphere and generate a powerful electrical field. The strength and effect of the EMP depends highly on the altitude at which the bomb detonates and the local strength of the Earth's magnetic field.

Note that it is also possible to produce less-powerful EMPs without nuclear weapons, for example, through *explosively pumped fluxed compression generators*, which are essentially normal electrical generators driven by an explosion using conventional fuel.

Electronic equipment can be protected from an EMP by placing it within a Faraday Cage, which is a metallic shield that can divert the electromagnetic energy directly to the ground.

SEE ALSO Compton Effect (1923), Little Boy Atomic Bomb (1945), Gamma-Ray Bursts (1967), HAARP (2007).

A side view of an E-4 advanced airborne command post (AABNCP) on the electromagnetic pulse (EMP) simulator for testing (Kirtland Air Force Base, New Mexico). The plane is designed to survive an EMP with systems intact.

UNITED STATES OF AMERICA

1963

Chaos Theory

Jacques Salomon Hadamard (1865–1963), **Jules Henri Poincaré** (1854–1912), **Edward Norton Lorenz** (1917–2008)

In Babylonian mythology, Tiamat was the goddess that personified the sea and was the frightening representation of primordial chaos. Chaos came to symbolize the unknown and the uncontrollable. Today, chaos theory is an exciting, growing field that involves the study of wide-ranging phenomena exhibiting a sensitive dependence on initial conditions. Although chaotic behavior often seems "random" and unpredictable, it often obeys strict mathematical rules derived from equations that can be formulated and studied. One important research tool to aid in the study of chaos is computer graphics. From chaotic toys with randomly blinking lights to wisps and eddies of cigarette smoke, chaotic behavior is generally irregular and disorderly; other examples include weather patterns, some neurological and cardiac activity, the stock market, and certain electrical networks of computers. Chaos theory has also often been applied to a wide range of visual art.

In science, certain famous and clear examples of chaotic physical systems exist, such as thermal convection in fluids, panel flutter in supersonic aircraft, oscillating chemical reactions, fluid dynamics, population growth, particles impacting on a periodically vibrating wall, various pendula and rotor motions, nonlinear electrical circuits, and buckled beams.

The early roots of chaos theory started around 1900 when mathematicians such as Jacques Hadamard and Henri Poincaré studied the complicated trajectories of moving bodies. In the early 1960s, Edward Lorenz, a research meteorologist at the Massachusetts Institute of Technology, used a system of equations to model convection in the atmosphere. Despite the simplicity of his formulas, he quickly found one of the hallmarks of chaos: extremely minute changes of the initial conditions led to unpredictable and different outcomes. In his 1963 paper, Lorenz explained that a butterfly flapping its wings in one part of the world could later affect the weather thousands of miles away. Today, we call this sensitivity the Butterfly Effect.

SEE ALSO Laplace's Demon (1814), Self-Organized Criticality (1987), Fastest Tornado Speed (1999).

LEFT: *According to Babylonian mythology, Tiamat gave birth to dragons and serpents.* RIGHT: *Chaos theory involves the study of wide-ranging phenomena exhibiting a sensitive dependence on initial conditions. Shown here is a portion of a Daniel White's Mandelbulb, a 3-dimensional analog of the Mandelbrot set, which represents the complicated behavior of a simple mathematical system.*

1963

Quasars

Maarten Schmidt (b. 1929)

"Quasars are among the most baffling objects in the universe because of their small size and prodigious energy output," write the scientists at hubblesite.org. "Quasars are not much bigger than Earth's solar system but pour out 100 to 1,000 times as much light as an *entire galaxy* containing a hundred billion stars."

Although a mystery for decades, today the majority of scientists believe that quasars are very energetic and distant galaxies with an extremely massive central **Black Hole** that spews energy as nearby galactic material spirals into the black hole. The first quasars were discovered with radio telescopes (instruments that receive radio waves from space) without any corresponding visible object. In the early 1960s, visually faint objects were finally associated with these strange sources that were termed *quasi-stellar radio sources*, or quasars for short. The spectrum of these objects, which shows the variation in the intensity of the object's radiation at different wavelengths, was initially puzzling. However, in 1963, Dutch-born American astronomer Maarten Schmidt made the exciting discovery that the spectral lines were simply coming from hydrogen, but that they were shifted far to the red end of the spectrum. This redshift, due to the expansion of the universe, implied that these quasars were part of galaxies that were extremely far away and ancient (see entries on **Hubble's Law** and **Doppler Effect**).

More than 200,000 quasars are known today, and most do not have detectable radio emissions. Although quasars appear dim because they are between about 780 million and 28 billion light-years away—they are actually the most luminous and energetic objects known in the universe. It is estimated that quasars can swallow 10 stars per year, or 600 Earths per minute, and then "turn off" when the surrounding gas and dust has been consumed. At this point, the galaxy hosting the quasar becomes an ordinary galaxy. Quasars may have been more common in the early universe because they had not yet had a chance to consume the surrounding material.

SEE ALSO Telescope (1608), Black Holes (1783), Doppler Effect (1842), Hubble's Law of Cosmic Expansion (1929), Gamma-Ray Bursts (1967).

A quasar, or growing black hole spewing energy, can be seen at the center of a galaxy in this artist's concept. Astronomers using NASA's Spitzer and Chandra space telescopes discovered similar quasars within a number of distant galaxies. X-ray emissions are illustrated by the white rays.

1963

Lava Lamp

Edward Craven Walker (1918–2000)

The Lava Lamp (U.S. Patent 3,387,396) is an ornamental illuminated vessel with floating globules, and it is included in this book because of its ubiquity and for the simple yet important principles it embodies. Many educators have used the Lava Lamp for classroom demonstrations, experiments, and discussions on topics that include thermal radiation, convection, and conduction.

The Lava Lamp was invented in 1963 by Englishman Edward Craven Walker. Author Ben Ikenson writes, "A World War II veteran, Walker adopted the lingo and lifestyle of the flower children. Part Thomas Edison, part Austin Powers, he was a nudist in the psychedelic days of the UK—and possessed some pretty savvy marketing skills to boot. 'If you buy my lamp, you won't need to buy drugs,' he was known to say."

To make a Lava Lamp, one must find two liquids that are immiscible, meaning that, like oil and water, they will not blend or mix. In one embodiment, the lamp consists of a 40-watt **Incandescent Light Bulb** at the bottom, which heats a tall tapered glass bottle containing water and globules made of a mixture of wax and carbon tetrachloride. The wax is slightly denser than the water at room temperature. As the base of the lamp is heated, the wax expands more than the water and becomes fluid. As the wax's *specific gravity* (density relative to water) decreases, the blobs rise to the top, and then the wax globules cool and sink. A metal coil at the base of the lamp serves to spread the heat and also to break the **Surface Tension** of the globules so that they may recombine when at the bottom.

The complex and unpredictable motions of the wax blobs within Lava Lamps have been used as a source of random numbers, and such a random-number generator is mentioned in U.S. Patent 5,732,138, issued in 1998.

Sadly, in 2004, a Lava Lamp killed Phillip Quinn when he attempted to heat it on his kitchen stove. The lamp exploded, and a piece of glass pierced his heart.

SEE ALSO Archimedes' Principle of Buoyancy (250 B.C.), Stokes' Law of Viscosity (1851), Surface Tension (1866), Incandescent Light Bulb (1878), Black Light (1903), Silly Putty (1943), Drinking Bird (1945).

Lava lamps demonstrate simple yet important physics principles, and many educators have used the Lava Lamp for classroom demonstrations, experiments, and discussions.

1964

The God Particle

Robert Brout (b. 1928), **Peter Ware Higgs** (b. 1929), **François Englert** (b. 1932)

"While walking in the Scottish Highlands in 1964," writes author Joanne Baker, "physicist Peter Higgs thought of a way to give particles their mass. He called this his 'one big idea.' Particles seemed more massive because they are slowed while swimming through a force field, now known as the Higgs field. It is carried by the Higgs boson, referred to as the 'God particle' by Nobel Laureate Leon Lederman."

Elementary particles are grouped into two classes: *bosons* (particles that transmit forces) and *fermions* (particles such as **Quarks**, **Electrons**, and **Neutrinos** that make up matter). The Higgs boson is a particle in the **Standard Model** that has not yet been observed, and scientists hope that the **Large Hadron Collider**—a high-energy particle accelerator in Europe—may provide experimental evidence relating to the particle's existence.

To help us visualize the Higgs field, imagine a lake of viscous honey that adheres to the otherwise massless fundamental particles that travel through the field. The field converts them into particles with mass. In the very early universe, theories suggest that all of the fundamental forces (i.e. strong, electromagnetic, weak, and gravitational) were united in one superforce, but as the universe cooled different forces emerged. Physicists have been able to combine the weak and electromagnetic forces into a unified "electroweak" force, and perhaps all of the forces may one day be unified. Moreover, physicists Peter Higgs, Robert Brout, and François Englert suggested that all particles had no mass soon after the **Big Bang**. As the Universe cooled, the Higgs boson and its associated field emerged. Some particles, such as massless photons of light, can travel through the sticky Higgs field without picking up mass. Others get bogged down like ants in molasses and become heavy.

The Higgs boson may be more than 100 times as massive as the proton. A large particle collider is required to find this boson because the higher the energy of collision, the more massive the particles in the debris.

SEE ALSO Standard Model (1961), Theory of Everything (1984), Large Hadron Collider (2009).

The Compact Muon Solenoid (CMS) is a particle detector located underground in a large cavern excavated at the site of the **Large Hadron Collider**. *This detector will assist in the search for the Higgs boson and in gaining insight into the nature of* **Dark Matter**.

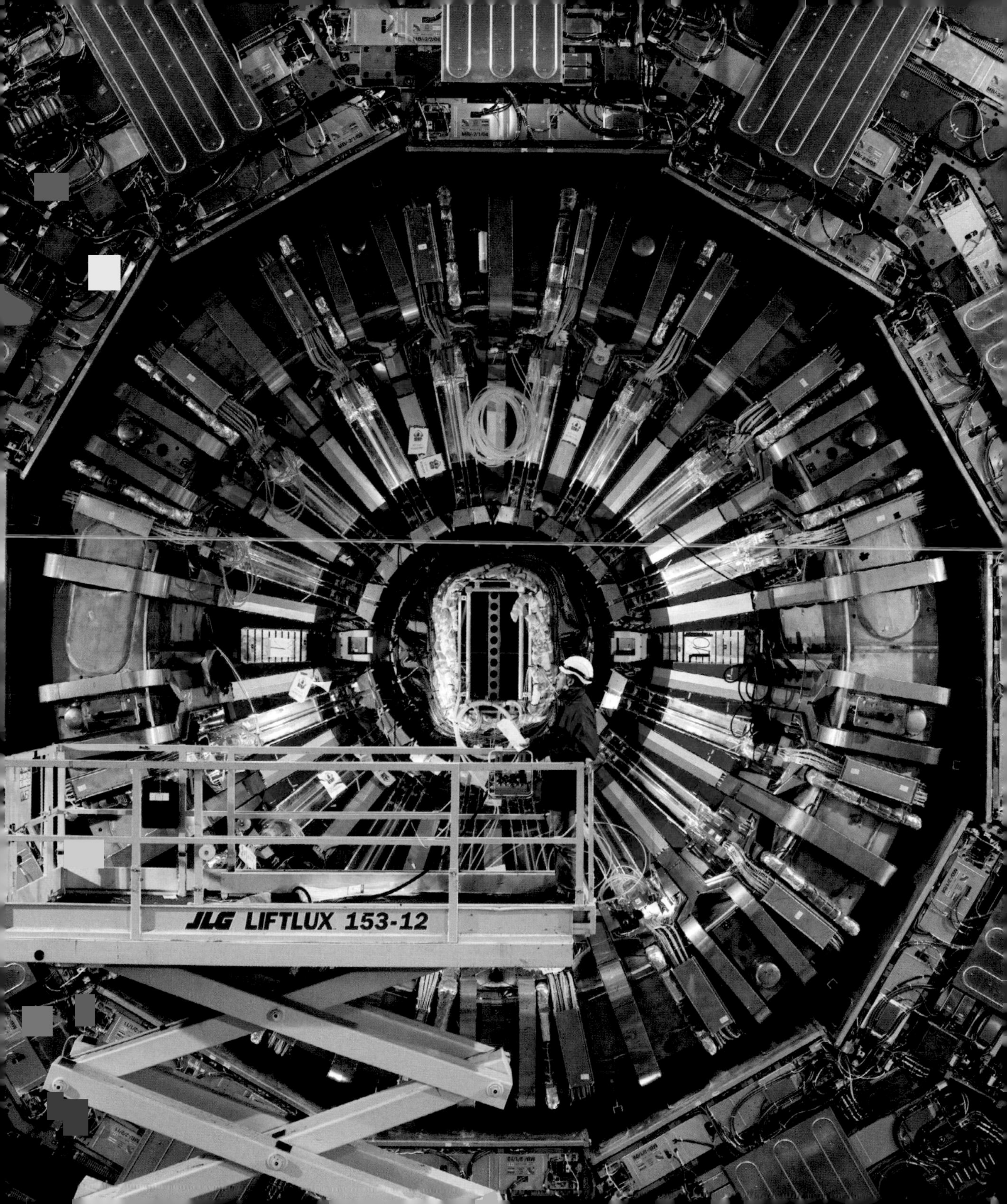
JLG LIFTLUX 153-12

1964

Quarks

Murray Gell-Mann (b. 1929), **George Zweig** (b. 1937)

Welcome to the particle zoo. In the 1960s, theorists realized that patterns in the relationships between various elementary particles, such as protons and neutrons, could be understood if these particles were not actually elementary but rather were composed of smaller particles called quarks.

Six types, or *flavors*, of quarks exist and are referred to as *up*, *down*, *charm*, *strange*, *top* and *bottom*. Only the up and down quarks are stable, and they are the most common in the universe. The other heavier quarks are produced in high-energy collisions. (Note that another class of particles called leptons, which include **Electrons**, are not composed of quarks.)

Quarks were independently proposed by physicists Murray Gell-Mann and George Zweig in 1964, and, by 1995, particle-accelerator experiments had yielded evidence for all six quarks. Quarks have fractional electric charge; for example, the up quark has a charge of +2/3, and the down quark has a charge of −1/3. **Neutrons** (which have no charge) are formed from two down quarks and one up quark, and the proton (which is positively charged) is composed of two up quarks and one down quark. The quarks are tightly bound together by a powerful short-range force called the color force, which is mediated by force-carrying particles called gluons. The theory that describes these strong interactions is called *quantum chromodynamics*. Gell-Mann coined the word *quark* for these particles after one of his perusals of the silly line in *Finnegans Wake*, "Three quarks for Muster mark."

Right after the **Big Bang**, the universe was filled with a quark-gluon **Plasma**, because the temperature was too high for hadrons (i.e. particles like protons and neutrons) to form. Authors Judy Jones and William Wilson write, "Quarks pack a mean intellectual wallop. They imply that nature is three-sided. . . . Specks of infinity on the one hand, building blocks of the universe on the other, quarks represent science at its most ambitious—also its coyest."

SEE ALSO Big Bang (13.7 Billion B.C.), Plasma (1879), Electron (1897), Neutron (1932), Quantum Electrodynamics (1948), Standard Model (1961).

Scientists used the photograph (left) of particle trails in a Brookhaven National Laboratory bubble chamber as evidence for the existence of a charmed baryon (a three-quark particle). A neutrino enters the picture from below (dashed line in right figure) and collides with a proton to produce additional particles that leave behind trails.

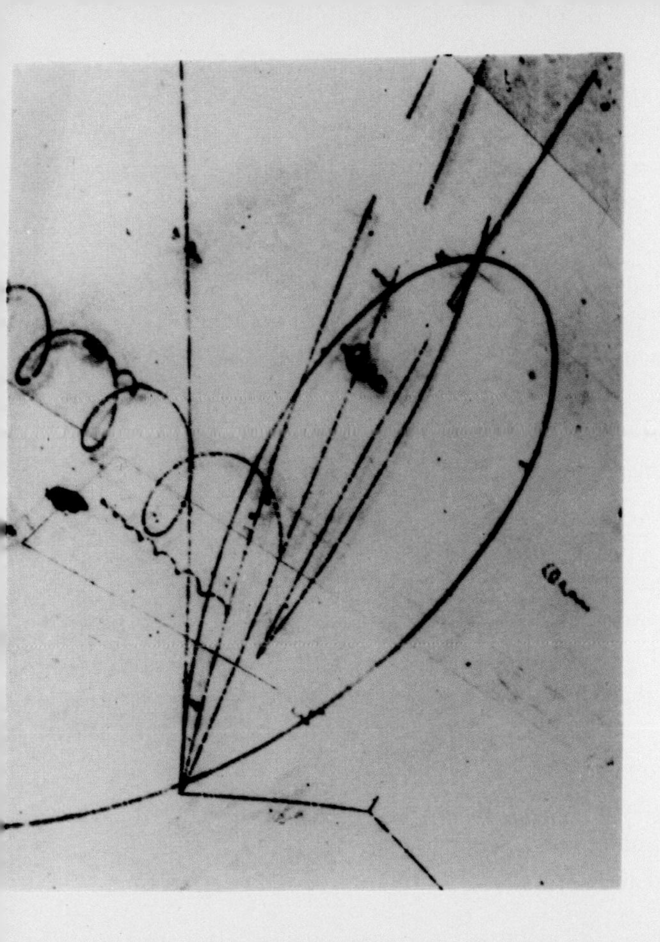

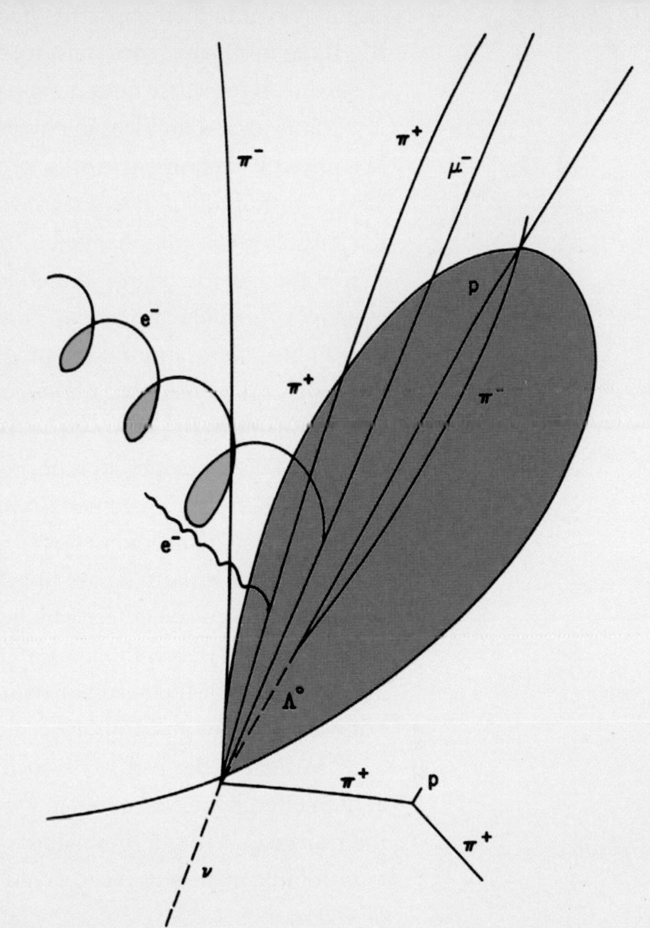

π⁻
π⁺
μ⁻
p
e⁻
π⁺
π⁻
e⁻
Λ°
π⁺
p
π⁺
ν

1964

CP Violation

James Watson Cronin (b. 1931), **Val Logsdon Fitch** (b. 1923)

You, me, the birds, and the bees are alive today due to CP violation and various laws of physics—and their apparent effect on the ratio of matter to **Antimatter** during the **Big Bang** from which our universe evolved. As a result of CP violation, asymmetries are created with respect to certain transformations in the subatomic realm.

Many important ideas in physics manifest themselves as symmetries, for example, in a physical experiment in which some characteristic is conserved, or remains constant. The C portion of *CP symmetry* suggests that the laws of physics should be the same if a particle were interchanged with its antiparticle, for example by changing the sign of the electric charge and other quantum aspects. (Technically, the C stands for charge *conjugation symmetry*.) The *P*, or parity, symmetry refers to a reversal of space coordinates, for example, swapping left and right, or, more accurately, changing all three space dimensions x, y, z to $-x$, $-y$, and $-z$. For instance, parity conservation would mean that the mirror images of a reaction occur at the same rate (e.g. the atomic nucleus emits decay products up as often as down).

In 1964, physicists James Cronin and Val Fitch discovered that certain particles, called neutral kaons, did not obey CP conservations whereby equal numbers of antiparticles and particles are formed. In short, they showed that nuclear reactions mediated by the weak force (which governs the radioactive decay of elements) violated the CP symmetry combination. In this case, neutral kaons can transform into their antiparticles (in which each quark is replaced with the others' antiquark) and vice versa, but with different probabilities.

During the Big Bang, CP violation and other as-yet unknown physical interactions at high energies played a role in the observed dominance of matter over antimatter in the universe. Without these kinds of interactions, nearly equal numbers of protons and antiprotons might have been created and annihilated each other, with no net creation of matter.

SEE ALSO Big Bang (13.7 Billion B.C.), Radioactivity (1896), Antimatter (1932), Quarks (1964), Theory of Everything (1984).

In the early 1960s, a beam from the Alternating Gradient Synchrotron at Brookhaven National Laboratory and the detectors shown here were used to prove the violation of conjugation (C) and parity (P)—winning the Nobel Prize in physics for James Cronin and Val Fitch.

1964

Bell's Theorem

John Stewart Bell (1928–1990)

In the entry **EPR Paradox**, we discussed *quantum entanglement*, which refers to an intimate connection between quantum particles, such as between two electrons or two photons. Neither particle has a definite spin before measurement. Once the pair of particles is entangled, a certain kind of change to one of them is reflected instantly in the other, even if, for example, one member of the pair is on the Earth while the other has traveled to the Moon. This entanglement is so counterintuitive that Albert Einstein thought it showed a flaw in quantum theory. One possibility considered was that such phenomena relied on some unknown "local hidden variables" outside of traditional quantum mechanical theory—and a particle was, in reality, still only influenced directly by its immediate surroundings. In short, Einstein did not accept that distant events could have an instantaneous or faster-than-light effect on local ones.

However, in 1964, physicist John Bell showed that no physical theory of local hidden variables can ever reproduce all of the predictions of quantum mechanics. In fact, the nonlocality of our physical world appears to follow from both Bell's Theorem and experimental results obtained since the early 1980s. In essence, Bell asks us first to suppose that the Earth particle and the Moon particle, in our example, have *determinate* values. Could such particles reproduce the results predicted by quantum mechanics for various ways scientists on the Earth and Moon might measure their particles? Bell proved mathematically that a statistical distribution of results would be produced that disagreed with that predicted by quantum mechanics. Thus, the particles may not carry determinate values. This is in contradiction to Einstein's conclusions, and the assumption that the universe is "local" is wrong.

Philosophers, physicists, and mystics have made extensive use of Bell's Theorem. Fritjof Capra writes, "Bell's theorem dealt a shattering blow to Einstein's position by showing that the conception of reality as consisting of separate parts, joined by local connections, is incompatible with quantum theory. . . . Bell's theorem demonstrates that the universe is fundamentally interconnected, interdependent, and inseparable."

SEE ALSO Complementarity Principle (1927), EPR Paradox (1935), Schrödinger's Cat (1935), Quantum Computers (1981).

Philosophers, physicists, and mystics have made extensive use of Bell's Theorem, which seemed to show Einstein was wrong and that the cosmos is fundamentally "interconnected, interdependent, and inseparable."

1965

Super Ball

"Bang, thump, bonk!" exclaimed the December 3, 1965 issue of *Life* magazine. "Willy-nilly the ball caroms down the hall as if it had a life of its own. This is Super Ball, surely the bouncingest spheroid ever, which has lept like a berserk grasshopper to the top of whatever charts psychologists may keep on U.S. fads."

In 1965, California chemist Norman Stingley, along with the Wham-O Manufacturing Company, developed the amazing Super Ball made from the elastic compound called Zectron. If dropped from shoulder height, it could bounce nearly 90% of that height and could continue bouncing for a minute on a hard surface (a tennis ball's bouncing lasts only ten seconds). In the language of physics, the coefficient of restitution, e, defined as the ratio of the velocity after collision to the velocity before collision, ranges from 0.8 to 0.9.

Released to the public in early summer of 1965, over six million Super Balls were bouncing around America by Fall. United States National Security Advisor McGeorge Bundy had five dozen shipped to the White House to amuse the White House staffers.

The secret of the Super Ball, often referred to as a bouncy ball, is polybutadiene, a rubber-like compound composed of long elastic chains of carbon atoms. When polybutadiene is heated at high pressure in the presence of sulfur, a chemical process called *vulcanization* converts these long chains to more durable material. Because the tiny sulfur bridges limit how much the Super Ball flexes, much of the bounce energy is returned to its motion. Other chemicals, such as di-ortho-tolylguanidine (DOTG), were added to increase the cross-linking of chains.

What would happen if one were to drop a Super Ball from the top of the Empire State Building? After the ball has dropped about 328 feet (around 100 meters, or 25–30 stories), it reaches a **Terminal Velocity** of about 70 miles per hour (113 kilometers/hour), for a ball of a radius of one inch (2.5 centimeters). Assuming $e = 0.85$, the rebound velocity will be about 60 miles per hour (97 kilometers/hour), corresponding to 80 feet (24 meters, or 7 stories).

SEE ALSO Baseball Curveball (1870), Golf Ball Dimples (1905), Terminal Velocity (1960).

If dropped from shoulder height, a Super Ball could bounce nearly 90 percent of that height and continue bouncing for a minute on a hard surface. Its coefficient of restitution ranges from 0.8 to 0.9.

1965

Cosmic Microwave Background

Arno Allan Penzias (b. 1933), **Robert Woodrow Wilson** (b. 1936)

The cosmic microwave background (CMB) is electromagnetic radiation filling the universe, a remnant of the dazzling "explosion" from which our universe evolved during the **Big Bang** 13.7 billion years ago. As the universe cooled and expanded, there was an increase in wavelengths of high-energy photons (such as in the gamma-ray and **X-ray** portion of the **Electromagnetic Spectrum**) and a shifting to lower-energy microwaves.

Around 1948, cosmologist George Gamow and colleagues suggested that this microwave background radiation might be detectable, and in 1965 physicists Arno Penzias and Robert Wilson of the Bell Telephone Laboratories in New Jersey measured a mysterious excess microwave noise that was associated with a thermal radiation field with a temperature of about −454 °F (3 K). After checking for various possible causes of this background "noise," including pigeon droppings in their large outdoor detector, it was determined that they were really observing the most ancient radiation in the universe and providing evidence for the Big Bang model. Note that because photons of energy take time to reach the Earth from distant parts of the universe; whenever we look outward in space, we are also looking back in time.

More precise measurements were made by the COBE (Cosmic Background Explorer) satellite, launched in 1989, which determined a temperature of -454.47 °F (2.735 K). COBE also allowed researchers to measure small fluctuations in the intensity of the background radiation, which corresponded to the beginning of structures, such as galaxies, in the universe.

Luck matters for scientific discoveries. Author Bill Bryson writes, "Although Penzias and Wilson had not been looking for the cosmic background radiation, didn't know what it was when they had found it, and hadn't described or interpreted its character in any paper, they received the 1978 Nobel Prize in physics." Connect an antenna to an analog TV; make sure it's not tuned to a TV broadcast and "about 1 percent of the dancing static you see is accounted for by this ancient remnant of the Big Bang. The next time you complain that there is nothing on, remember you can always watch the birth of the universe."

SEE ALSO Big Bang (13.7 Billion B.C.), Telescope (1608), Electromagnetic Spectrum (1864), X-rays (1895), Hubble's Law of Cosmic Expansion (1929), Gamma-Ray Bursts (1967), Cosmic Inflation (1980).

The Horn reflector antenna at Bell Telephone Laboratories in Holmdel, New Jersey, was built in 1959 for pioneering work related to communication satellites. Penzias and Wilson discovered the cosmic microwave background using this instrument.

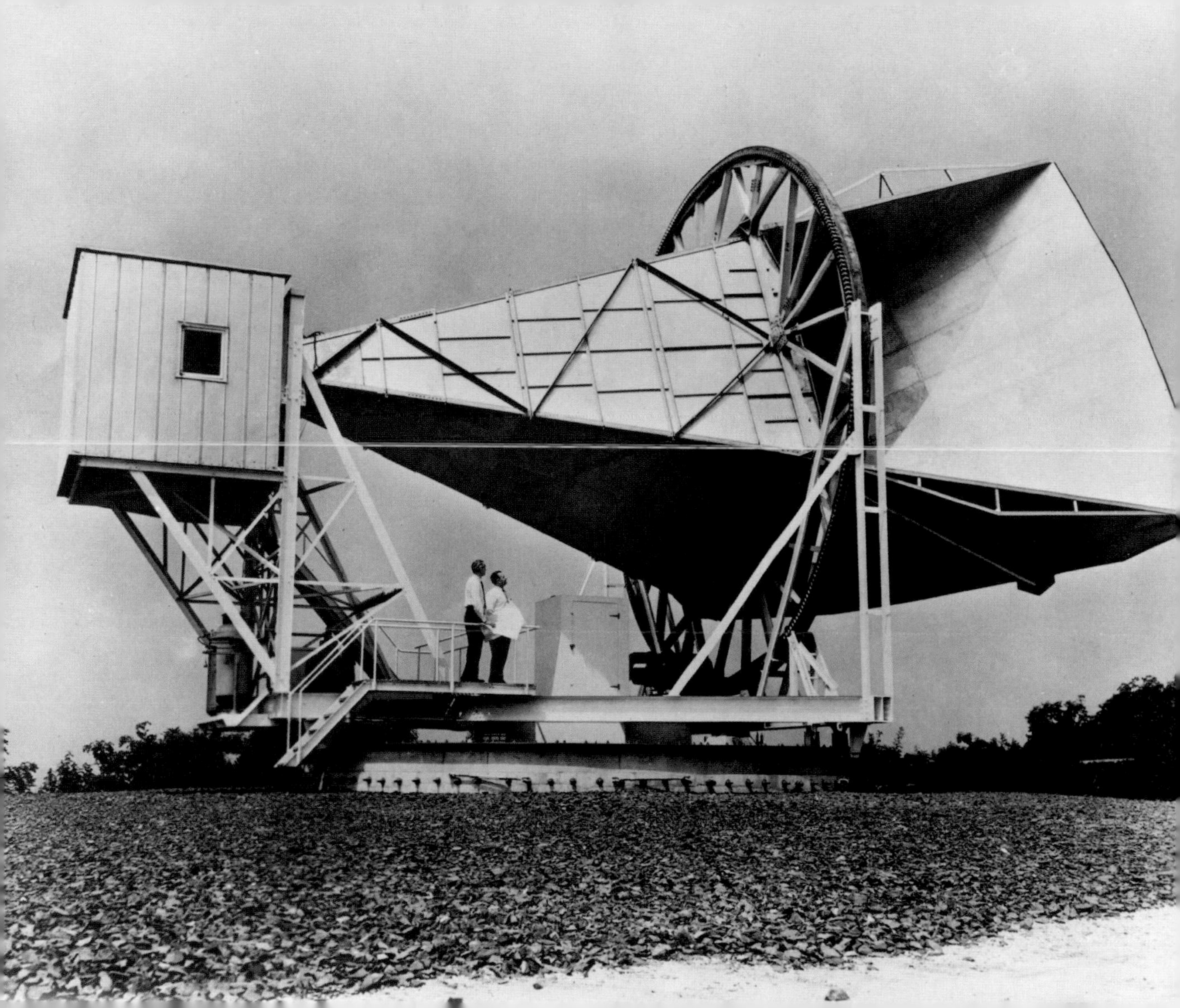

1967

Gamma-Ray Bursts

Paul Ulrich Villard (1860–1934)

Gamma-ray bursts (GRBs) are sudden, intense bursts of gamma rays, which are an extremely energetic form of light. "If you could see gamma rays with your eyes," write authors Peter Ward and Donald Brownlee, "you would see the sky light up about once a night, but these distant events go unnoticed by our natural senses." However, if a GRB would ever flash closer to the Earth, then "one minute you exist, and the next you are either dead or dying from radiation poisoning." In fact, researchers have suggested that the mass extinction of life 440 million years ago in the late Ordovician period was caused by a GRB.

Until recently, GRBs were one of the biggest enigmas in high-energy astronomy. They were discovered accidentally in 1967 by U.S. military satellites that scanned for Soviet nuclear tests in violation of the atmospheric nuclear test-ban treaty. After the typically few-second burst, the initial event is usually followed by a longer-lived afterglow at longer wavelengths. Today, physicists believe that most GRBs come from a narrow beam of intense radiation released during a supernova explosion, as a rotating high-mass star collapses to form a black hole. So far, all observed GRBs appear to originate outside our Milky Way galaxy.

Scientists are not certain as to the precise mechanism that could cause the release of as much energy in a few seconds as the Sun produces in its entire lifetime. Scientists at NASA suggest that when a star collapses, an explosion sends a blast wave that moves through the star at close to the speed of light. The gamma rays are created when the blast wave collides with material still inside the star.

In 1900, chemist Paul Villard discovered gamma rays while studying the **Radioactivity** of radium. In 2009, astronomers detected a GRB from an exploding megastar that existed a mere 630 million years after the **Big Bang** kicked the Universe into operation some 13.7 billion years ago, making this GRB the most distant object ever seen and an inhabitant of a relatively unexplored epoch of our universe.

SEE ALSO Big Bang (13.7 Billion B.C.), Electromagnetic Spectrum (1864), Radioactivity (1896), Cosmic Rays (1910), Quasars (1963).

Hubble Space Telescope image of Wolf-Rayet star WR-124 and its surrounding nebula. These kinds of stars may be generators of long-duration GRBs. These stars are large stars that rapidly lose mass via strong stellar winds.

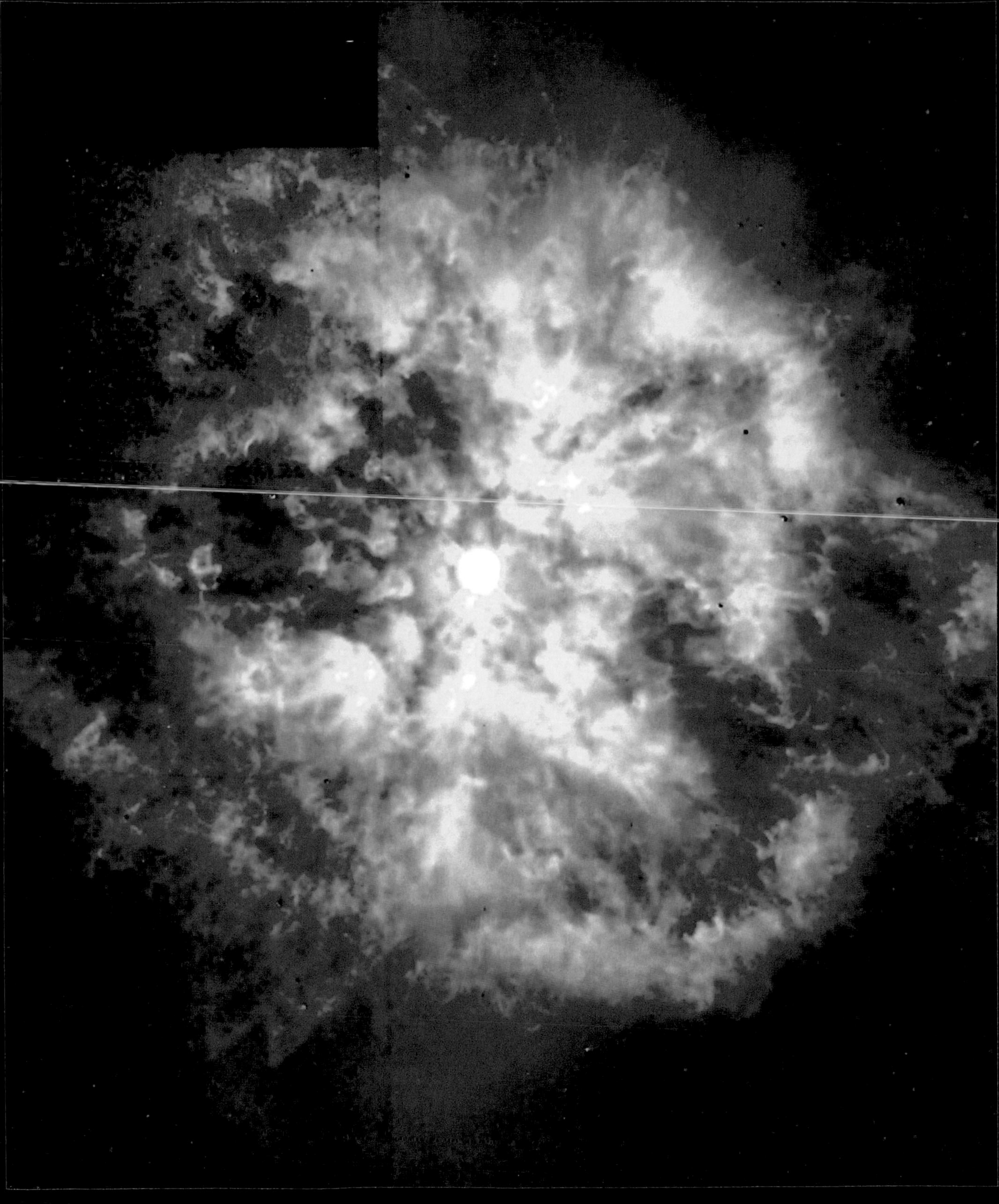

1967

Living in a Simulation

Konrad Zuse (1910–1995), **Edward Fredkin** (b. 1934), **Stephen Wolfram** (b. 1959), **Max Tegmark** (b. 1967)

As we learn more about the universe and are able to simulate complex worlds using computers, even serious scientists begin to question the nature of reality. Could we be living in a computer simulation?

In our own small pocket of the universe, we have already developed computers with the ability to simulate lifelike behaviors using software and mathematical rules. One day, we may create thinking beings that live in simulated spaces as complex and vibrant as a rain forest. Perhaps we'll be able to simulate reality itself, and it is possible that more advanced beings are already doing this elsewhere in the universe.

What if the number of these simulations is larger than the number of universes? Astronomer Martin Rees suggests that if the simulations outnumber the universes, "as they would if one universe contained many computers making many simulations," then it is likely that *we* are artificial life. Rees writes, "Once you accept the idea of the multiverse . . . , it's a logical consequence that in some of those universes there will be the potential to simulate parts of themselves, and you may get a sort of infinite regress, so we don't know where reality stops . . . , and we don't know what our place is in this grand ensemble of universes and simulated universes."

Astronomer Paul Davies has also noted, "Eventually, entire virtual worlds will be created inside computers, their conscious inhabitants unaware that they are the simulated products of somebody else's technology. For every original world, there will be a stupendous number of available virtual worlds—some of which would even include machines simulating virtual worlds of their own, and so on ad infinitum."

Other researchers, such as Konrad Zuse, Ed Fredkin, Stephen Wolfram, and Max Tegmark, have suggested that the physical universe may be running on a cellular automaton or discrete computing machinery—or be a purely mathematical construct. The hypothesis that the universe is a digital computer was pioneered by German engineer Zuse in 1967.

SEE ALSO Fermi Paradox (1950), Parallel Universes (1956), Anthropic Principle (1961).

As computers become more powerful, perhaps someday we will be able to simulate entire worlds and reality itself, and it is possible that more advanced beings are already doing this elsewhere in the universe.

1967

Tachyons

Gerald Feinberg (1933–1992)

Tachyons are hypothetical subatomic particles that travel faster than the speed of light (FTL). "Although most physicists today place the probability of the existence of tachyons only slightly higher than the existence of unicorns," writes physicist Nick Herbert, "research into the properties of these hypothetical FTL particles has not been entirely fruitless." Because such particles might travel backward in time, author Paul Nahin humorously writes, "If tachyons are one day discovered, the day before the momentous occasion a notice from the discoverers should appear in newspapers announcing 'Tachyons have been discovered tomorrow'."

Albert Einstein's theory of relativity doesn't preclude objects from going faster than light speed; rather it says that nothing traveling slower than the speed of light (SL) can ever travel faster than 186,000 miles per second (about 299,000 kilometers/second), which is the speed of light in a vacuum. However, FTL objects may exist so long as they have never traveled slower than light. Using this framework of thought, we might place all things in the universe into three classes: those always traveling less than SL, those traveling exactly at SL (photons), and those always traveling faster than SL. In 1967, the American physicist Gerald Feinberg coined the word *tachyon* for such hypothetical FTL particles, from the Greek word *tachys* for fast.

One reason that objects cannot start at a speed less than light and go faster than SL is that **Special Relativity** states that an object's mass would become infinite in the process. This relativistic mass increase is a well-tested phenomenon by high-energy physicists. Tachyons don't produce this contradiction because they never existed at sublight speeds.

Perhaps tachyons were created at the moment of the **Big Bang** from which our Universe evolved. However, in minutes these tachyons would have plunged backward in time to the universe's origin and been lost again in its primordial chaos. If tachyons are being created today, physicists feel they might detect them in **Cosmic Ray** showers or in records of particle collisions in the lab.

SEE ALSO Lorentz Transformation (1904), Special Theory of Relativity (1905), Cosmic Rays (1910), Time Travel (1949).

Tachyons are used in science fiction. If an alien, made of tachyons, approached you from his ship, you might see him arrive before you saw him leave his ship. The image of him leaving his ship would take longer to reach you than his actual FTL body.

1967

Newton's Cradle

Edme Mariotte (c. 1620–1684), **Willem Gravesande** (1688–1742), **Simon Prebble** (b. 1942)

Newton's Cradle has fascinated physics teachers and students ever since it became well known in the late 1960s. Designed by English actor Simon Prebble, who coined the phrase *Newton's Cradle* for the wooden-framed version sold by his company in 1967, the most common versions available today usually consist of five or seven metal balls suspended by wires so that they may oscillate along a single plane of motion. The balls are the same size and just touch when at rest. If one ball is pulled away and released, it collides with the stationary balls, stops, and a single ball at the other end swings upwards. The motions conserve both momentum and energy, although a detailed analysis involves more complicated considerations of the ball interactions.

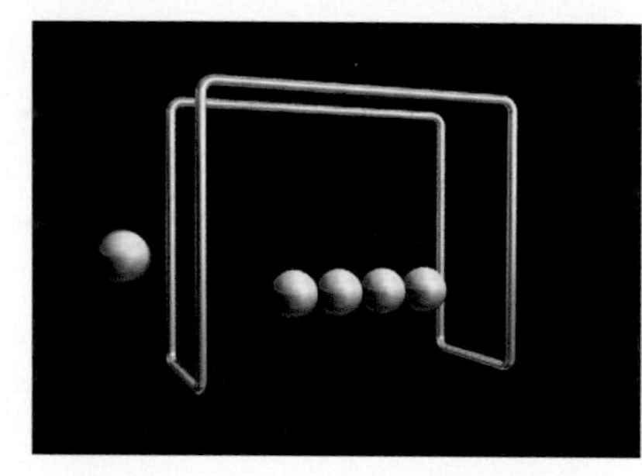

When the released ball makes an impact with the other balls, a shock wave is produced that propagates through the balls. These kinds of impacts were demonstrated in the seventeenth century by French physicist Edme Mariotte. Dutch philosopher and mathematician Willem Gravesande also performed collision experiments with devices similar to Newton's Cradle.

Today, discussions of Newton's Cradle span a range of sizes. For example, one of the largest cradles ever made holds 20 bowling balls (15 pounds [6.9 kilograms] each), suspended using cables with a length of 20 feet (6.1 meters). The other end of the size scale is described in a 2006 paper published in nature titled "A Quantum Newton's Cradle," by physicists from Pennsylvania State University who constructed a quantum version of Newton's Cradle. The authors write, "Generalization of Newton's cradle to quantum mechanical particles lends it a ghostly air. Rather than just reflecting off each other, colliding particles can also transmit through each other."

Numerous references to Newton's Cradle appear in the *American Journal of Physics*, which focuses on physics teachers, suggesting that the cradle continues to be of interest for teaching purposes.

SEE ALSO Conservation of Momentum (1644), Newton's Laws of Motion and Gravitation (1687), Conservation of Energy (1843), Foucault's Pendulum (1851).

The motions of the spheres in Newton's Cradle conserve both momentum and energy, although a detailed analysis involves more complicated considerations of the ball interactions.

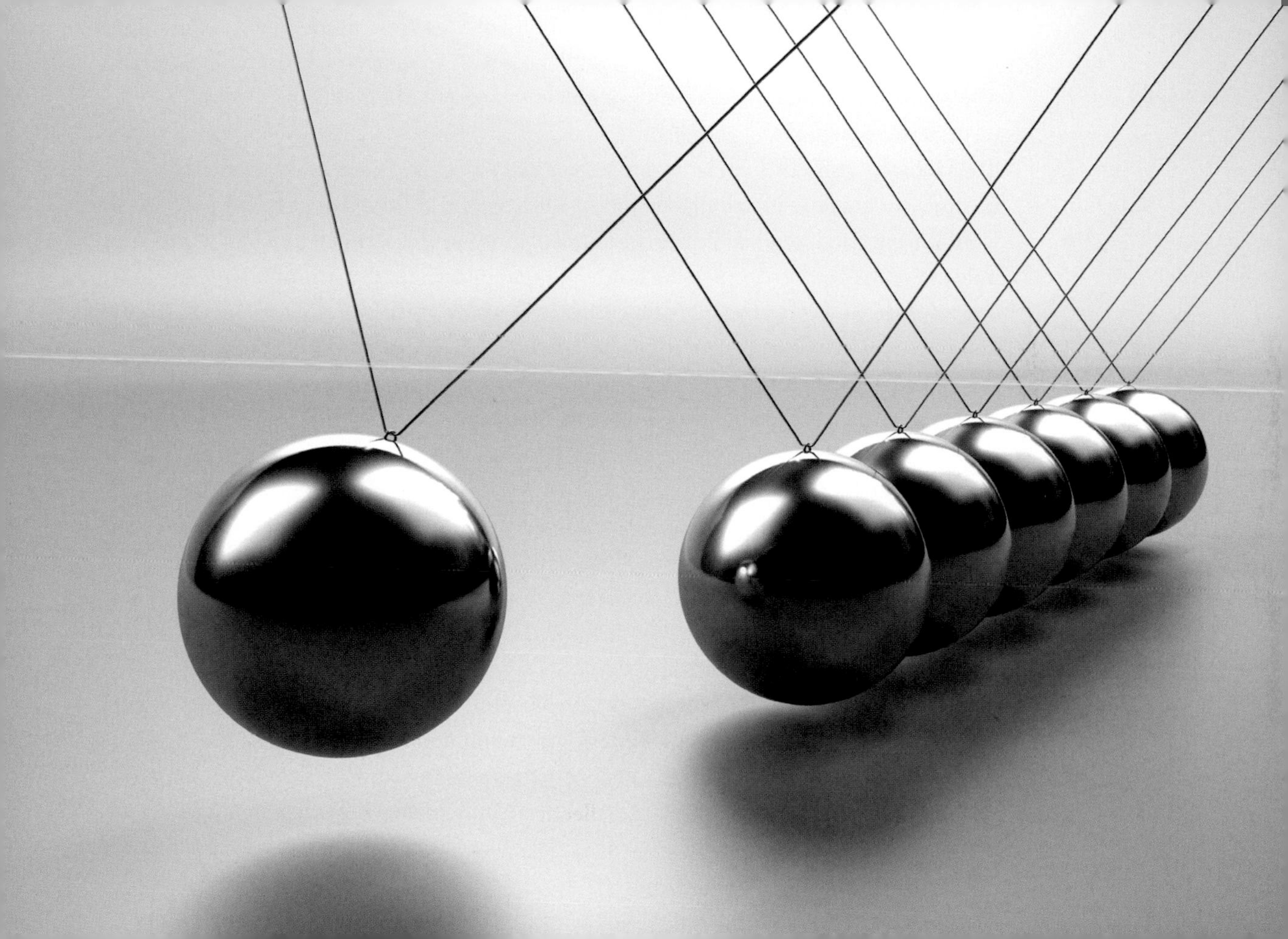

1967

Metamaterials

Victor Georgievich Veselago (b. 1929)

Will scientists ever be able to create an invisibility cloak, such as the one used by the alien Romulans to render their warships invisible in *Star Trek*? Some of the earliest steps have already been taken toward this difficult goal with *metamaterials*, artificial materials with small-scale structures and patterns that are designed to manipulate electromagnetic waves in unusual ways.

Until the year 2001, all known materials had a positive index of refraction that controls the bending of light. However, in 2001, scientists from the University of California at San Diego described an unusual composite material that had a negative index, essentially reversing **Snell's Law**. This odd material was a mix of fiberglass, copper rings, and wires capable of focusing light in novel ways. Early tests revealed that microwaves emerged from the material in the exact opposite direction from that predicted by Snell's Law. More than a physical curiosity, these materials may one day lead to the development of new kinds of antennas and other electromagnetic devices. In theory, a sheet of negative-index material could act as a super-lens to create images of exceptional detail.

Although most early experiments were performed with microwaves, in 2007 a team led by physicist Henri Lezec achieved negative refraction for visible light. In order to create an object that acted as if it were made of negatively refracting material, Lezec's team built a prism of layered metals perforated by a maze of nanoscale channels. This was the first time that physicists had devised a way to make visible light travel in a direction opposite from the way it traditionally bends when passing from one material to another. Some physicists suggest that the phenomenon may someday lead to optical microscopes for imaging objects as small as molecules and for creating cloaking devices that render objects invisible. Metamaterials were first theorized by Soviet physicist Victor Veslago in 1967. In 2008, scientists described a fishnet structure that had a negative refractive index for near-infrared light.

SEE ALSO Snell's Law of Refraction (1621), Newton's Prism (1672), Explaining the Rainbow (1304), Blackest Black (2008).

Artistic rendition of light-bending metamaterials developed by researchers working with the National Science Foundation. A layered material can cause light to refract, or bend, in a manner not seen with natural materials.

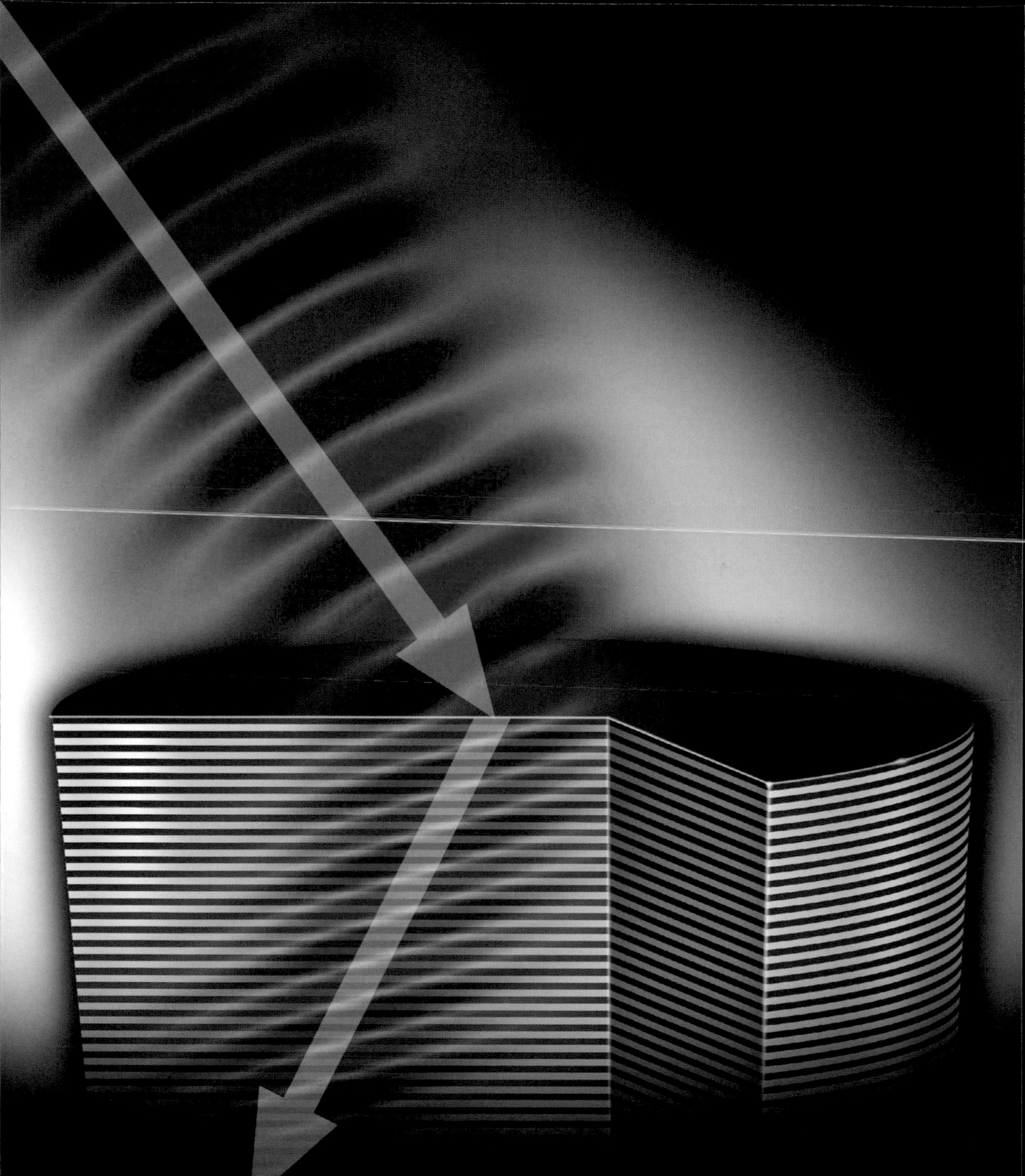

1969

Unilluminable Rooms

Ernst Gabor Straus (1922–1983), **Victor L. Klee, Jr.** (1925–2007), **George Tokarsky** (b. 1946)

American novelist Edith Wharton once wrote, "There are two ways of spreading light: to be the candle or the mirror that reflects it." In physics, the law of *reflection* states that for mirror-like reflections, the angle at which the wave is incident on a surface is equal to the angle at which it is reflected. Imagine that we are in a dark room with flat walls covered with mirrors. The room has several turns and side-passages. If I light a candle somewhere in the room, would you be able to see it no matter where you stand in the room, and no matter what the room shape or in which side-passage you stand? Stated in terms of billiard balls, must there be a pool shot between any two points on a polygonal pool table?

If we happen to be trapped in an L-shaped room, you'd be able to see my candle no matter where you and I stand because the light ray can bounce off various walls to get to your eye. But can we imagine a perplexing polygonal room that is so complicated that a point exists that light never reaches? (For our problem, we consider a person and candle to be transparent and the candle to be a point source.)

This conundrum was first presented in print by mathematician Victor Klee in 1969, although it dates back to the 1950s when mathematician Ernst Straus pondered such problems. It seems incredible that no one knew the answer until 1995, when mathematician George Tokarsky of the University of Alberta discovered such a room that is not completely illuminable. His published floor plan of the room had 26 sides. Subsequently, Tokarsky found an example with 24 sides, and this strange room is the least-sided unilluminable polygonal room currently known. Physicists and mathematicians do not know if unilluminable polygonal rooms with fewer sides actually exist.

Other similar light-reflection problems exist. In 1958, mathematical physicist Roger Penrose and his colleague showed that unlit regions can exist in certain rooms with curved sides.

SEE ALSO Archimedes' Burning Mirrors (212 B.C.), Snell's Law of Refraction (1621), Brewster's Optics (1815).

In 1995, mathematician George Tokarsky discovered this unilluminable 26-sided polygonal room. The room contains a location at which a candle can be held that leaves another point in the room in the dark.

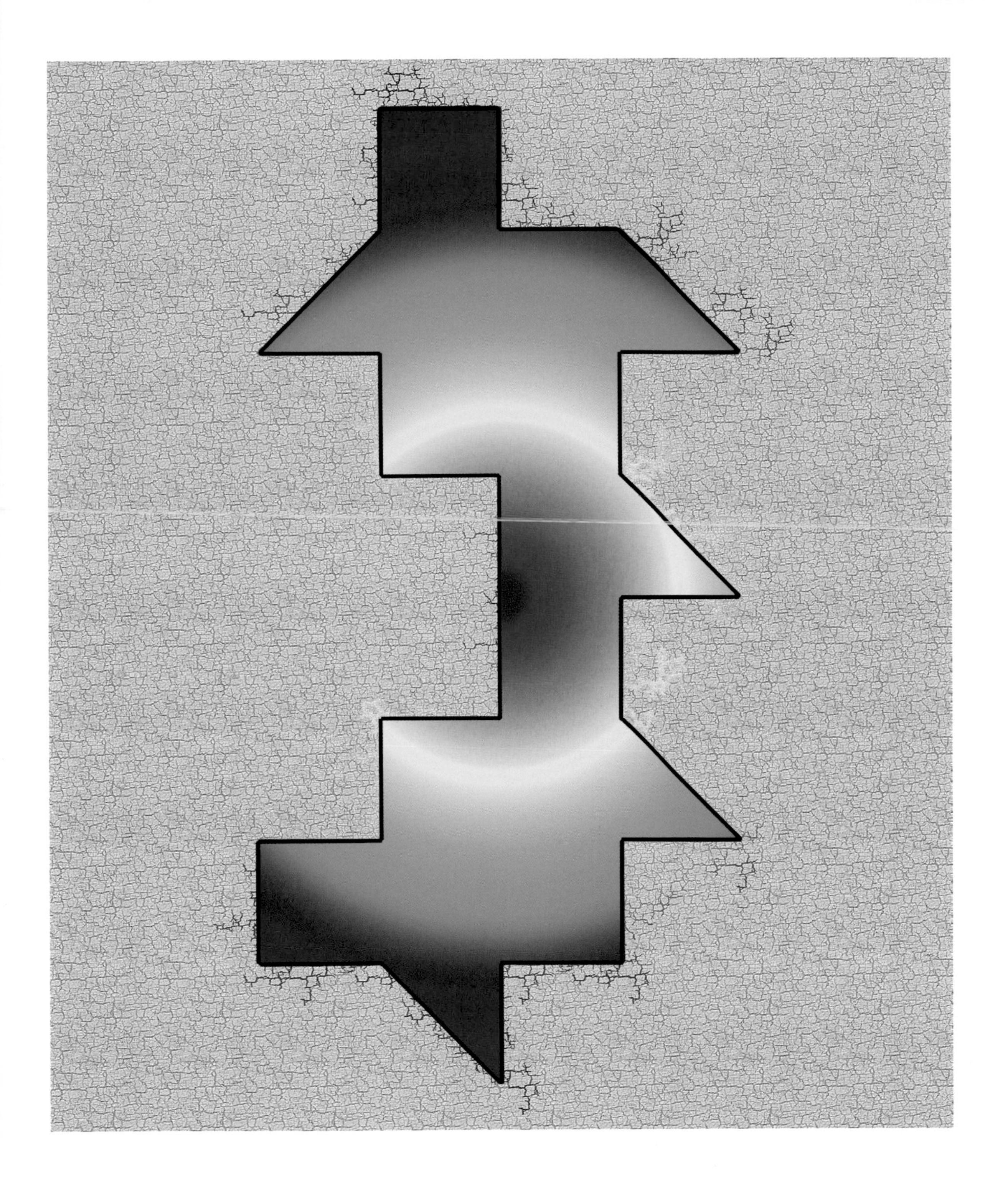

1971

Supersymmetry

Bruno Zumino (b. 1923), **Bunji Sakita** (1930–2002), **Julius Wess** (1934–2007)

"Physicists have conjured a theory about matter that sounds as if were straight out of a *Star Trek* plot," writes journalist Charles Seife. "It proposes that every particle has an as-yet-undiscovered doppelganger, a shadowy twin *superpartner* that has vastly different properties from the particles we know. . . . If supersymmetry is correct, these . . . particles are probably the source of exotic **Dark Matter** . . . that makes up almost all of the mass in the cosmos."

According to the theory of supersymmetry (SUSY), every particle in the **Standard Model** has a supersymmetric heavier twin. For example, **Quarks** (the tiny particles that combine to form other subatomic particles such as protons and neutrons) would have a heavier partner particle called a squark, which is short for supersymmetric quark. The supersymmetric partner of an electron is called a selectron. SUSY pioneers include physicists B. Sakita, J. Wess, and B. Zumino.

Part of the motivation of SUSY is the sheer aesthetics of the theory, since it adds a satisfying symmetry with respect to the properties of the known particles. If SUSY did not exist, writes Brian Greene, "It would be as if Bach, after developing numerous intertwining voices to fill out an ingenious pattern of musical symmetry, left out the final resolving measure." SUSY is also an important feature of **String Theory** in which some of the most basic particles, like quarks and electrons, can be modeled by inconceivably tiny, essentially one-dimensional entities called strings.

Science-journalist Anil Ananthaswamy writes, "The key to the theory is that in the high-energy soup of the early universe, particles and their super-partners were indistinguishable. Each pair co-existed as single massless entities. As the universe expanded and cooled, though, this supersymmetry broke down. Partners and super-partners went their separate ways, becoming individual particles with a distinct mass all their own."

Seife concludes, "If these shadowy partners stay undetectable, then the theory of supersymmetry would be merely a mathematical toy. Like the Ptolemaic universe, it would appear to explain the workings of the cosmos, yet it would not reflect reality."

SEE ALSO String Theory (1919), Standard Model (1961), Dark Matter (1933), Large Hadron Collider (2009).

According to the theory of supersymmetry (SUSY), every particle in the Standard Model *has a massive "shadow" particle partner. In the high-energy conditions of the early universe, particles and their super-partners were indistinguishable.*

1980

Cosmic Inflation

Alan Harvey Guth (b. 1947)

The **Big Bang** theory states that our universe was in an extremely dense and hot state 13.7 billion years ago, and space has been expanding ever since. However, the theory is incomplete because it does not explain several observed features in the universe. In 1980, physicist Alan Guth proposed that 10^{-35} seconds (a 100 billion trillion trillionths of a second) after the Big Bang, the universe expanded (or inflated) in a mere 10^{-32} seconds from a size smaller than a proton to the size of a grapefruit—an increase in size of 50 orders of magnitude. Today, the observed temperature of the background radiation of the universe seems relatively constant even though the distant parts of our visible universe are so far apart that they do not appear to have been connected, unless we invoke inflation that explains how these regions were originally in close proximity (and had reached the same temperature) and then separated faster than the speed of light.

Additionally, inflation explains why the universe appears to be, on the whole, quite "flat"—in essence why parallel light rays remain parallel, except for deviations near bodies with high gravitation. Any curvature in the early universe would have been smoothed away, like stretching the surface of a ball until it is flat. Inflation ended 10^{-30} seconds after the Big Bang, allowing the universe to continue its expansion at a more leisurely rate.

Quantum fluctuations in the microscopic inflationary realm, magnified to cosmic size, become the seeds for larger structures in the universe. Science-journalist George Musser writes, "The process of inflation never ceases to amaze cosmologists. It implies that giant bodies such as galaxies originated in teensy-weensy random fluctuations. Telescopes become microscopes, letting physicists see down to the roots of nature by looking up into the heavens." Alan Guth writes that inflationary theory allows us to "consider such fascinating questions as whether other big bangs are continuing to happen far away, and whether it is possible in principle for a super-advanced civilization to recreate the big bang."

SEE ALSO Big Bang (13.7 Billion B.C.), Comic Microwave Background (1965), Hubble's Law of Cosmic Expansion (1929), Parallel Universes (1956), Dark Energy (1998), Cosmological Big Rip (36 Billion), Cosmic Isolation (100 Billion).

A map produced by the Wilkinson Microwave Anisotropy Probe (WMAP) showing a relatively uniform distribution of cosmic background radiation, produced by an early universe more than 13 billion years ago. Inflation theory suggests that the irregularities seen here are the seeds that became galaxies.

1981

Quantum Computers

Richard Phillips Feynman (1918–1988), **David Elieser Deutsch** (b. 1953)

One of the first scientists to consider the possibility of a quantum computer was physicist Richard Feynman who, in 1981, wondered just how small computers could become. He knew that when computers finally reached the size of sets of atoms, the computer would be making use of the strange laws of quantum mechanics. Physicist David Deutsch in 1985 envisioned how such a computer would actually work, and he realized that calculations that took virtually an infinite time on a traditional computer could be performed quickly on a quantum computer.

Instead of using the usual binary code, which represents information as either a 0 or 1, a quantum computer uses qubits, which essentially are simultaneously both 0 and 1. Qubits are formed by the quantum states of particles, for example, the spin state of individual electrons. This superposition of states allows a quantum computer to effectively test every possible combination of qubits at the same time. A thousand-qubit system could test $2^{1,000}$ potential solutions in the blink of an eye, thus vastly outperforming a conventional computer. To get a sense for the magnitude of $2^{1,000}$ (which is approximately 10^{301}), note that there are only about 10^{80} atoms in the visible universe.

Physicists Michael Nielsen and Isaac Chuang write, "It is tempting to dismiss quantum computation as yet another technological fad in the evolution of the computer that will pass in time. . . . This is a mistake, since quantum computation is an *abstract* paradigm for information processing that may have many *different* implementations in technology."

Of course, many challenges still exist for creating a practical quantum computer. The slightest interaction or impurity from the surroundings of the computer could disrupt its operation. "These quantum engineers . . . will have to get information into the system in the first place," writes author Brian Clegg, "then trigger the operation of the computer, and, finally, get the result out. None of these stages is trivial. . . . It's as if you were trying to do a complex jigsaw puzzle in the dark with your hands tied behind your back."

SEE ALSO Complementarity Principle (1927), EPR Paradox (1935), Parallel Universes (1956), Integrated Circuit (1958), Bell's Theorem (1964), Quantum Teleportation (1993).

In 2009, physicists at the National Institute of Standards and Technology demonstrated reliable quantum information processing in the ion trap at the left center of this photograph. The ions are trapped inside the dark slit. By altering the voltages applied to each of the gold electrodes, scientists can move the ions between the six zones of the trap.

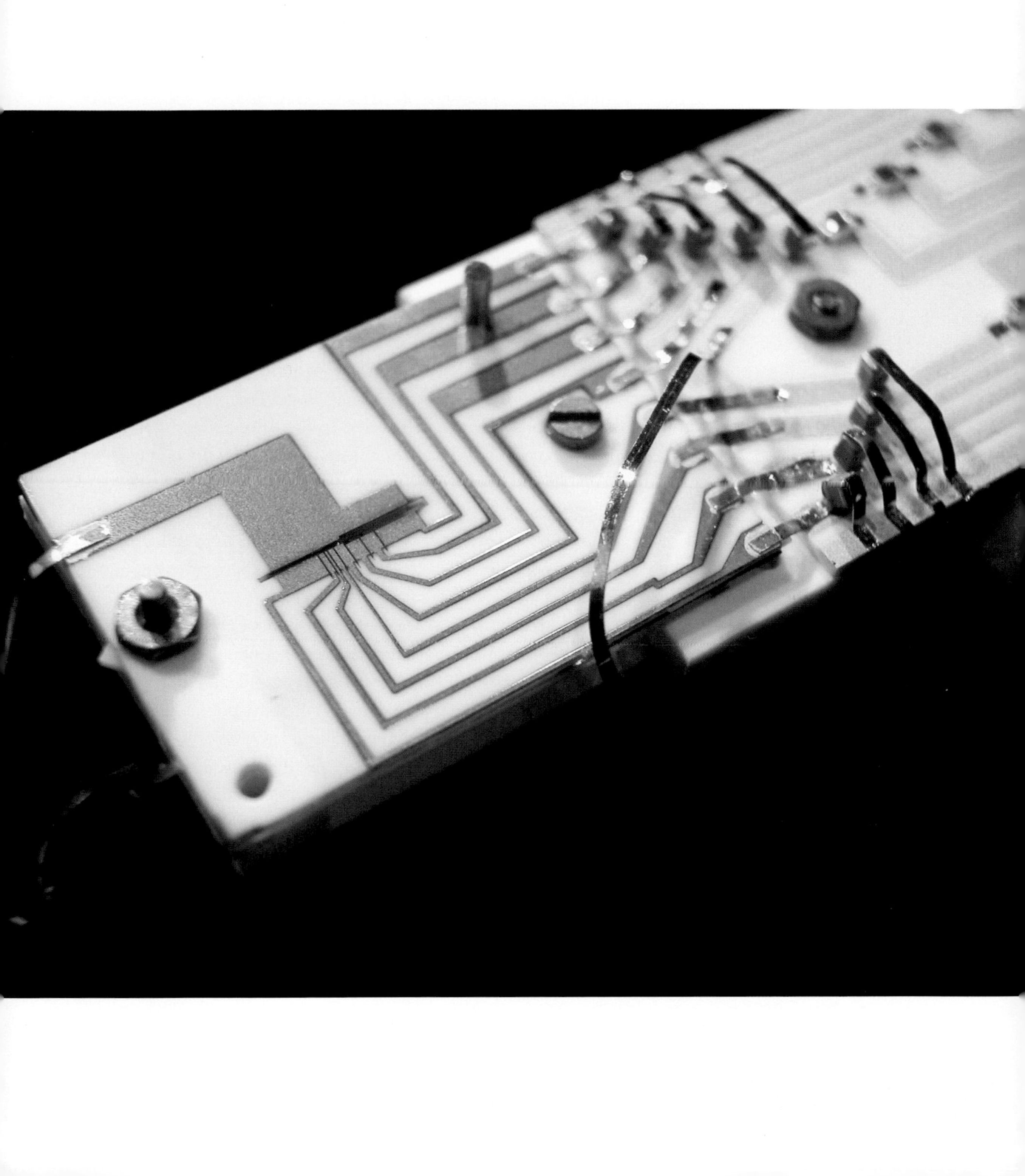

1982

Quasicrystals

Sir Roger Penrose (b. 1931), **Dan Shechtman** (b. 1941)

I am often reminded of exotic quasicrystals when I read the Biblical description from Ezekiel 1:22 of an "awe-inspiring" or "terrible" crystal that is spread over the heads of living creatures. In the 1980s, quasicrystals shocked physicists with a surprising mixture of order and *nonperiodicity*, which means that they lack *translational symmetry*, and a shifted copy of the pattern will never match with its original.

Our story begins with Penrose tiles—two simple geometric shapes that, when put side by side, can cover a plane in a pattern with no gaps or overlaps, and the pattern does not repeat periodically like the simple hexagonal tile patterns on some bathroom floors. Penrose tilings, named after mathematical physicist Roger Penrose, have fivefold rotational symmetry, the same kind of a symmetry exhibited by a five-pointed star. If you rotate the entire tile pattern by 72 degrees, it looks the same as the original. Author Martin Gardner writes, "Although it is possible to construct Penrose patterns with a high degree of symmetry . . . , most patterns, like the universe, are a mystifying mixture of order and unexpected deviations from order. As the patterns expand, they seem to be always striving to repeat themselves but never quite managing it."

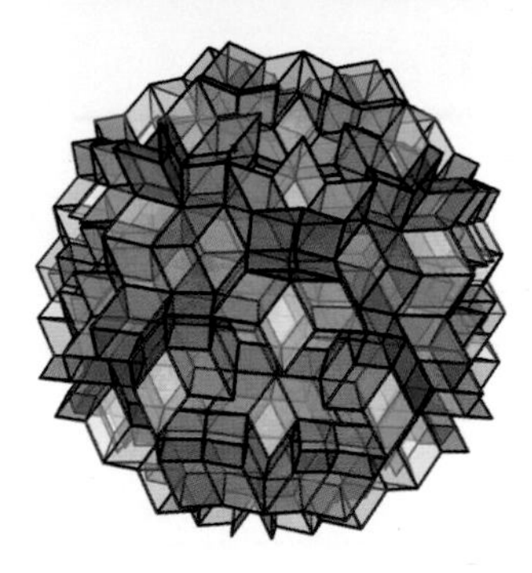

Before Penrose's discovery, most scientists believed that crystals based on fivefold symmetry would be impossible to construct, but *quasicrystals* resembling Penrose tile patterns have since been discovered, and they have remarkable properties. For example, metal quasicrystals are poor conductors of heat, and quasicrystals can be used as slippery nonstick coatings.

In the early 1980s, scientists had speculated about the possibility that the atomic structure of some crystals might be based on a nonperiodic lattice. In 1982, materials scientist Dan Shechtman discovered a nonperiodic structure in the electron micrographs of an aluminum-manganese alloy with an obvious fivefold symmetry reminiscent of a Penrose tiling. At the time, this finding was so startling that some said it was as shocking as finding a five-sided snowflake.

SEE ALSO Kepler's "Six-Cornered Snowflake" (1611), Bragg's Law of Crystal Diffraction (1912).

LEFT (UPPER): *The hexagonal symmetry of a beehive is periodic.* LEFT (LOWER): *A generalization of Penrose tiling and a possible model for quasicrystals based on an icosahedral tiling made from two rhombohedra. (Courtesy of Edmund Harriss.)* RIGHT: *Penrose tiling with two simple geometric shapes that, when put side-by-side, can cover a plane in a pattern with no gaps or overlaps and that does not repeat periodically.*

1984

Theory of Everything

Michael Boris Green (b. 1946), **John Henry Schwarz** (b. 1941)

"My ambition is to live to see all of physics reduced to a formula so elegant and simple that it will fit easily on the front of a T-shirt," wrote physicist Leon Lederman. "For the first time in the history of physics," writes physicist Brian Greene, we "have a framework with the capacity to explain every fundamental feature upon which the universe is constructed [and that may] explain the properties of the fundamental particles and the properties of the forces by which they interact and influence one another."

The theory of everything (TOE) would conceptually unite the four fundamental forces of nature, which are, in decreasing order of strengths: 1) the *strong nuclear force*—which holds the nucleus of the atom together, binds quarks into elementary particles, and makes the stars shine, 2) the *electromagnetic force*—between electric charges and between magnets, 3) the *weak nuclear force*—which governs the radioactive decay of elements, and 4) the *gravitational force*—which holds the Earth to the Sun. Around 1967, physicists showed how electromagnetism and the weak forces could be unified as the *electroweak* force.

Although not without controversy, one candidate for a possible TOE is M-theory, which postulates that the universe has ten dimensions of space and one of time. The notion of extra dimensions also may help resolve the *hierarchy problem* concerning why gravity is so much weaker than the other forces. One solution is that gravity leaks away into dimensions beyond our ordinary three spatial dimensions. If humanity did find the TOE, summarizing the four forces in a short equation, this would help physicists determine if time machines are possible and what happens at the center of black holes, and, as astrophysicist Stephen Hawking said, it gives us the ability to "read the mind of God."

The entry is arbitrarily dated as 1984, the date of an important breakthrough in superstring theory by physicists Michael Green and John Schwarz. M-theory, an extension of **String Theory**, was developed in the 1990s.

SEE ALSO Big Bang (13.7 Billion B.C.), Maxwell's Equations (1861), String Theory (1919), Randall-Sundrum Branes (1999), Standard Model (1961), Quantum Electrodynamics (1948), God Particle (1964).

Particle accelerators provide information on subatomic particles to help physicists develop a Theory of Everything. Shown here is the Cockroft-Walton generator, once used at Brookhaven National Laboratory to provide the initial acceleration to protons prior to injection into a linear accelerator and then a synchrotron.

1985

Buckyballs

Richard Buckminster "Bucky" Fuller (1895–1983), **Robert Floyd Curl, Jr.** (b. 1933), **Harold (Harry) Walter Kroto** (b. 1939), **Richard Errett Smalley** (1943–2005)

Whenever I think about buckyballs, I humorously imagine a team of microscopic soccer players kicking these rugged soccer-shaped carbon molecules and scoring goals in a range of scientific fields. Buckminsterfullerene (or buckyball, or C_{60}, for short) is composed of 60 carbon atoms and was made in 1985 by chemists Robert Curl, Harold Kroto, and Richard Smalley. Every carbon atom is at the corner of one pentagon and two hexagons. The name derives from inventor Buckminster Fuller who created cage-like structures, like the geodesic dome, that reminded the C_{60} discovers of the buckyball. C_{60} was subsequently discovered in everything from candle soot to meteorites, and researchers have been able to place selected atoms within the C_{60} structure, like a bird in a cage. Because C_{60} readily accepts and donates electrons, it may one day be used in batteries and electronic devices. The first cylindrical *nanotubes* made of carbon were obtained in 1991. These tubes are quite sturdy and may one day serve as molecular-scale electrical wires.

Buckyballs always seem to be in the news. Researchers have studied C_{60} derivatives for drug delivery and for inhibiting HIV (human immunodeficiency virus). C_{60} is of interest theoretically for various quantum mechanical and superconducting characteristics. In 2009, chemist Junfeng Geng and colleagues discovered convenient ways to form *buckywires* on an industrial scale by joining buckyballs like a string of pearls. According to *Technology Review*, "Buckywires ought to be handy for all kinds of biological, electrical, optical, and magnetic applications. . . . These buckywires look as if they could be hugely efficient light harvesters because of their great surface area and the way they can conduct photon-liberated electrons. [They may have] electronic applications in wiring up molecular circuit boards."

Also in 2009, researchers developed a new highly conductive material consisting of a crystalline network of negatively charged buckyballs with positively charged lithium ions moving through the structure. Experiments continue on these and related structures to determine if they may one day serve as "superionic" materials for batteries of the future.

SEE ALSO Battery (1800), De Broglie Relation (1924), Transistor (1947).

Buckminsterfullerne (or buckyball, or C_{60}, for short) is composed of 60 carbon atoms. Every carbon atom is at the corner of one pentagon and two hexagons.

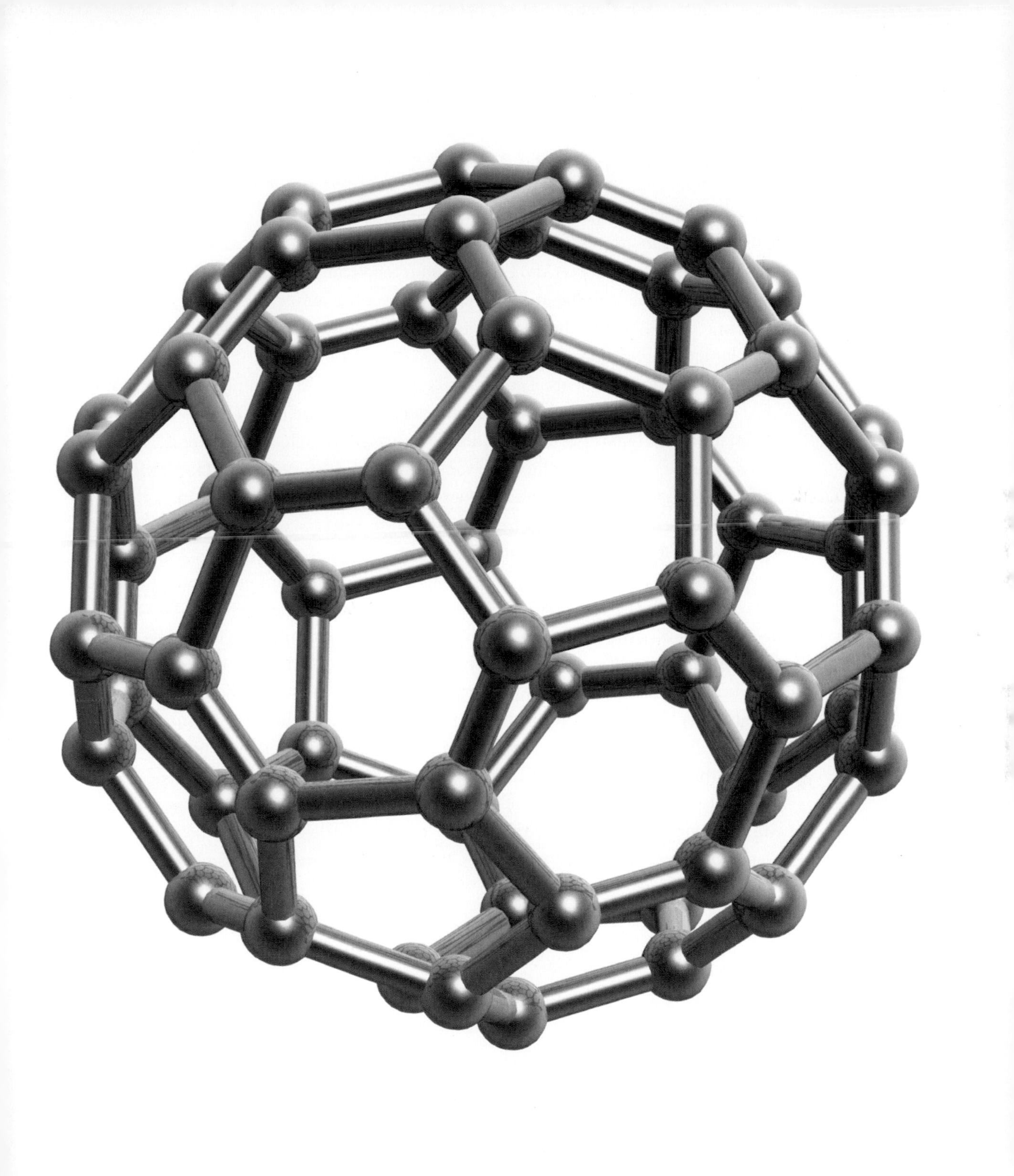

1987

Quantum Immortality

Hans Moravec (b. 1948), **Max Tegmark** (b. 1967)

The mind-boggling concept of quantum immortality, and related concepts discussed by technologist Hans Moravec in 1987 and later by physicist Max Tegmark, relies on the many-worlds interpretation (MWI) of quantum mechanics discussed in the entry on **Parallel Universes**. This theory holds that whenever the universe ("world") is confronted by a choice of paths at the quantum level, it actually follows the various possibilities, splitting into multiple universes.

According to proponents of *quantum immortality*, the MWI implies that we may be able to live virtually forever. For example, suppose you are in an electric chair. In almost all parallel universes, the electric chair will kill you. However, there is a small set of alternate universes in which you somehow survive—for example an electrical component may fail when the executioner pulls the switch. You are alive in, and thus able to experience, one of the universes in which the electric chair malfunctions. From your own point of view, you live virtually forever.

Consider a thought experiment. Don't try this at home, but imagine that you are in your basement next to a hammer that is triggered or not triggered based on the decay of a radioactive atom. With each run of the experiment, a 50-50 chance exists that the hammer will smash your skull, and you will die. If the MWI is correct, then each time you conduct the experiment, you will be split into one universe in which the hammer smashes and kills you and another universe in which the hammer does not move. Perform the experiment for a thousand times, and you may find yourself to be surprisingly alive. In the universe in which the hammer falls, you are dead. However, from the point of view of the living version of you, the hammer experiment will continue running and you will be alive, because at each branch in the multiverse there exists a version of you that survives. If the MWI is correct, you may slowly begin to notice that you never seem to die!

SEE ALSO Schrödinger's Cat (1935), Parallel Universes (1956), Quantum Resurrection (100 Trillion).

According to proponents of quantum immortality, *we can avoid the lurking specter of death virtually forever. Perhaps there is a small set of alternate universes in which you continue to survive, and thus from your own point of view, you live for an eternity.*

1987

Self-Organized Criticality

Per Bak (1948–2002)

"Consider a collection of electrons, or a pile of sand grains, a bucket of fluid, an elastic network of springs, an ecosystem, or the community of stock-market dealers," writes mathematical physicist Henrik Jensen. "Each of these systems consists of many components that interact through some kind of exchange of forces or information. . . . Is there some simplifying mechanism that produces a typical behavior shared by large classes of systems . . .?"

In 1987, physicists Per Bak, Chao Tang, and Kurt Wiesenfel published their concept of self-organized criticality (SOC), partly in response to this kind of question. SOC is often illustrated with avalanches in a pile of sand grains. One by one, grains are dropped onto a pile until the pile reaches a stationary critical state in which its slope fluctuates about a constant angle. At this point, each new grain is capable of inducing a sudden avalanche at various possible size scales. Although some numerical models of sand piles exhibit SOC, the behavior of real sand piles has sometimes been ambiguous. In the famous 1995 Oslo Rice-pile Experiment, performed at the University of Oslo in Norway, if the rice grains have a large aspect ratio, the piles exhibited SOC; however, for less-elongated grains, SOC was not found. Thus, SOC may be sensitive to the details of the system. When Sara Grumbacher and colleagues used tiny iron and glass spheres to study avalanche models, SOC was found in all cases.

SOC has been looked for in fields ranging from geophysics to evolutionary biology, economics, and cosmology, and may link many complex phenomena in which small changes result in sudden chain-reactions through the system. One key element of SOC involves power-law distributions. For a sandpile, this would imply that there will be far fewer large avalanches than small avalanches. For example, we might expect one avalanche a day involving 1000 grains but 100 avalanches involving 10 grains, and so on. In a wide variety of contexts, apparently complex structures or behaviors emerge in systems that can be characterized by simple rules.

SEE ALSO Rogue Waves (1826), Soliton (1834), Chaos Theory (1963).

LEFT: *Past research in SOC has involved the stability of rice-pile formations.* RIGHT: *Studies have shown that snow avalanches may exhibit self-organized criticality. The relationships between frequency and size of avalanches may be helpful for quantifying the risk of avalanches.*

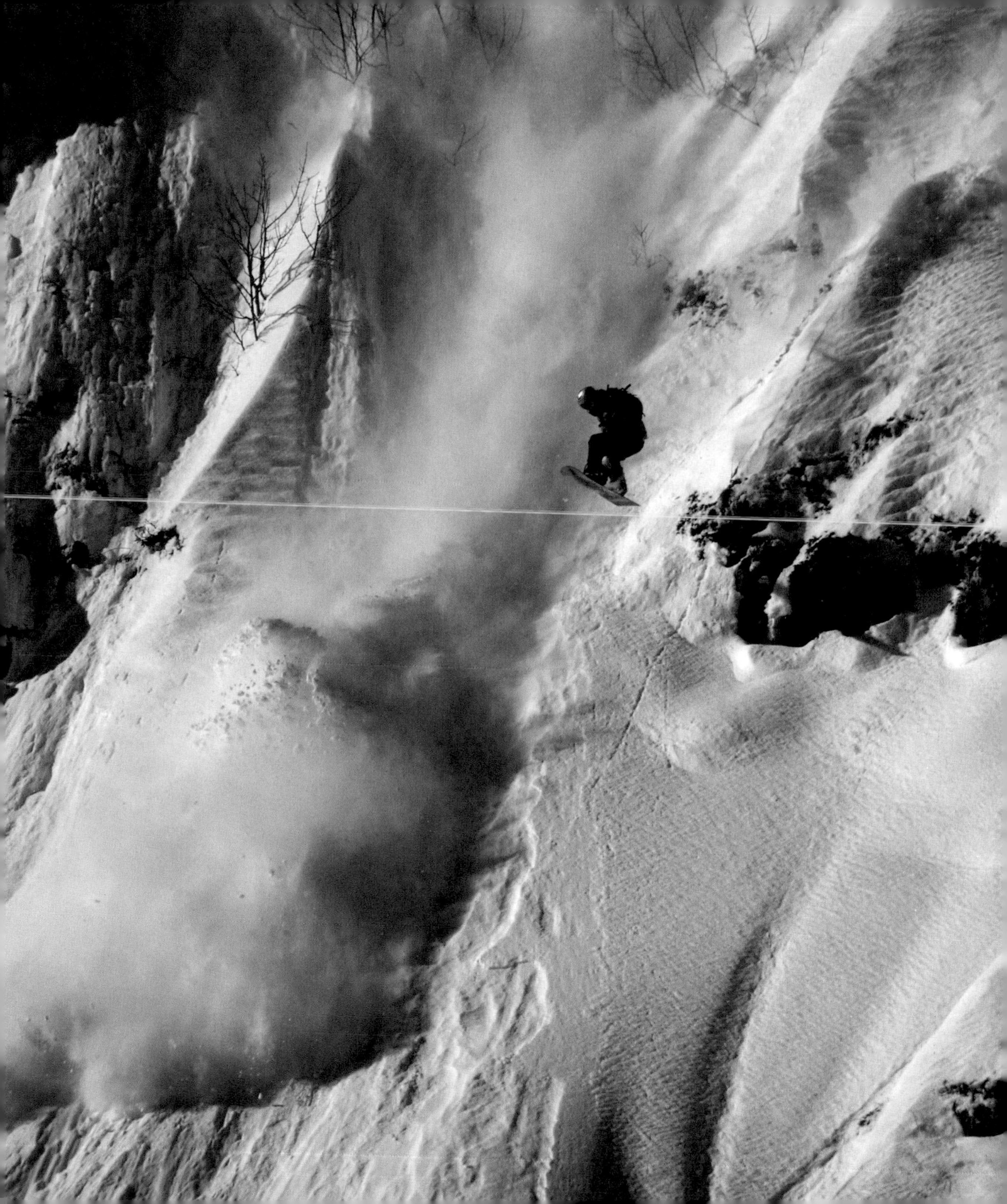

1988

Wormhole Time Machine

Kip Stephen Thorne (b. 1940)

As discussed in the entry on **Time Travel**, Kurt Gödel's time machine, proposed in 1949, worked on huge size scales—the entire universe had to rotate to make it function. At the other extreme of time-travel devices are cosmic wormholes created from subatomic quantum foam as proposed by Kip Thorne and colleagues in 1988 in their prestigious *Physical Review Letters* article. In their paper, they describe a wormhole connecting two regions that exit in different time periods. Thus, the wormhole may connect the past to the present. Since travel through the wormhole is nearly instantaneous, one could use the wormhole for backward time travel. Unlike the time machine in H. G. Wells' *The Time Machine*, the Thorne machine requires vast amounts of energy to use—energy that our civilization cannot possibly produce for many years to come. Nevertheless, Thorne optimistically writes in his paper: "From a single wormhole, an arbitrarily advanced civilization can construct a machine for backward time travel."

The Thorne traversable wormhole might be created by enlarging submicroscopic wormholes that exist in the quantum foam that pervades all of space. Once enlarged, one end of the wormhole is accelerated to extremely high speeds and then returned. Another approach involves placing a wormhole mouth near a very high gravity body and then returning it. In both cases, time dilation (slowing) causes the end of the wormhole that has been moved to have aged less than the end that has not moved with respect to your laboratory. For example, a clock on the accelerated end might read 2012 while a clock on the stationary end could read 2020. If you leaped into the 2020 end, you would arrive back in the year 2012. However, you could not go back in time to a date before the wormhole time machine was created. One difficulty for creating the wormhole time machine is that in order to keep the throat of the wormhole open, a significant amount of *negative energy* (e.g. associated with so-called *exotic matter*) would be required—something not technologically feasible to create today.

SEE ALSO Time Travel (1949), Casimir Effect (1948), Chronology Protection Conjecture (1992).

Artistic depiction of a wormhole in space. The wormhole may function as both a shortcut through space and a time machine. The two mouths (openings) of the wormhole are the yellow and blue regions.

1990

Hubble Telescope

Lyman Strong Spitzer, Jr. (1914–1997)

"Since the earliest days of astronomy," write the folks at the Space Telescope Science Institute, "since the time of Galileo, astronomers have shared a single goal—to see more, see farther, see deeper. The Hubble Space Telescope's launch in 1990 sped humanity to one of its greatest advances in that journey." Unfortunately, ground-based telescope observations are distorted by the Earth's atmosphere that makes stars seem to twinkle and that partially absorbs a range of electromagnetic radiation. Because the Hubble Space Telescope (HST) orbits outside of the atmosphere, it can capture high-quality images.

Incoming light from the heavens is reflected from the telescope's concave main mirror (7.8 feet [2.4 meters] in diameter) into a smaller mirror that then focuses the light through a hole in the center of the main mirror. The light then travels toward various scientific instruments for recording visible, ultraviolet, and infrared light. Deployed by NASA using a space shuttle, the HST is the size of a Greyhound bus, powered by solar arrays, and uses **Gyroscopes** to stabilize its orbit and point at targets in space.

Numerous HST observations have led to breakthroughs in astrophysics. Using the HST, scientists were able to determine the age of the universe much more accurately than ever before by allowing scientists to carefully measure distance to Cepheid variable stars. The HST has revealed protoplanetary disks that are likely to be the birthplaces of new planets, galaxies in various stages of evolution, optical counterparts of **Gamma-Ray Bursts** in distance galaxies, the identity of **Quasars**, the occurrence of extrasolar planets around other stars, and the existence of **Dark Energy** that appears to be causing the universe to expand at an accelerating rate. HST data established the prevalence of giant black holes at the centers of galaxies and the fact that the masses of these black holes are correlated with other galactic properties.

In 1946, American astrophysicist Lyman Spitzer, Jr. justified and promoted the idea of a space observatory. His dreams were realized in his lifetime.

SEE ALSO Big Bang (13.7 Billion B.C.), Telescope (1608), Nebular Hypothesis (1796), Gyroscope (1852), Cepheid Variables Measure the Universe (1912), Hubble's Law of Cosmic Expansion (1929), Quasars (1963), Gamma-Ray Bursts (1967), Dark Energy (1998).

Astronauts Steven L. Smith and John M. Grunsfeld appear as small figures as they replace gyroscopes inside the Hubble Space Telescope (1999).

NASA
GODDARD
SPACE
FLIGHT
CENTER

1992

Chronology Protection Conjecture

Stephen William Hawking (b. 1942)

If time travel to the past is possible, how can various paradoxes be avoided, such as your traveling back in time and killing your grandmother, thus preventing your birth in the first place? Travel to the past may not be ruled out by known physical laws and may be permitted by hypothetical techniques that employ wormholes (shortcuts through space and time) or high gravities (see **Time Travel**). If time travel is possible, why don't we see evidence of such time travelers? Novelist Robert Silverberg eloquently stated the potential problem of time-traveling tourists: "Taken to its ultimate, the cumulative audience paradox yields us the picture of an audience of billions of time-travelers piled up in the past to witness the Crucifixion, filling all the holy land and spreading out into Turkey, into Arabia, even to India and Iran. . . . Yet at the original occurrence of that event, no such hordes were present. . . . A time is coming when we will throng the past to the choking point. We will fill all our yesterdays with ourselves and crowd out our own ancestors."

Partly due to the fact that we have never seen a time traveler from the future, physicist Stephen Hawking formulated the Chronology Projection Conjecture, which proposes that the laws of physics prevent the creation of a time machine, particularly on macroscopic size scales. Today, debate continues as to the precise nature of the conjecture, or if it is actually valid. Could paradoxes be avoided simply through a string of coincidences that prevented you from killing your grandmother even if you could go back in time—or would backward time travel be prohibited by some fundamental law of nature such as a law concerning quantum mechanical aspects of gravity?

Perhaps if backward time travel was possible, our past would not be altered, because the moment someone traveled back in time, the time traveler would enter a parallel universe the instant the past is entered. The original universe would remain intact, but the new one would include whatever acts the time traveler made.

SEE ALSO Time Travel (1949), Fermi Paradox (1950), Parallel Universes (1956), Wormhole Time Machine (1988), Stephen Hawking on *Star Trek* (1993).

Stephen Hawking formulated the Chronology Projection Conjecture, which proposes that the laws of physics prevent the creation of a time machine, particularly on macroscopic size scales. Today, debate continues as to the precise nature of the conjecture.

1993

Quantum Teleportation

Charles H. Bennett (b. 1943)

In *Star Trek*, when the captain had to escape from a dangerous situation on a planet, he asked a transporter engineer on the starship to "beam me up." In seconds, the captain would disappear from the planet and reappear on the ship. Until recently, teleportation of matter was pure speculation.

In 1993, computer-scientist Charles Bennett and colleagues proposed an approach in which a particle's quantum state might be transmitted over a distance using quantum entanglement (discussed in **EPR Paradox**). Once a pair of particles (like photons) is *entangled*, a certain kind of change to one of them is reflected instantly in the other, and it doesn't matter if the pair is separated by inches or by interplanetary distances. Bennett proposed a method for scanning and transmitting part of the information of a particle's quantum state to its distant partner. The partner's state is subsequently modified using the scanned information so that it is in the state of the original particle. In the end, the first particle is no longer in its original state. Although we transfer a particle state, we can think of this as if the original particle magically jumped to the new location. If two particles of the same kind share identical quantum properties, they are indistinguishable. Because this method of teleportation has a step in which information is sent to the receiver by conventional means (such as a laser beam), teleportation does not occur at faster than light speed.

In 1997, researchers teleported a photon, and, in 2009, teleported the state of one ytterbium ion to another ytterbium ion in unconnected enclosures a meter apart. Currently, it is far beyond our technical capacity to perform quantum teleportation for people or even viruses.

Quantum teleportation may one day be useful for facilitating long-range quantum communications in quantum computers that perform certain tasks, such as encryption calculations and searches of information, much faster than traditional computers. Such computers may use quantum bits that exist in a superposition of states, like a coin that is simultaneously both tails and heads.

SEE ALSO Schrödinger's Cat (1935), EPR Paradox (1935), Bell's Theorem (1964), Quantum Computers (1981).

Researchers have been able to teleport a photon and also teleport information between two separate atoms (ytterbium ions) in unconnected enclosures. Centuries from now, will humans be teleported?

1993

Stephen Hawking on *Star Trek*

Stephen William Hawking (b. 1942)

According to surveys, astrophysicist Stephen Hawking is considered to be "the most famous scientist" at the start of the twenty-first century. Because of his inspiration, he is included in this book as a special entry. Like Einstein, Hawking also crossed over into popular culture, and he has appeared on many TV shows as himself, including *Star Trek: The Next Generation*. Because it is extremely rare for a top scientist to become a cultural icon, the title of this entry celebrates this aspect of his importance.

Many principles that concern **Black Holes** have been attributed to Stephen Hawking. Consider, for example, that the rate of evaporation of a Schwarzschild black hole of mass M can be formulated as $dM/dt = -C/M^2$, where C is a constant, and t is time. Another law of Hawking states the temperature of a black hole is inversely proportional to its mass. Physicist Lee Smolin writes, "A black hole the mass of Mount Everest would be no larger than a single atomic nucleus, but would glow with a temperature greater than the center of a star."

In 1974, Hawking determined that black holes should thermally create and emit subatomic particles, a process known as Hawking radiation, and in the same year he was elected as one of the youngest fellows of the Royal Society in London. Black holes emit this radiation and eventually evaporate and disappear. From 1979 until 2009, Hawking was the Lucasian Professor of Mathematics at the University of Cambridge, a post once held by Sir Isaac Newton. Hawking has also conjectured that the universe has no edge or boundary in imaginary time, which suggests that "the way the universe began was completely determined by the laws of science." Hawking wrote in the October 17, 1988 *Der Spiegel* that because "it is possible for the way the universe began to be determined by the laws of science . . . , it would not be necessary to appeal to God to decide how the universe began. This doesn't prove that there is no God, only that God is not necessary."

SEE ALSO Newton as Inspiration (1687), Black Holes (1783), Einstein as Inspiration (1921), Chronology Protection Conjecture (1992).

LEFT: *On* Star Trek, *Stephen Hawking plays poker with holographic representations of Isaac Newton and Albert Einstein.* RIGHT: *U.S. President Barack Obama talks with Stephen Hawking in the White House before a ceremony presenting Hawking with the Presidential Medal of Freedom (2009). Hawking has a motor neuron disease that leaves him almost completely paralyzed.*

1995

Bose-Einstein Condensate

Satyendra Nath Bose (1894–1974), **Albert Einstein** (1879–1955), **Eric Allin Cornell** (b. 1961), **Carl Edwin Wieman** (b. 1951)

The cold matter in a Bose-Einstein condensate (BEC) exhibits an exotic property in which atoms lose their identity and merge into a mysterious collective. To help visualize the process, imagine an ant colony with 100 ants. You lower the temperature to a frigid 170 billionths of a kelvin—colder than the deep reaches of interstellar space—and each ant morphs into an eerie cloud that spreads through the colony. Each ant cloud overlaps with every other one, so the colony is filled with a single dense cloud. No longer can you see individual insects; however, if you raise the temperature, the ant cloud differentiates and returns to the 100 individuals who continue to go about their ant business as if nothing untoward has happened.

A BEC is a state of matter of a very cold gas composed of bosons, particles that can occupy the same quantum state. At low temperatures, their wave functions can overlap, and interesting quantum effects can be observed on much larger size scales. First predicted by physicists Satyendra Nath Bose and Albert Einstein around 1925, BECs were not created in the laboratory until 1995 by physicists Eric Cornell and Carl Wieman, using a gas of rubidium-87 atoms (which are bosons), cooled to near absolute zero. Driven by the **Heisenberg Uncertainty Principle**—which dictates that as the velocity of the gas atoms decreases, the positions become uncertain—the atoms condense into one giant "superatom" behaving as a single entity, a quantum ice cube of sorts. Unlike an actual ice cube, the BEC is very fragile and disrupts easily to form a normal gas. Despite this, the BEC is being increasingly studied in numerous areas of physics including quantum theory, superfluidity, the slowing of light pulses, and even in the modeling of **Black Holes**.

Researchers can create such ultracold temperatures using **Lasers** and magnetic fields to slow and trap atoms. The laser beam actually can exert pressure against the atoms, slowing and cooling them at the same time.

SEE ALSO Heisenberg Uncertainty Principle (1927), Superfluids (1937), Laser (1960).

In the July 14, 1995, issue of Science *magazine, researchers from JILA (formerly known as the Joint Institute for Laboratory Astrophysics) reported the creation of a BEC. The graphic shows successive representations of the condensation (represented as a blue peak). JILA is operated by NIST (National Institute of Standards and Technology) and the University of Colorado at Boulder.*

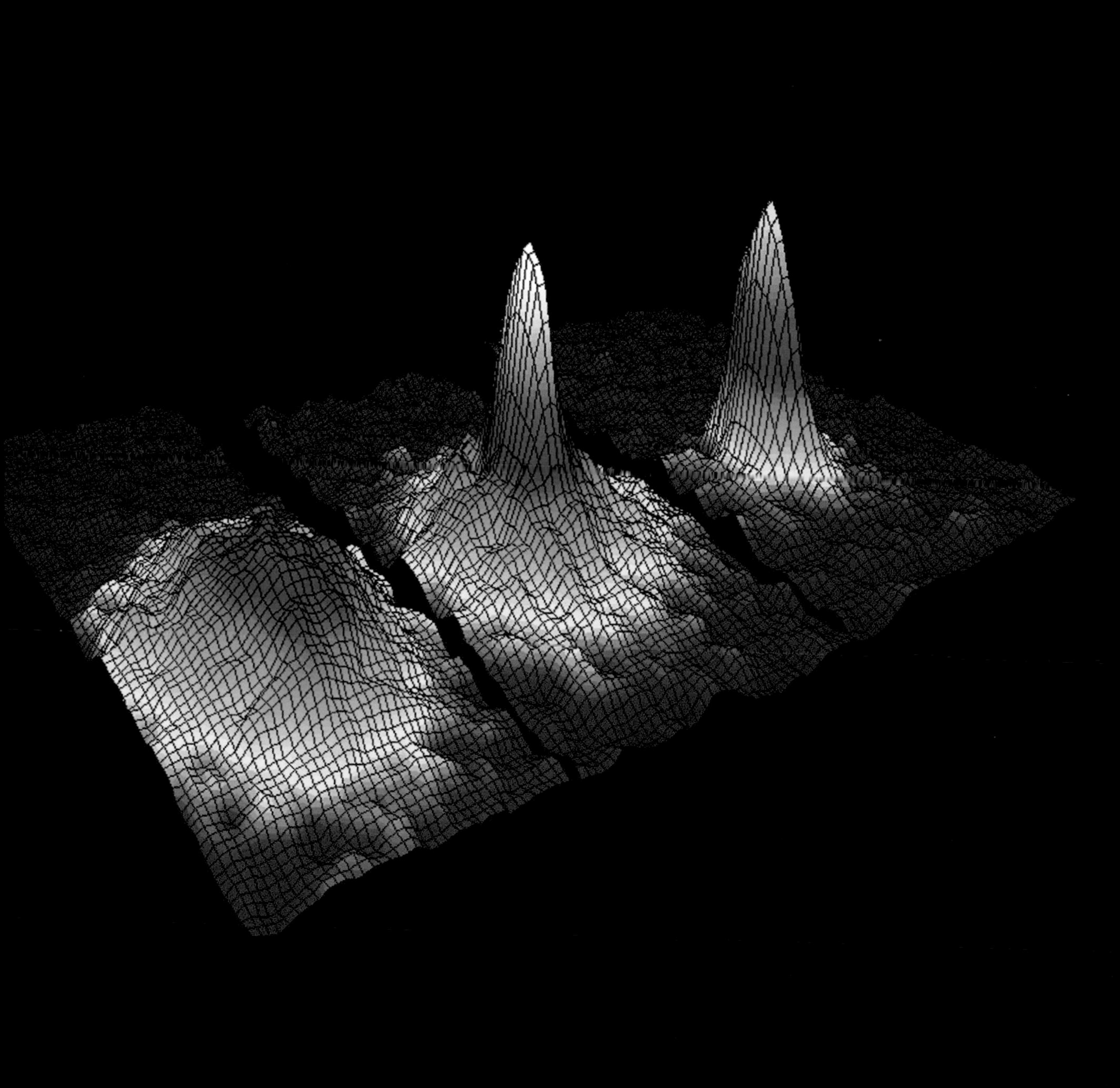

1998

Dark Energy

"A strange thing happened to the universe five billion years ago," writes science-journalist Dennis Overbye. "As if God had turned on an antigravity machine, the expansion of the cosmos speeded up, and galaxies began moving away from one another at an ever faster pace." The cause appears to be dark energy—a form of energy that may permeate all of space and that is causing the universe to accelerate its expansion. Dark energy is so abundant that it accounts for nearly three-quarters of the total mass-energy of the universe. According to astrophysicist Neil deGrasse Tyson and astronomer Donald Goldsmith, "If cosmologists could only explain where the dark energy comes from . . . they could claim to have uncovered a fundamental secret of the universe."

Evidence of the existence of dark energy came in 1998, during astrophysical observations of certain kinds of distant supernovae (exploding stars) that are receding from us at an accelerating rate. In the same year, American cosmologist Michael Turner coined the term *dark energy*.

If the acceleration of the universe continues, galaxies outside our local supercluster of galaxies will no longer be visible, because their recessional velocity will be greater than the speed of light. According to some scenarios, dark energy may eventually exterminate the universe in a **Cosmological Big Rip** as matter (in forms that range from atoms to planets) is torn apart. However, even without a Big Rip, the universe may become a lonely place (see **Cosmic Isolation**). Tyson writes, "Dark energy . . . will, in the end, undermine the ability of later generations to comprehend their universe. Unless contemporary astrophysicists across the galaxy keep remarkable records . . . future astrophysicists will know nothing of external galaxies. . . . Dark energy will deny them access to entire chapters from the book of the universe. . . . [Today] are we, too, missing some basic pieces of the universe that once was, [thus] leaving us groping for answers we may never find?"

SEE ALSO Hubble's Law of Cosmic Expansion (1929), Dark Matter (1933), Cosmic Microwave Background (1965), Cosmic Inflation (1980), Cosmological Big Rip (36 Billion), Cosmic Isolation (100 Billion).

SNAP (which stands for Supernova Acceleration Probe, a cooperative venture between NASA and the U.S. Department of Energy) is a proposed space observatory for measuring the expansion of the Universe and for elucidating the nature of dark energy.

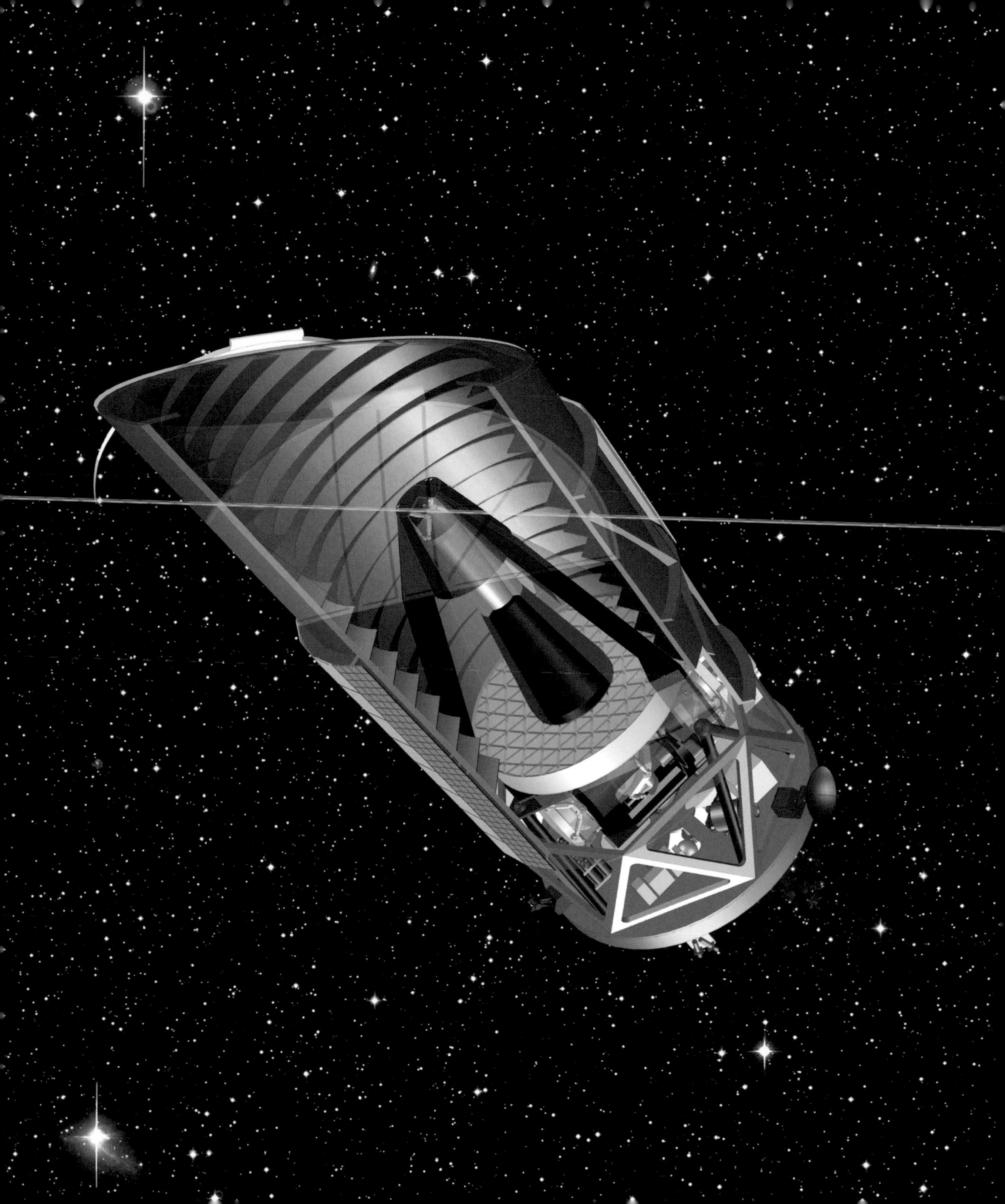

1999

Randall-Sundrum Branes

Lisa Randall (b. 1962), **Raman Sundrum** (b. 1964)

The Randall-Sundrum (RS) brane theory attempts to address the *hierarchy problem* in physics that concerns questions as to why the force of gravity appears to be so much weaker than other fundamental forces such as the electromagnetic force and the strong and weak nuclear force. Although gravity may seem strong, just remember that the electrostatic forces on a rubbed balloon are sufficient to hold it to a wall and defeat the gravity of the entire planet. According to the RS theory, gravity may be weak because it is concentrated in another dimension.

As evidence of the worldwide interest in the 1999 paper "A Large Mass Hierarchy from a Small Extra Dimension" by physicists Lisa Randall and Raman Sundrum, note that from 1999 to 2004, Dr. Randall was the most-cited theoretical physicist in the world for this and other works. Randall is also notable as she was the first tenured woman in the Princeton University physics department. One way to visualize the RS theory is to imagine that our ordinary world with its three obvious dimensions of space and one dimension of time is like a vast shower curtain, which physicists call a *brane*. You and I are like water droplets, spending our lives attached to the curtain, unaware that another brane may reside a short distance away in another spatial dimension. It is on this other *hidden brane* that gravitons, the elementary particles that give rise to gravity, may primarily reside. The other kinds of particles in the **Standard Model**, like **Electrons** and protons, are on the *visible brane* in which our visible universe resides. Gravity is actually as strong as the other forces, but it is diluted as it "leaks" into our visible brane. Photons that are responsible for our eyesight are stuck to the visible brain, and thus we are not able to see the hidden brane.

So far, no one has actually discovered a graviton. However, it is possible that high-energy particle accelerators may be able to allow scientists to identify this particle, which may also provide some evidence for the existence of additional dimensions.

SEE ALSO General Theory of Relativity (1915), String Theory (1919), Parallel Universes (1956), Standard Model (1961), Theory of Everything (1984), Dark Matter (1933), Large Hadron Collider (2009).

ATLAS is a particle detector at the site of the **Large Hadron Collider**. *ATLAS is being used to search for possible evidence related to the origins of mass and the existence of extra dimensions.*

1999

Fastest Tornado Speed

Joshua Michael Aaron Ryder Wurman (b. October 1, 1960)

Dorothy's fictional journey in *The Wizard of Oz* was not pure fantasy. Tornadoes are one of nature's most destructive forces. When the early American pioneers traveled to the Central Plains and encountered tornadoes for the first time, some witnessed adult buffalos being carried away into the air. The relatively low pressure within a tornado's swirling vortex causes cooling and condensation, making the storm visible as a funnel.

On May 3, 1999, scientists recorded the fastest tornado wind speed near the ground—roughly 318 miles (512 kilometers) per hour. Led by atmospheric scientist Joshua Wurman, a team began to follow a developing *supercell thunderstorm*—that is, a thunderstorm with an accompanying mesocyclone, which is a deep, continuously rotating updraft located a few miles up in the atmosphere. Using truck-mounted **Doppler** radar equipment, Wurman fired pulses of microwaves toward the Oklahoma storm. The waves bounced off rain and other particles, changing their frequency and providing the researchers with an accurate estimation of wind speed at about 100 feet (30 meters) above the ground.

Thunderstorms are generally characterized by rising air, called updrafts. Scientists continue to study why these updrafts become twisting whirlwinds in some thunderstorms but not others. Updraft air rises from the ground and interacts with higher-altitude winds blowing from a different direction. The tornado funnel that forms is associated with the low-pressure region as air and dust rush into the vortex. Although air is rising in a tornado, the funnel itself starts from the storm cloud and grows to the ground as a tornado forms.

Most tornadoes occur in Tornado Alley of the middle United States. Tornadoes may be created by heated air close to the ground trapped under localized colder air higher in the atmosphere. The heavier colder air spills around the warmer air region, and the warmer, lighter area rises rapidly to replace the cold air. Tornadoes sometimes form in the U.S. when warm, moist air from the Gulf of Mexico collides with cool, dry air from the Rocky Mountains.

SEE ALSO Barometer (1643), Rogue Waves (1826), Doppler Effect (1842), Buys-Ballot's Weather Law (1857).

A tornado observed by the VORTEX-99 team on May 3, 1999, in central Oklahoma.

2007

HAARP

Oliver Heaviside (1850–1925), **Arthur Edwin Kennelly** (1861–1939), **Marchese Guglielmo Marconi** (1874–1937)

If one follows the writings of conspiracy theorists, the high-frequency active auroral research program, or HAARP, is the ultimate secret missile defense tool, a means for disrupting the weather and communications around the world, or a method for controlling the minds of millions of people. However, the truth is somewhat less frightening but nonetheless fascinating.

HAARP is an experimental project funded, in part, by the U.S. Air Force, U.S. Navy, and the Defense Advanced Research Projects Agency (DARPA). Its purpose is to facilitate study of the ionosphere, one of the outermost layers of the atmosphere. Its 180-antenna array, located on a 35-acre (4,000 meters^2) plot in Alaska, became fully operational in 2007. HAARP employs a high-frequency transmitter system that beams 3.6 million watts of radio waves into the ionosphere, which starts about 50 miles (80 kilometers) above ground. The effects of heating the ionosphere can then be studied with sensitive instruments on the ground at the facilities of HAARP.

Scientists are interested in studying the ionosphere because of its effect on both civilian and military communications systems. In this region of the atmosphere, sunlight creates charged particles (see **Plasma**). The Alaska location was chosen partly because it exhibits a wide variety of ionosphere conditions for study, including aurora emissions (see **Aurora Borealis**). Scientists can adjust the signal of HAARP to stimulate reactions in the lower ionosphere, causing radiation auroral currents that send low-frequency waves back to the Earth. Such waves reach deep into the ocean and might be used by the Navy to direct its submarine fleet—no matter how deeply submerged the sub.

In 1901, Guglielmo Marconi demonstrated transatlantic communication, and people wondered precisely how radio waves were able to bend around the Earth's curvature. In 1902, engineers Oliver Heaviside and Arthur Kennelly independently suggested that a conducting layer existed in the upper atmosphere that would reflect radio waves back to the Earth. Today, the ionosphere facilitates long-range communications, and it can also lead to communication blackouts arising from the effect of a solar flare on the ionosphere.

SEE ALSO Plasma (1879), Aurora Borealis (1621), Green Flash (1882), Electromagnetic Pulse (1962).

LEFT: *The HAARP high-frequency antenna array.* RIGHT: *HAARP research may lead to improved methods for allowing the U.S. Navy to more easily communicate with submarines deep beneath the ocean's surface.*

2008

Blackest Black

All manmade materials, even asphalt and charcoal, reflect some amount of light—but this has not prevented futurists from dreaming of a perfect black material that absorbs all the colors of light while reflecting nothing back. In 2008, reports began to circulate about a group of U.S. scientists who had made the "blackest black," a superblack—the "darkest ever" substance known to science. The exotic material was created from carbon nanotubes that resemble sheets of carbon, only an atom thick, curled into a cylindrical shape. Theoretically, a perfect black material would absorb light of any wavelength shined on it at all angles.

Researchers at Rensselaer Polytechnic Institute and Rice University had constructed and studied a microscopic carpet of the nanotubes. In some sense, we can think of the "roughness" of this carpet as being adjusted to minimize the reflectance of light.

The black carpet contained tiny nanotubes that reflect only 0.045 percent of all light shined upon the substance. This black is more than 100 times darker than black paint! This "ultimate black" may one day be used to more efficiently capture energy from the Sun or to design more sensitive optical instruments. To limit reflection of light shining upon the superblack material, the researchers made the surface of the nanotube carpet irregular and rough. A significant portion of light is "trapped" in the tiny gaps between the loosely packed carpet strands.

Early tests of the superblack material were conducted using visible light. However, materials that block or highly absorb other wavelengths of electromagnetic radiation may one day be used in defense applications for which the military seeks to make objects difficult to detect.

The quest to produce the blackest black never ends. In 2009, researchers from Leiden University demonstrated that a thin layer of niobium nitride (NbN) is ultra-absorbent, with a light absorption of almost 100% at certain viewing angles. Also in 2009, Japanese researchers described a sheet of carbon nanotubes that absorbed nearly every photon of a wide range of tested wavelengths.

SEE ALSO Electromagnetic Spectrum (1864), Metamaterials (1967), Buckyballs (1985).

In 2008, scientists created the darkest material known at that time, a carpet of carbon nanotubes more than 100 times darker than the paint on a black sports car. The quest for the blackest black continues.

2009

Large Hadron Collider

According to Britain's *The Guardian* newspaper, "Particle physics is the unbelievable in pursuit of the unimaginable. To pinpoint the smallest fragments of the universe you have to build the biggest machine in the world. To recreate the first millionths of a second of creation you have to focus energy on an awesome scale." Author Bill Bryson writes, "Particle physicists divine the secrets of the Universe in a startlingly straightforward way: by flinging particles together with violence and seeing what flies off. The process has been likened to firing two Swiss watches into each other and deducing how they work by examining their debris."

Built by the European Organization for Nuclear Research (usually referred to as CERN), the Large Hadron Collider (LHC) is the world's largest and highest-energy particle accelerator, designed primarily to create collisions between opposing beams of protons (which are one kind of hadron). The beams circulate around the circular LHC ring inside a continuous vacuum guided by powerful electromagnets, the particles gaining energy with every lap. The magnets exhibit **Superconductivity** and are cooled by a large liquid-helium cooling system. When in their superconduction states, the wiring and joints conduct current with very little resistance.

The LHC resides within a tunnel 17 miles (27 kilometers) in circumference across the Franco-Swiss border and may potentially allow physicists to gain a better understanding of the Higgs boson (also called the **God Particle**), a hypothetical particle that may explain why particles have mass. The LHC may also be used to find particles predicted by **Supersymmetry**, which suggests the existence of heavier partner particles for elementary particles (for example, selectrons are the predicted partners of electrons). Additionally, the LHC may be able to provide evidence for the existence of spatial dimensions beyond the three obvious spatial dimensions. In some sense, by colliding the two beams, the LHC is re-creating some of the kinds of conditions present just after the **Big Bang**. Teams of physicists analyze the particles created in the collisions using special detectors. In 2009, the first proton–proton collisions were recorded at the LHC.

SEE ALSO Superconductivity (1911), String Theory (1919), Cyclotron (1929), Standard Model (1961), God Particle (1964), Supersymmetry (1971), Randall-Sundrum Branes (1999).

Installing the ATLAS calorimeter for the LHC. The eight toroid magnets can be seen surrounding the calorimeter that is subsequently moved into the middle of the detector. This calorimeter measures the energies of particles produced when protons collide in the center of the detector.

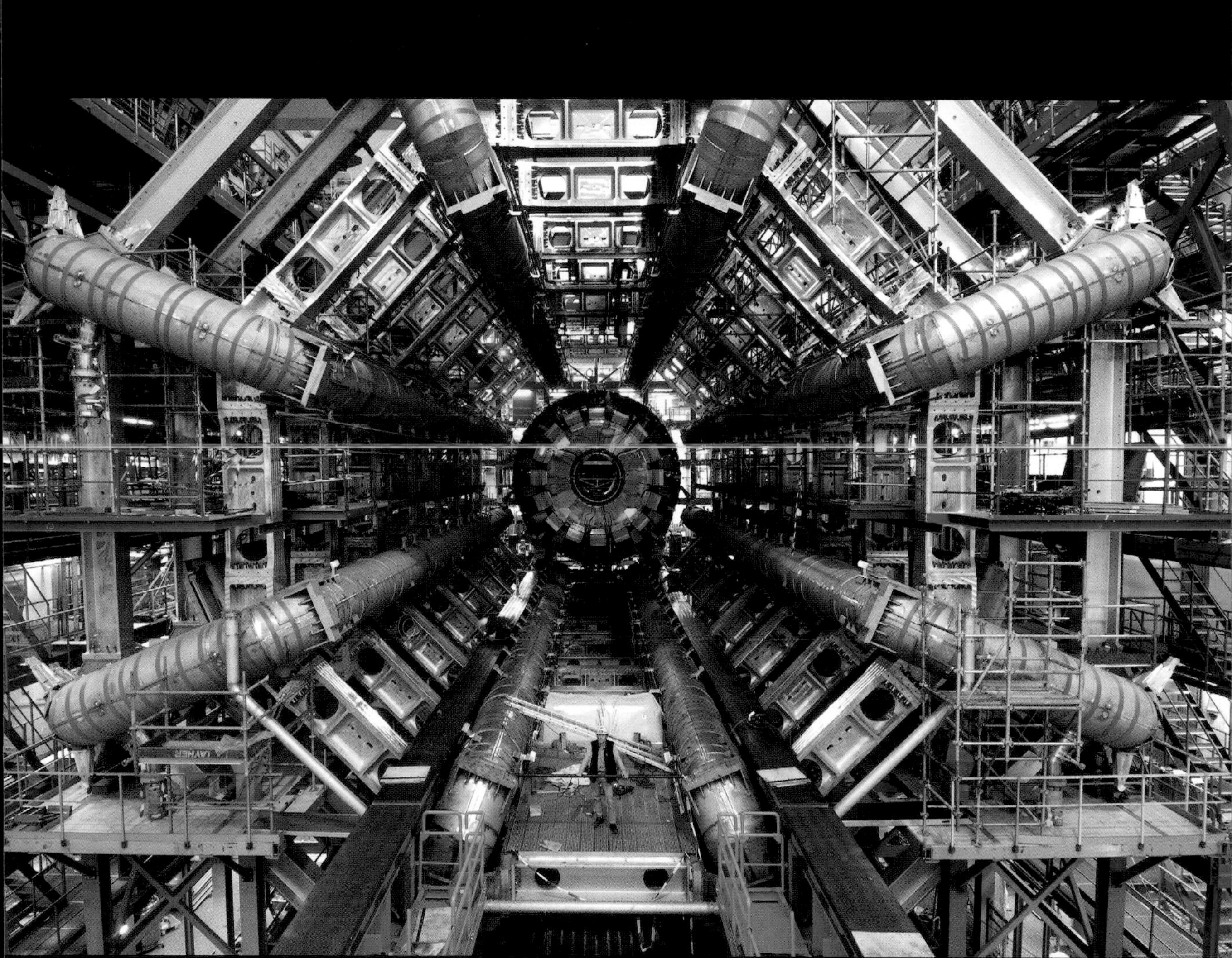

36 Billion

Cosmological Big Rip

Robert R. Caldwell (b. 1965)

The final fate of the universe is determined by many factors, including the degree to which **Dark Energy** drives the expansion of the universe. One possibility is that the acceleration will proceed at a constant rate, like a car that moves 1 mile per hour faster with each mile traveled. All galaxies will eventually recede from one another at speeds that can be greater than the speed of light, leaving each galaxy alone in a dark universe (see **Cosmic Isolation**). Eventually the stars all go out, like candles slowly burning away on a birthday cake. However, in other scenarios, the candles on the cake are ripped apart, as dark energy eventually destroys the universe in a "Big Rip" in which matter—ranging from subatomic particles to planets and stars—is torn apart. If the repulsive effect of dark energy were to somehow turn off, gravity would predominate in the cosmos, and the universe would collapse into a Big Crunch.

Physicist Robert Caldwell of Dartmouth College and colleagues first published the Big Rip hypotheses in 2003 in which the universe expands at an ever-increasing rate. At the same time, the size of our *observable* universe shrinks and eventually becomes subatomic in size. Although the precise date of this cosmic demise is uncertain, one example in Caldwell's paper is worked out for a universe that ends roughly 22 billion years from now.

If the Big Rip eventually occurs, about 60 million years before the end of the universe, gravity would be too weak to hold individual galaxies together. Roughly three months before the final rip, the Solar system will be gravitationally unbound. The Earth explodes 30 minutes before the end. Atoms tear apart 10^{-19} seconds before everything ends. The nuclear force that binds the **Quarks** in **Neutrons** and protons has finally been overcome.

Note that in 1917, Albert Einstein actually suggested the idea of an anti-gravitational repulsion in the form of a *cosmological constant* to explain why the gravity of bodies in the universe did not cause the universe to contract.

SEE ALSO Big Bang (13.7 Billion B.C.), Hubble's Law of Cosmic Expansion (1929), Cosmic Inflation (1980), Dark Energy (1998), Cosmic Isolation (100 Billion).

During the Big Rip, planets, stars, and all matter are torn apart.

100 Billion

Cosmic Isolation

Clive Staples "Jack" Lewis (1898–1963), **Gerrit L. Verschuur** (b. 1937), **Lawrence M. Krauss** (b. 1954)

The chances of an extraterrestrial race making physical contact with us may be quite small. Astronomer Gerrit Verschuur believes that if extraterrestrial civilizations are, like ours, in their infancy, then no more than 10 or 20 of them exist at this moment in our visible universe, and each such civilization is a lonely 2,000 light years apart from another. "We are," says Vershuur, "effectively alone in the Galaxy." In fact, C. S. Lewis, the Anglican lay theologian, proposed that the great distances separating intelligent life in the universe is a form of divine quarantine to "prevent the spiritual infection of a fallen species from spreading."

Contact with other galaxies will be even more difficult in the future. Even if the **Cosmological Big Rip** does not occur, the expansion of our universe may be pulling galaxies away from each other faster than the speed of light and causing them to become invisible to us. Our descendants will observe that they live in a blob of stars, which results from gravity pulling together a few nearby galaxies into one supergalaxy. This blob may then sit in an endless and seemingly static blackness. This sky will not be totally black, because the stars in this supergalaxy will be visible, but **Telescopes** that peer beyond will see nothing. Physicists Lawrence Krauss and Robert Scherrer write that in 100 billion years a dead Earth may "float forlornly" through the supergalaxy, an "island of stars embedded in a vast emptiness." Eventually, the supergalaxy itself disappears as it collapses into a **Black Hole**.

If we never encounter alien visitors, perhaps space-faring life is extremely rare and interstellar flight is extremely difficult. Another possibility is that there are signs of alien life all around us of which we are unaware. In 1973, radio astronomer John A. Ball proposed the zoo hypothesis about which he wrote that "the perfect zoo (or wilderness area or sanctuary) would be one in which the fauna do not interact with, and are unaware, of their zoo-keepers."

SEE ALSO Black Holes (1783), Black Eye Galaxy (1779), Dark Matter (1933), Fermi Paradox (1950), Dark Energy (1998), Universe Fades (100 Trillion), Cosmological Big Rip (36 Billion).

This **Hubble Telescope** *image of the Antennae Galaxies beautifully illustrates a pair of galaxies undergoing a collision. Our descendants may observe that they live in a blob of stars, which results from gravity pulling together a few nearby galaxies into one supergalaxy.*

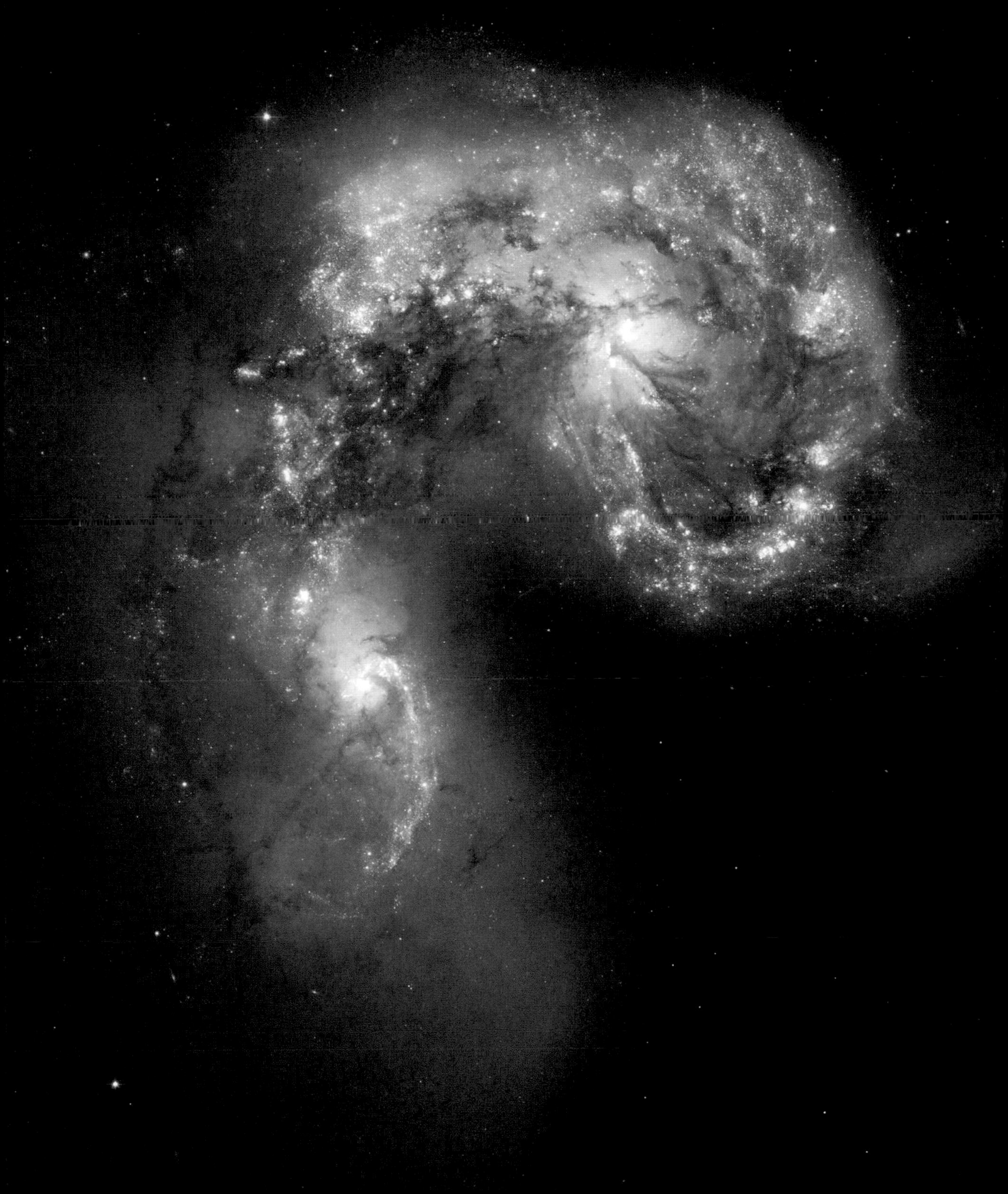

Universe Fades

Fred Adams (b. 1961), **Stephen William Hawking** (b. 1942)

The poet Robert Frost wrote, "Some say the world will end in fire, some say in ice." The ultimate destiny of our universe depends on its geometrical shape, the behavior of **Dark Energy**, the amount of matter, and other factors. Astrophysicists Fred Adams and Gregory Laughlin have described the dark ending as our current star-filled cosmos eventually evolves to a vast sea of subatomic particles while stars, galaxies, and even **Black Holes** fade.

In one scenario, the death of the universe unfolds in several acts. In our *current* era, the energy generated by stars drives astrophysical processes. Even though our universe is about 13.7 billion years old, the vast majority of stars have barely begun to shine. Alas, all stars will die after 100 trillion years, and star formation will be halted because galaxies will have run out of gas—the raw material for making new stars. At this point, the stelliferous, or star-filled, era draws to a close.

During the *second era*, the universe continues to expand while energy reserves and galaxies shrink—and material clusters at galactic centers. Brown dwarfs, objects that don't have sufficient mass to shine as stars do, linger on. By this point in time, gravity will have already drawn together the burned-out remains of dead stars, and these shrunken objects will have formed super-dense objects such as **White Dwarfs**, **Neutron Stars**, and black holes. Eventually even these white dwarfs and neutron stars disintegrate due to the decay of protons.

The *third era*—the era of black holes—is one in which gravity has turned entire galaxies into invisible, supermassive black holes. Through a process of energy radiation described by astrophysicist Stephen Hawking in the 1970s, black holes eventually dissipate their tremendous mass. This means a black hole with the mass of a large galaxy will evaporate completely in 10^{98} to 10^{100} years.

What is left as the curtain closes on the black hole era? What fills the lonely cosmic void? Could any creatures survive? In the end, our universe may consist of a diffuse sea of **Electrons**.

SEE ALSO Black Holes (1783), Stephen Hawking on *Star Trek* (1993), Dark Energy (1998), Cosmological Big Rip (36 Billion).

Artistic view of gravitationally linked brown dwarfs, discovered in 2006.

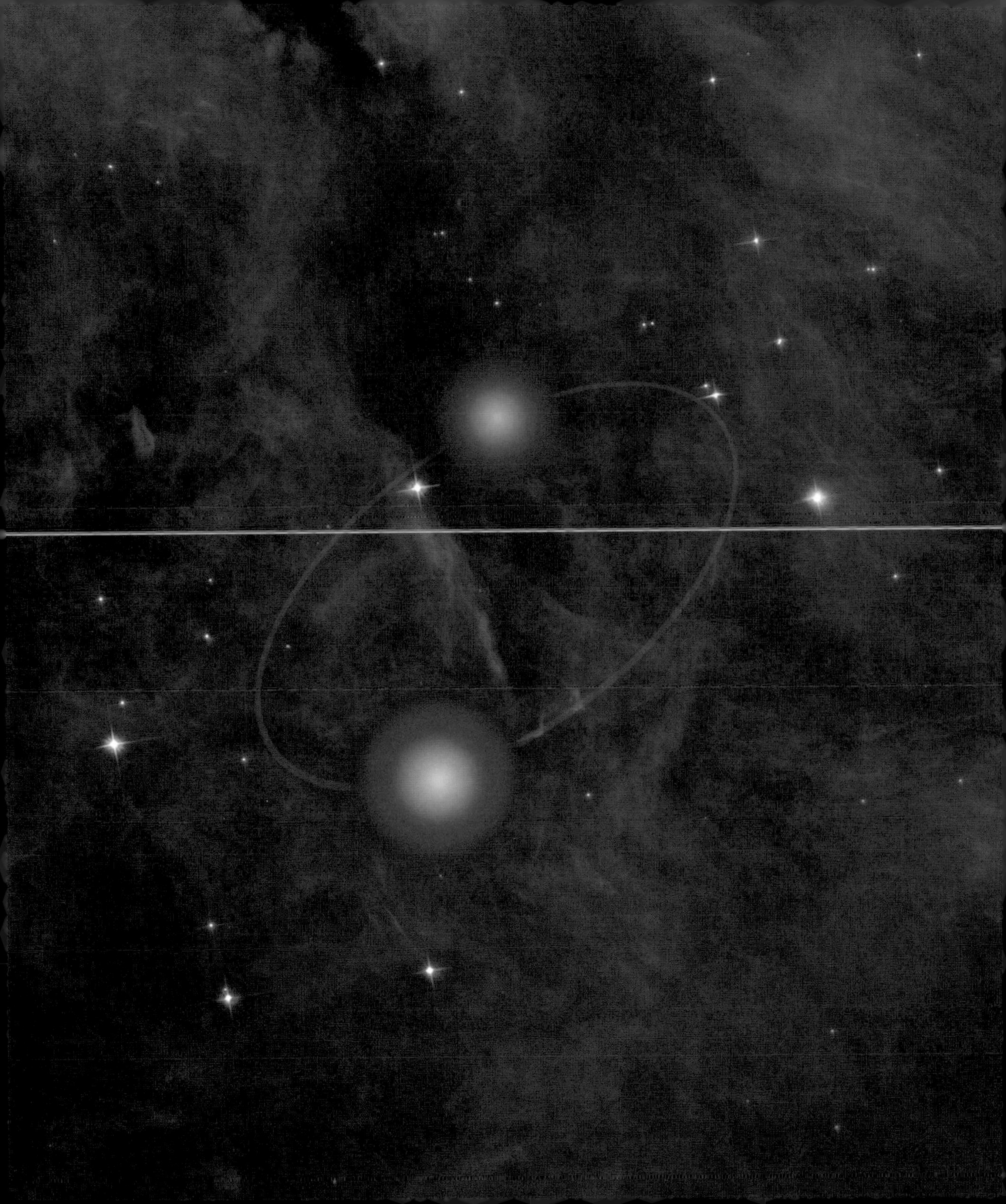

Quantum Resurrection

Ludwig Eduard Boltzmann (1844–1906)

As discussed in the preceding few entries, the fate of the universe is unknown, and some theories posit the continual creation of universes that "bud" from our own. However, let's focus on our own universe. One possibility is that our universe will continue to expand forever, and particles will become increasingly sparse. This seems like a sad end, doesn't it? However, even in this empty universe, quantum mechanics tells us that residual energy fields will have random fluctuations. Particles will spring out of the vacuum as if out of nowhere. Usually, this activity is small, and large fluctuations are rare. But particles *do* emerge, and given a long amount of time, something big is bound to appear, for example, a hydrogen atom, or even a small molecule like ethylene, $H_2C=CH_2$. This may seem unimpressive, but if our future is infinite, we can wait a long time, and almost anything could pop into existence. Most of the gunk that emerges will be an amorphous mess, but every now and then, a tiny number of ants, planets, people, or Jupiter-sized brains made from gold will emerge. Given an *infinite* amount of time, you *will* reappear, according to physicist Katherine Freese. Quantum resurrection may await all of us. Be happy.

Today, serious researchers even contemplate the universe being overrun by *Boltzmann Brains*—naked, free-floating brains in outer space. Of course, the Boltzmann Brains are highly improbable objects, and there is virtually no chance that one has appeared in the 13.7 billion years our universe has existed. According to one calculation by physicist Tom Banks, the probability of thermal fluctuations producing a brain is *e* to the power of -10^{25}. However, given an infinitely large space existing for an infinitely long time, these spooky conscious observers spring into existence. Today, there is a growing literature on the implications of Boltzmann Brains, kick-started by a 2002 publication by researchers Lisa Dyson, Matthew Kleban, and Leonard Susskind that seemed to imply that the *typical* intelligent observer may arise through thermal fluctuations, rather than cosmology and evolution.

SEE ALSO Casimir Effect (1948), Quantum Immortality (1987).

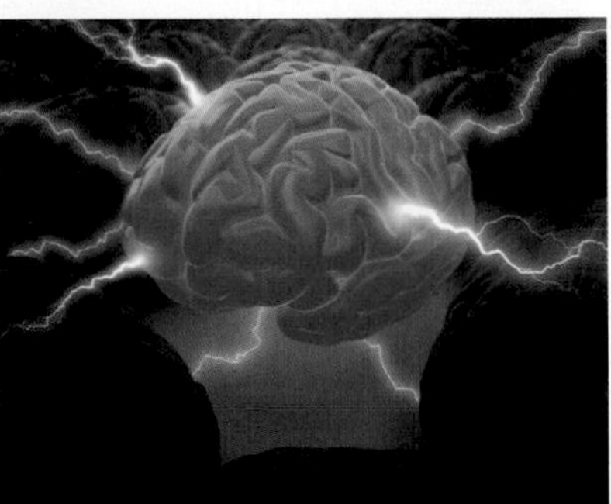

Boltzmann Brains, or thermally produced disembodied intelligences, may someday dominate our universe and outnumber all the naturally evolved intelligences that had ever existed before them.

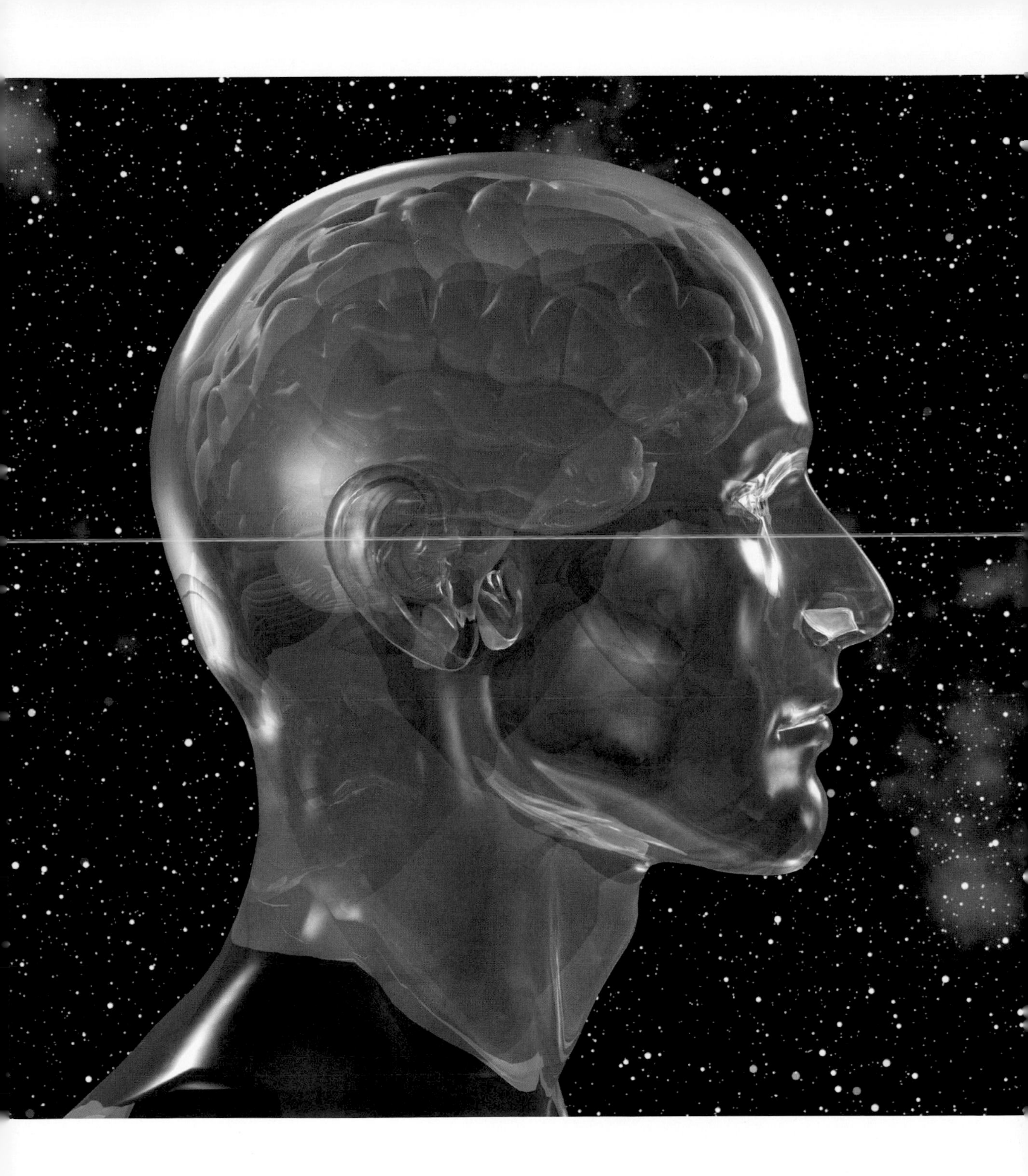

Notes and Further Reading

I've compiled the following list that identifies some of the material I used to research and write this book, along with sources for quotations. As many readers are aware, Internet websites come and go. Sometimes they change addresses or completely disappear. The website addresses listed here provided valuable background information when this book was written.

If I have overlooked an interesting or pivotal moment in physics that you feel has not been fully appreciated, please let me know about it. Just visit my website *pickover.com* and send me an e-mail explaining the idea and how you feel it influenced physics. Perhaps future editions of the book will provide more detail with respect to physics marvels such as exoplanets, geosynchronous satellites, caustics, thermometers, the Zeeman Effect, the Stark Effect, the wave equation $\partial^2 u/\partial t^2 = c^2\nabla^2 u$ (studied b y Jean le Rond d'Alembert, Leonhard Euler, and others), matrix mechanics, the discovery of the meson (a particle composed of one quark and one antiquark), the Lamb shift, the quantum eraser, Noether's (first) theorem, and the heat equation $\partial u/\partial t = \alpha\nabla^2 u$.

General Reading

Baker, J., *50 Physics Ideas You Really Need to Know*, London: Quercus, 2007.
Tallack, P., ed., *The Science Book*, London: Weidenfeld & Nicolson, 2001.
Trefil, J., *The Nature of Science*, NY: Houghton Mifflin, 2003.
Nave, R., "Hyperphysics," *tinyurl.com/lq5r*.
Wikipedia Encyclopedia, *www.wikipedia.org*.

Pickover Books

I have frequently made use of my own books for background information for different entries; however, in order to save space, I do not usually relist them for any of the entries that follow. As mentioned in the Acknowledgments, *Archimedes to Hawking* provided background information for many of entries concerning physical laws, and the reader is urged to consult this book for a fuller treatment.

Pickover, C., *Black Holes: A Traveler's Guide*, Hoboken, NJ: Wiley, 1996.
Pickover, C., *The Science of Aliens*, NY: Basic Books, 1998.
Pickover, C., *Time: A Traveler's Guide*, NY: Oxford University Press, 1998.
Pickover, C., *Surfing Through Hyperspace*, NY: Oxford University Press, 1999.
Pickover, C., *The Stars of Heaven*, NY: Oxford University Press, 2001.
Pickover, C., *From Archimedes to Hawking*, NY: Oxford University Press, 2008.
Pickover, C., *The Math Book*, NY: Sterling, 2009.

Introduction

American Physical Society, *tinyurl.com/ycmemtm*.
Simmons, J., *Doctors and Discoveries*, Boston, MA: Houghton Mifflin, 2002.

13.7 Billion B.C., Big Bang

Chown, M., *The Magic Furnace*, NY: Oxford University Press, 2001.
NASA, *tinyurl.com/ycuyjj8*.
Hawking, S., *A Brief History of Time*, NY: Bantam Books, 1988.

3 Billion B.C., Black Diamonds

Ideas on the origin of carbonados are controversial, and many hypotheses still exist.

Garai1, J., et al., *Astrophysical J.* 653: L153; 2006.
Tyson, P., *tinyurl.com/3jbk8*.

30,000 B.C., Atlatl

Elpel, T., *tinyurl.com/ydyjhsb*.

2500 B.C., Truss

Moffett, M., Fazio, M., Wodehouse, L., *A World History of Architecture*, London: Lawrence King Publishing, 2003.

1000 B.C., Olmec Hematite Compass

Carlson, J., *Science*, 189: 753; 1975.

341 B.C., Crossbow

Gedgaud, J., *tinyurl.com/y989tq8*.
Wilson, T., *tinyurl.com/lcc6mw*.

250 B.C., Baghdad Battery

BBC, *tinyurl.com/6oal*.

250 B.C., Archimedean Screw

Hasan, H., *Archimedes*, NY: Rosen Publishing Group, 2006.
Rorres, C., *tinyurl.com/yfe6hlg*.

240 B.C., Eratosthenes Measures the Earth

Hubbard, D., *How to Measure Anything*, Hoboken, NJ, 2007.

230 B.C., Pulley

Haven, K., *100 Greatest Science Inventions of All Time*, Westport, CT: Libraries Unlimited, 2005.

125 B.C., Antikythera Mechanism

Marchant, J., *tinyurl.com/ca8ory*.

50, Hero's Jet Engine

Quart. J. Sci., *tinyurl.com/yamqwjc*.

78, St. Elmo's Fire

Callahan, P., *Tuning in to Nature*, London: Routledge and Kegan Paul, 1977.

1132, Cannon

Bernal, J., *Science in History*, Cambridge, MA: MIT Press, 1971.
Kelly, J., *Gunpowder*, NY: Basic Books, 2004.

1304, Explaining the Rainbow

Lee, R., Fraser, A., *The Rainbow Bridge*, University Park: PA, 2001.

1338, Hourglass

Mills, A., et al., *Eur. J. Phys.* 17: 97; 1996.

1543, Sun-Centered Universe

O'Connor, J., Robertson, E., *tinyurl.com/yhmuks4*.

1596, *Mysterium Cosmographicum*

Gingerich, O., *Dictionary of Scientific Biography*, Gillispie, C., ed., NY: Scribner, 1970.

1600, *De Magnete*

Keithley, J., *The Story of Electrical and Magnetic Measurements*, Hoboken, NJ: Wiley-IEEE Press, 1999.

Reynolds J., Tanford, C., in Tallack, P., ed., *The Science Book*, London: Weidenfeld & Nicolson, 2001.

1608, Telescope

Lang, K., *tinyurl.com/yad22mv*.
Brockman, J., ed., *The Greatest Inventions of the Past 2000 Years*, NY: Simon & Schuster, 2000.

1609, Kepler's Laws of Planetary Motion

Gingerich, O., *Dictionary of Scientific Biography*, Gillispie, C., ed., NY: Scribner, 1970.

1610, Discovery of Saturn's Rings

Courtland, R., *tinyurl.com/yf4dfdb*.
Sagan, C., *Pale Blue Dot*, NY: Ballantine, 1997.

1611, Kepler's "Six-Cornered Snowflake"

Other symmetries can arise in snowflakes, such as three-fold symmetry.

1621, Aurora Borealis

Angot, A., *The Aurora Borealis*, London: Kegan Paul, 1896.
Hall, C., Pederson, D., Bryson, G., *Northern Lights*, Seattle, WA: Sasquatch Books, 2001.

1638, Acceleration of Falling Objects

Other earlier researchers into the acceleration of falling bodies include Nicole Oresme and Domingo de Soto.

Cohen, I., *The Birth of a New Physics*, NY: Norton, 1985.

1660, Von Guericke's Electrostatic Generator

Brockman, J., ed., *The Greatest Inventions of the Past 2000 Years*, NY: Simon & Schuster, 2000.
Gurstelle, W., *Adventures from the Technology Underground*, NY: Three Rivers Press, 2006.

1662, Boyle's Gas Law

Boyle's Law is sometimes called the Boyle-Mariotte Law, because French physicist Edme Mariotte (1620–1684) discovered the same law independently of Boyle but did not publish it until 1676.

1665, *Micrographia*

Westfall, R., *Dictionary of Scientific Biography*, Gillispie, C., ed., NY: Scribner, 1970.

1669, Amontons Friction

The true area of contact between an object and another surface is a small percentage of the apparent contact area. The true contact area, formed by tiny asperities, grows with increasing load. Leonardo da Vinci did not publish his findings and thus did not initially receive credit for his ideas.

1672, Measuring the Solar System

Haven, K., 100 *Greatest Science Inventions of All Time*, Westport, CT: Libraries Unlimited, 2005.

1672, Newton's Prism

Douma, M., *http://tinyurl.com/ybu2k7j*.

1673, Tautochrone Ramp

Darling, D., *The Universal Book of Mathematics*, Hoboken, NJ: Wiley, 2004.
Pickover, C., *The Math Book*, NY: Sterling, 2009.

1687, Newton as Inspiration

Cropper, W., *Great Physicists*, NY: Oxford University Press, 2001.
Gleick, J., *Isaac Newton*, NY: Vintage, 2004.
Koch, R., Smith, C., *New Scientist*, **190**: 25; 2006.
Hawking, S., *Black Holes and Baby Universes*, NY: Bantam, 1993.

1744, Leyden Jar

The original form of the Leyden jar was partially filled with water. Later it was discovered that the glass jar was not required and that two plates could be separated by a vacuum and store charge. Georg Matthias Bose (1710–1761) also invented early versions of the Leyden jar. Because the jar was so effective at storing charge, it was considered to be a condenser or electric charge.

McNichol, T., *AC/DC*, San Francisco, CA: Jossey-Bass, 2006.

1752, Ben Franklin's Kite

For details of any controversies surrounding Franklin's experiments, and for information on experiments that predated his, see I. B. Cohen and W. Isaacson.

Chaplin, J., *The First Scientific American*, NY: Basic Books, 2006.
Hindle, B., Foreword in I. B. Cohen's *Ben Franklin's Science*, Cambridge: MA, Harvard University Press, 1996.
Isaacson, W., *Benjamin Franklin*, NY: Simon & Schuster, 2003.

1761, Black Drop Effect

Because Mercury also exhibits a black-drop effect, and Mercury has virtually no atmosphere, we know that the black-drop effect does not require an atmosphere on the observed body.

Bergman, T., Philos. Trans., **52**: 227;1761.
Pasachoff, J., Schneider, G., Golub, L., *Proc. IAU Colloquium*, **196**: 242; 2004.
Shiga, D., *tinyurl.com/yjjbtte*.

1766, Bode's Law of Planetary Distances

Neptune and Pluto have mean orbital distances of 30.07 and 39.5, and these two bodies provide large discrepancies from the predicted values of 38.8 and 77.2, respectively. (Curiously, 38.8 is close to the actual value of 39.5 for Pluto, as if Bode's Law skips the location of Neptune.)

1777, Lichtenberg Figures

Hickman, B., *tinyurl.com/ybneddf*.

1779, Black Eye Galaxy

O'Meara, S., *Deep-Sky Companions*, NY: Cambridge University Press, 2007.
Darling, D., *tinyurl.com/yhulfpd*.

1783, Black Holes

Quantum physics suggests that short-lived pairs of particles are created in space, and these pairs flicker in and out of existence in very short time scales. The process by which a black hole emits particles involves the creation of virtual particle pairs right on the edge of the black hole's horizon. The black hole's tidal gravity pulls the pair of virtual photons apart, thereby feeding energy into them. One member of the pair is swallowed by the black hole, and the leftover particle scoots away out into the universe.

1796, Nebular Hypothesis

Earlier theories related to the nebular hypothesis include the ideas of theologian and Christian mystic Emanuel Swedenborg (1734).

1800, Battery

Brain, M., Bryant, C., *tinyurl.com/a2vpe*.
Guillen, M., *Five Equations that Changed the World*, NY: Hyperion, 1995.

1801, Wave Nature of Light

Tallack, P., ed., *The Science Book*, London: Weidenfeld & Nicolson, 2001.
Moring, G., *The Complete Idiot's Guide to Understanding Einstein*, NY: Alpha, 2004.

1807, Fourier Analysis
Pickover, C., *The Math Book*, NY: Sterling, 2009.
Ravetz, J., Grattan-Guiness, I., *Dictionary of Scientific Biography*, Gillispie, C., ed., NY: Scribner, 1970.
Hassani, S., *Mathematical Physics*, NY: Springer, 1999.

1814, Laplace's Demon
Markus, M., *Charts for Prediction and Chance*, London: Imperial College Press, 2007.

1815, Brewster's Optics
Baker, C., *Kaleidoscopes*, Concord, CA: C&T Publishing, 1999.
Land, E., *J. Opt. Soc. Am.* **41**: 957; 1951.

1816, Stethoscope
In the early 1850s, A. Leared and G. P. Camman invented the binaural stethoscope (with two earpieces). Note that the stiffer a membrane, the higher its natural frequency of oscillation and the more efficiently it operates at higher frequencies. The ability of a chestpiece to collect sound is roughly proportional to its diameter. Larger diameters are also more efficient at capturing lower frequencies.

The skin can be turned into a taut diaphragm by pressing hard on the bell. Higher frequencies are diminished by tubing that is too long. Other physics principles of stethoscopes are discussed in Constant, J., Bedside Cardiology, NY: Lippincott Williams & Wilkins, 1999.

Porter, R., *The Greatest Benefit to Mankind*, NY: Norton, 1999.

1823, Olbers Paradox
Olbers was not the first to pose this problem. Note that Lord Kelvin provided a satisfactory resolution of the paradox in 1901.

Note that the sky would be very bright in an infinite visible universe, but not infinitely bright. Beyond a certain distance, the surface areas of the stars would appear to touch and make a "shield." Also, stars themselves have a finite lifetime, which should be considered.

1824, Greenhouse Effect
Friedman, T., Hot, *Flat and Crowded*, NY: Farrar, Straus and Giroux, 2008.
Gonzalez, J., Werthman, T., Sherer, T., *The Complete Idiot's Guide to Geography*, NY: Alpha, 2007.
Sagan, C., *Billions and Billions*, NY: Ballantine, 1998.
Tallack, P., ed., *The Science Book*, London: Weidenfeld & Nicolson, 2001.

1826, Rogue Waves
Lehner, S., Foreword to Smith, C., *Extreme Waves*, Washington DC: Joseph Henry Press, 2006.

1827, Brownian Motion
Other early researchers who worked in the area of Brownian motion include T. N. Thiele, L. Bachelier, and J. Ingenhousz.

1834, Soliton
Girvan, R., *tinyurl.com/ychluer*.

1835, Gauss and the Magnetic Monopole
The traditional form of Maxwell's Equations allows for an electric charge, but not a magnetic charge; however, it is possible to extend such equations to include magnetic charges. Some modern approaches assume the existence of monopoles.

In 2009, spin-ice magnetic monopoles were detected but have a different origin from those predicted by Dirac's work and are unlikely to contribute to the development of unified theories of particle physics.

1839, Fuel Cell
In 1959, engineer F. Bacon refined the fuel cell, demonstrating how it could create significant amounts of electricity. Note also that in a car powered by a fuel cell, hydrogen gas is not actually burned. Hydrogen fuel cells make use of a proton-exchange membrane (PEM) that permits the passage of protons from hydrogen.

Solid-oxide fuel cells generate power from heated air and hydrocarbons. No expensive catalyst such as platinum is required. Oxidation of the fuel forms water and electrons. The electrolyte is the solid oxide (ceramic).

1840, Poiseuille's Law of Fluid Flow
The law is also sometimes called the Hagen-Poiseuille Law after the German physicist G. Hagen for his similar findings in 1839.

1840, Joule's Law of Electric Heating
German physicist J. von Mayer (1814–1878) actually discovered Joule's mechanical equivalent of heat before Joule; however, von Mayer's manuscript on the subject was poorly written and initially unnoticed.

1841, Anniversary Clock
Hack, R., *Hughes*, Beverly Hills, CA: Phoenix Books, 2007.

1842, Doppler Effect
Seife, C., *Alpha and Omega*, NY: Viking, 2003.

1843, Conservation of Energy
Angier, N., *The Canon*, NY: Houghton Mifflin, 2007.
Trefil, J., *The Nature of Science*, NY: Houghton Mifflin, 2003.

1844, I-Beams
Around 1750, the Euler-Bernoulli beam equation was introduced to quantify the relationship between the applied load and a beam's bending. In 1848, F. Zores seemed to have been rolling wrought-iron I-beams in France.

Peterson, C., *tinyurl.com/y9mv4vo*.
Gayle, M., Gayle, C., *Cast-Iron Architecture in America*, NY: Norton, 1998.
Kohlmaier, G., von Sartory, B., Harvey, J., *Houses of Glass*, Cambridge, MA: MIT Press, 1991.

1846, Discovery of Neptune
In 2003, documents were discovered that suggested that the traditional story of Adams' wonderful prediction of Neptune's location was exaggerated, and his "predictions" were frequently changing.

Kaler, J., *The Ever-Changing Sky*, NY: Cambridge University Press, 2002.

1850, Second Law of Thermodynamics
French physicist S. Carnot in 1824 realized that the efficiency of converting heat to mechanical work depended on the difference of temperature between hot and cold objects. Other scientists, such as C. Shannon and R. Landauer, have shown how the Second Law and the notion of entropy also apply to communications and information theory. In 2010, scientists from the University of Twente conducted an experiment by bouncing beads against the vanes of a windmill-like device. One side of each vane was softer than the other, and a net windmill movement occurred. Of course, the machine does not violate the Second Law—most of the beads' energy is lost through heat and sound.

1850, Ice Slipperiness
Michael Faraday's work was published in 1859, but many researchers had their doubts about his explanation.

1851, Foucault's Pendulum
Davis, H., *Philosophy and Modern Science*, Bloomington, IN: Principia Press, 1931.

1852, Gyroscope
Today there exist tiny gyroscopes that make use of a vibrating element to indicate the directionality and movements of an object. A vibrating element tends to continue vibrating in the same plane while its support is rotated.

Hoffmann, L., *Every Boy's Book of Sport and Pastime*, London: Routledge, 1897.

1861, Maxwell's Equations
Mathematical theories and formulas have predicted phenomena that were only confirmed years after the theory was proposed. For example, Maxwell's Equations predicted radio waves. All four equations in the set of equations are found, with slightly different notation, in his 1861 paper "On Physical Lines of Force." Note that it is possible to extend Maxwell's equations to allow for the possibility of "magnetic charges," or magnetic monopoles, analogous to electric charges, e.g. $\nabla \cdot \mathbf{B} = 4\pi\rho_m$, where ρ_m is the density of these magnetic charges.

Crease, R., *tinyurl.com/dxstsw*.
Feynman, R., *The Feynman Lectures on Physics*, Reading, MA: Addison Wesley, 1970.

1867, Dynamite
Bown, S., A *Most Damnable Invention*, NY: St. Martin's, 2005.
Bookrags, *tinyurl.com/ybr78me*.

1867, Maxwell's Demon
Maxwell formulated his idea in 1867 and presented it to the public in his book *Theory of Heat* in 1871. In 1929, Leo Szilard also helped to banish the demon by various arguments involving energy expenditure required for obtaining information about the molecules.

Leff, H., Rex, A. *Maxwell's Demon 2*, Boca Raton, FL: CRC Press, 2003.

1868, Discovery of Helium
Janssen and Lochyer were initially ridiculed because no other element was found based solely on extraterrestrial evidence. Helium can be generated by radioactive decay.

Garfinkle, D., Garfinkle, R., *Three Steps to the Universe*, Chicago, IL: University of Chicago Press, 2008.

1870, Baseball Curveball
In 1959, L. Briggs also conducted wind tunnel tests. The rough seams of a baseball help create the whirlpool of air. The dropping of the ball follows a continuous flight curve, but the vertical component of the speed is much faster near the plate due to gravity and spin. The first known curveball thrown in a game may be C. Cummings' 1867 pitch in Worcester, Massachusetts.

Adair, R., *The Physics of Baseball*, NY: HarperCollins, 2002.

1878, Incandescent Light Bulb
Note that incandescent lights have many advantages, such as being able to operate at low voltages in flashlights.

1879, Hall Effect
Electricity traveling through a conductor produces a magnetic field, which can be measured with the Hall sensor without interrupting the current flow. Hall Effect sensors are used in electronic compasses.

1880, Piezoelectric Effect
In 1881, G. Lippmann predicted the reverse piezoelectric effect in which an electric field applied to the crystal produced a stress in the crystal. Aside from crystals, the piezoelectric effect may occur in ceramics, non-metallic solids prepared through heating and cooling.

McCarthy, W., *Hacking Matter*, NY: Basic Book, 2003.

1880, War Tubas
Self, D., *tinyurl.com/yfbgv97*.
Stockbridge, F., *Pop. Sci.*, **93**: 39; 1918.

1882, Galvanometer
Moving-coil galvanometers were also developed by J. Schweigger in 1825 and C. L. Nobilli in 1828. Other important names in the history of galvanometers are M. Deprez and A. Ampère.

Wilson, J., *Memoirs of George Wilson*, London: Macmillan, 1861.

1887, Michelson-Morley Experiment
Trefil, J., *The Nature of Science*, NY: Houghton Mifflin, 2003.

1889, Birth of the Kilogram
Brumfiel, G., *tinyurl.com/n5xajq*.

1889, Birth of the Meter
Galison, P., *Einstein's Clocks, Poincaré's Maps*, NY: Norton, 2003.

1890, Eötvös' Gravitational Gradiometry
Király, P., *tinyurl.com/ybxumh9*.

1891, Tesla Coil
PBS, *tinyurl.com/ybs88x2*.
Warren, J., *How to Hunt Ghosts*, NY: Fireside, 2003.

1892, Thermos
Levy, J., *Really Useful*, NY: Firefly, 2002.

1895, X-rays
In 2009, physicists turned on the world's first X-ray laser. It was capable of producing pulses of X-rays as brief as 2 millionths of a nanosecond. Prior to Röntgen's work, N. Tesla began his observations of X-rays (at that time still unknown and unnamed).

Haven, K., *100 Greatest Science Inventions of All Time*, Westport, CT: Libraries Unlimited, 2005.

1895, Curie's Magnetism Law
Curie's Law holds only for a limited range of values of B_{ext}. For iron, the Curie temperature is 1043 K.

1896, Radioactivity
Hazen, R., Trefil, J., *Science Matters*, NY: Anchor, 1992.
Battersby, S., in Tallack, P., ed., *The Science Book*, London: Weidenfeld & Nicolson, 2001.

1897, Electron
AIP, *tinyurl.com/42snq*.
Sherman, J., *J. J. Thomson and the Discovery of Electrons*, Hockessin, DE: Mitchell Lane, 2005.

1898, Mass Spectrometer
By changing the magnetic field strength, ions with different *m/z* values can be directed toward the window leading to the detector. In 1918, A. Dempster developed the first modern mass spectrometer, with much greater accuracy than previous designs.

Davies, S., in *Defining Moments in Science*, Steer, M., Birch, H., and Impney, A., eds., NY: Sterling, 2008.

1903, Black Light
Rielly, E., *The 1960s*, Westport, CT: Greenwood Press, 2003.
Stover, L., *Bohemian Manifesto*, NY: Bulfinch, 2004.

1903, Tsiolkovsky Rocket Equation
Kitson, D., in *Defining Moments in Science*, Steer, M., Birch, H., and Impney, A., eds., NY: Sterling, 2008.

1904, Lorentz Transformation
In 1905, Einstein derived the Lorentz transformation using assumptions of his special theory of relativity, along with the fact that the speed of light is constant in a vacuum in any inertial reference frame. Other physicists who considered the transformation include G. FitzGerald, J. Larmor, and W. Voigt.

1905, $E = mc^2$
Farmelo, G., *It Must be Beautiful*, London: Granta, 2002.
Bodanis, D., $E = mc^2$, NY: Walker, 2005.

1905, Photoelectric Effect
Lamb, W., Scully, M., *Jubilee Volume in Honor of Alfred Kastler* (Paris: Presses Universitaires de France, 1969.
Kimble, J., et al., *Phys. Rev. Lett.* **39**: 691; 1977.

1905, Third Law of Thermodynamics
Trefil, J., *The Nature of Science*, NY: Houghton Mifflin, 2003.

1908, Geiger Counter
Bookrags, *tinyurl.com/y89j2yz*.

1910, Cosmic Rays
When cosmic rays interact with interstellar gas and radiation within a galaxy, the cosmic rays produce gamma rays, which can reach detectors on the Earth.

Pierre Auger Observatory, *tinyurl.com/y8eleez*.

1911, Superconductivity
Baker, J., *50 Physics Ideas You Really Need to Know*, London: Quercus, 2007.
How Stuff Works, *tinyurl.com/anb2ht*.

1911, Atomic Nucleus
Rutherford conducted the gold foil experiment with H. Geiger and E. Marsden in 1909.

Gribbin, J., *Almost Everyone's Guide to Science*, New Haven, CT: Yale University Press, 1999.

1911, Kármán Vortex Street
Chang, I., *Thread of the Silkworm*, NY: Basic Books, 1996.
Hargittai, I., *The Martians of Science*, NY: Oxford University Press, 2006.

1912, Cepheid Variables Measure the Universe
Leavitt, H., Pickering, E., *Harvard College Observatory Circular*, **173**; 1, 1912.

1913, Bohr Atom
In 1925, matrix mechanics (a formulation of quantum mechanics) was created by Max Born, Werner Heisenberg, and Pascual Jordan.

Goswami, A., *The Physicists' View of Nature*, Vol. 2, NY: Springer, 2002.
Trefil, J., *The Nature of Science*, NY: Houghton Mifflin, 2003.

1913, Millikan Oil Drop Experiment
Millikan also used X-rays to irradiate droplets to change the ionization of molecules in the droplet and, thus, the total charge.

Tipler, P., Llewellyn, R., *Modern Physics*, NY: Freeman, 2002.

1919, String Theory
See notes for **Theory of Everything.**

Atiyah, M., *Nature*, **438**, 1081; 2005.

1921, Einstein as Inspiration
Levenson, T., *Discover*, **25**: 48; 2004.
Ferren, B., *Discover*, **25**: 82; 2004.

1922, Stern-Gerlach Experiment
A neutral atom was used in the Stern-Gerlach experiment because if actual ions or free electrons were used, the deflections under study with respect to atomic magnetic dipoles would be obscured by the larger deflections due to electric charge.

Gilder, L., *The Age of Entanglement*, NY: Knopf, 2008.

1923, Neon Signs
Hughes, H., West, L., *Frommer's 500 Places to See Before They Disappear*. Hoboken, NJ: Wiley, 2009.
Kaszynski, W., *The American Highway*, Jefferson, NC: McFarland, 2000.

1924, De Broglie Relation
Baker, J., 50 *Physics Ideas You Really Need to Know*, London: Quercus, 2007.

1925, Pauli Exclusion Principle
Massimi, M., *Pauli's Exclusion Principle*, NY: Cambridge University Press, 2005.
Watson, A., *The Quantum Quark*, NY: Cambridge University Press, 2005.

1926, Schrödinger's Wave Equation
Max Born interpreted ψ as probability amplitude.

Miller, A., in Farmelo, G., *It Must be Beautiful*, London: *Granta*, 2002.
Trefil, J., *The Nature of Science*, NY: Houghton Mifflin, 2003.

1927, Complementarity Principle
Cole, K., *First You Build a Cloud, NY: Harvest*, 1999
Gilder, L., *The Age of Entanglement*, NY: Knopf, 2008.
Wheeler, J., *Physics Today*, **16: 30; 1963.**

1927, Hypersonic Whipcracking
McMillen, T, Goriely, A., *Phys. Rev. Lett.*, **88**: 244301-1; 2002.
McMillen, T, Goriely, A., *Physica D*, **184**: 192; 2003.

1928, Dirac Equation
Wilczek, F., in Farmelo, G., *It Must Be Beautiful*, NY: Granata, 2003.
Freeman, D., in Cornwell, J., *Nature's Imagination*, NY: Oxford University Press, 1995.

1929, Hubble's Law of Cosmic Expansion
Huchra, J., *tinyurl.com/yc2vy*38.

1929, Cyclotron
Dennison, N., in *Defining Moments in Science*, Steer, M., Birch, H., and Impney, A., eds., NY: Sterling, 2008.
Herken, G., *Brotherhood of the Bomb*, NY: Henry Holt, 2002.

1931, White Dwarfs and Chandrasekhar Limit
White dwarfs may accrete material from nearby stars to increase their mass. As the mass approaches the Chandrasekhar Limit, the star exhibits increased rates of fusion, which may lead to a type 1a supernova explosion that destroys the star.

1932, Neutron
During the process of beta decay of the free neutron, the free neutron becomes a proton and emits an electron and an antineutrino in the process.

Cropper, W., *Great Physicists*, NY: Oxford University Press, 2001.
Oliphant, M., Bull. *Atomic Scientists*, **38**: 14; 1982.

1932, Antimatter
In 2009, researchers detected positrons in lightning storms.

Baker, J., 50 *Physics Ideas You Really Need to Know*, London: Quercus, 2007.
Kaku, M., *Visions*, NY: Oxford University Press, 1999.

1933, Dark Matter
Dark matter is also suggested by astronomical observations of the ways in which galactic clusters cause gravitational lensing of background objects.

McNamara, G., Freeman, K., *In Search of Dark Matter*. NY: Springer, 2006.

1933, Neutron Stars
The neutrons in the neutron star are created during the crushing process when protons and electrons form neutrons.

1935, EPR Paradox
Although we have used spin in this example, other observable quantities, such as photon polarization, can be used to demonstrate the paradox.

1935, Schrödinger's Cat
Moring, G., *The Complete Idiot's Guide to Understanding Einstein*, NY: Alpha, 2004.

1937, Superfluids
Superfluidity has been achieved with two isotopes of helium, one isotope of rubidium, and one isotope of lithium. Helium-3 becomes a superfluid at a different lambda point temperature and for different reasons than helium-4. Both isotopes never turn solid at the lowest temperatures achieved (at ordinary pressures).

1938, Nuclear Magnetic Resonance
Ernst, R., Foreword to *NMR in Biological Systems*, Chary, K., Govil, G., eds., NY: Springer, 2008.

1942, Energy from the Nucleus
Weisman, A., *The World Without US*, NY: Macmillan, 2007.

1943, Silly Putty
Credit for the invention of Silly Putty is sometimes disputed and includes other individuals such as E. Warrick of the Dow Corning Corporation.

Fleckner, J., *tinyurl.com/ydmubcy*.

1945, Drinking Bird
Sobey, E., Sobey, W., *The Way Toys Work*, Chicago, IL: Chicago Review Press, 2008.

1945, Little Boy Atomic Bomb
The second atomic bomb, the "Fat Man," was dropped three days later on Nagasaki. Fat Man made use of plutonium-239 and an implosion device—similar to the Trinity bomb tested in New Mexico. Six days after the Nagasaki bombing, Japan surrendered.

1947, Transistor
Riordan, M., Hoddeson, L., *Crystal Fire*, NY: Norton, 1998.

1947, Sonic Booms
The plane's speed affects the cone width of the shock waves.

1947, Hologram
Gabor's hologram theories predated the availability of laser light sources.

One can create the illusion of movement in a hologram by exposing a holographic film multiple times using an object in different positions. Interestingly, a hologram film can be broken into small pieces, and the original object can still be reconstructed and seen from each small piece. The hologram is a record of the phase and amplitude information of light reflected from the object.

Kasper, J., Feller, S., *The Complete Book of Holograms*, Hoboken, NJ: Wiley, 1987.

1948, Quantum Electrodynamics
The Lamb shift is a small difference in energy between two states of the hydrogen atom caused by the interaction between the electron and the vacuum. This observed shift led to renormalization theory and a modern theory of QED.

Greene, B., *The Elegant Universe*, NY: Norton, 2003.
QED, Britannica, *tinyurl.com/yaf6uuu*.

1948, Tensegrity
In 1949, B. Fuller developed an icosahedron based on the tensegrity. Early work in tensegrity-like structures was also explored by K. Ioganson. Ancient sailing ships may have relied on tensegrities, using ropes interacting with vertical struts. Fuller also filed many key patents on tensegrities. The cell's cytoskeleton is composed of molecular struts and cables that connect receptors on the cell surface to connection points on the cell's central nucleus.

1948, Casimir Effect
The effect was accurately measured in 1996 by physicist S. Lamoreaux. One of the earliest experimental tests was conducted by M. Sparnaay, in 1958.

Reucroft, S., Swain, J. *tinyurl.com/yajqouc*.
Klimchitskaya, G., et al., *tinyurl.com/yc2eq4q*.

1949, Radiocarbon Dating
Other methods such as potassium-argon dating are employed for dating very old rocks.

Bryson, B., *A Short History of Everything*, NY: Broadway, 2003.

1954, Solar Cells
The process of adding impurities (such as phosphorous) to silicon is called doping. The phosphorous-doped silicon is an N-type (for negative). Silicon is a semiconductor, and its conductivity can be drastically changed by doping. Other materials used for solar cells include gallium arsenide.

1955, Leaning Tower of Lire
Walker, J., *The Flying Circus of Physics*, Hoboken, NJ: Wiley, 2007.

1955, Seeing the Single Atom
See the entry on **Quantum Tunneling** for the scanning tunneling microscope. Note also that atomic force microscopy (AFM), developed in the 1980s, is another form of high-resolution scanning probe microscopy. Here, various forces between the sample surface and probe cause a deflection in a cantilever according to Hooke's Law.

Crewe, A., Wall, J., Langmore, J., *Science*, **168**: 1338; 1970.
Markoff, J., *tinyurl.com/ya2vg4q*.
Nellist, P., *tinyurl.com/ydvj5qr*.

1955, Atomic Clocks
In 2010, "optical clocks" that employ atoms (e.g. aluminum-27) that oscillate at the frequencies of light rather than in the microwave range were among the most precise timekeepers.

1956, Neutrinos
Aside from the Sun, other sources of neutrinos include nuclear reactors, atmospheric neutrinos produced by interactions between atomic nuclei and cosmic rays, supernovae, and neutrinos

produced during the Big Bang. The Ice Cube Neutrino Observatory is a recently built neutrino telescope in the deep Antarctic ice.

Lederman, L., Teresi, D., *The God Particle* (Boston, MA: Mariner, 2006).

1956, Tokamak
Deuterium and tritium (isotopes of hydrogen) are used as input to the fusion reaction. Neutrons, created by the reaction, leave the tokamak and are absorbed in the surrounding walls, creating heat. The waste products are not useable to create nuclear weapons. One of the magnetic fields is produced by external electric currents flowing in coils wrapped around the "doughnut." A smaller magnetic field is generated by an electric current flowing around the torus in the plasma.

1958, Integrated Circuit
Bellis, M., *tinyurl.com/y93fp7u*.
Miller, M., *tinyurl.com/nab2ch*.

1960, Laser
Numerous researchers are associated with the development of laser theory, including Nikolay Basov, Aleksandr Prokhorov, Gordon Gould, and many more.

1961, Anthropic Principle
Hawking, S., *A Brief History of Time*, NY: Bantam Books, 1988.
Trefil, J., *The Nature of Science*, NY: Houghton Mifflin, 2003.

1961, Standard Model
S. Glashow's discovery of a way to combine the electromagnetic and weak interactions (the latter of which is also referred to as weak nuclear force) provided one of the earlier steps toward the Standard Model. Other key individuals include S. Weinberg and A. Salam.

While on the subject of subatomic particles, note that in 1935, Hideki Yukawa predicted the existence of the meson (later called the pion) as a carrier of the strong nuclear force that binds atomic nuclei. Gluons are involved in interactions among quarks, and are indirectly involved with the binding of protons and neutrons.

Battersby, S., in Tallack, P., ed., *The Science Book*, London: Weidenfeld & Nicolson, 2001.

1962, Electromagnetic Pulse
Publishers Weekly, *tinyurl.com/62ht4pp*.

1963, Chaos Theory
Pickover, C., *The Math Book*, NY: Sterling, 2009.

1963, Quasars
Hubblesite, *tinyurl.com/yadj3by*.

1963, Lava Lamp
Ikenson, B., *Patents*, NY: Black Dog & Leventhal, 2004.

1964, God Particle
The Higgs mechanism has many other important physicists associated with its development, including F. Englert, R. Brout, P. Anderson, G. Guralnik, C. R. Hagen, T. Kibble, S. Weinberg, and A. Salam.

Baker, J., *50 Physics Ideas You Really Need to Know*, London: Quercus, 2007.

1964, Quarks
Jones, J., Wilson, W., *An Incomplete Education*, NY: Ballantine, 1995.

1964, CP Violation
CP violation appears to be insufficient to account for the matter-antimatter asymmetry during the Big Bang. Accelerator experiments may reveal sources of CP violation beyond the sources discovered in 1964.

1964, Bell's Theorem
Albert, D., Galchen, R., *tinyurl.com/yec7zoe*.
Kapra, F., *The Tao of Physics*, Boston, MA: Shambhala, 2000.

1965, Super Ball
Smith, W., *tinyurl.com/ykay4hh*.

1965, Cosmic Microwave Background
In 1965, R. Dicke, P. J. E. Peebles, P. G. Roll, and D. T. Wilkinson interpreted the results of A. Penzias and R. Wilson and declared the background radiation as a signature of the big bang. The WMAP satellite, launched in 2001, provided additional detail on these fluctuations. The HIGH-altitude BOOMERANG balloon, flown over Antarctica in 1997, 1998, and 2003, also provided observations of the CMB.

Bryson, B., *A Short History of Everything*, NY: Broadway, 2003.

1967, Gamma-Ray Bursts
Ward, P., Brownlee, D., *The Life and Death of Planet Earth*. NY: Macmillan, 2003.
Melott, A., et al., *Intl. J. Astrobiol.* **3**: 55;2004.
NASA, *tinyurl.com/4prwz*.

1967, Living in a Simulation
Davies, P., *tinyurl.com/yfyap9t*.
Reese, M., *tinyurl.com/yke5b7w*.

1967, Tachyons
Other individuals associated with the idea of tachyons include A. Sommerfeld, G. Sudarshan, O.-M. Bilaniuk, and V. Deshpande. Strangely, as a tachyon increases its speed, it loses energy. Once a tachyon has lost all its energy, it travels at infinite velocity and simultaneously occupies every point along its trajectory. Particles that live in this bizarre state of omnipresence are called "transcendent" particles.

Herbert, N., *Faster Than Light*, NY: New American Library, 1989.
Nahin, P., *Time Machines*, NY: Springer, 1993.

1967, Newton's Cradle
Kinoshita1, T., et al., *Nature*, **440**: 900; 2006.

1969, Unilluminable Rooms
Darling, D., *UBM*, Wiley, 2004.
Pickover, C., *The Math Book*, NY: Sterling, 2009.
Stewart, I., *Sci. Am.* **275**: 100; 1996.
Stewart, I., *Math Hysteria*, OUP, 2004.

1971, Supersymmetry
Other early contributors to supersymmetry include H. Miyazawa, J. L. Gervais, Y. Golfand, E. P. Likhtman, D. V. Volkov, V. P. Akulov, and J. Wess. The lightest SUSY particle is a possible candidate for dark matter. SUSY suggests that for every known subatomic matter particle, the universe should have a corresponding force-carrying particle, and vice versa.

Seife, C., *Alpha and Omega*, NY: Viking, 2003.
Greene, B., *The Elegant Universe*, NY: Norton, 2003.
Ananthaswamy, A., *tinyurl.com/yh3fjr2*.

1980, Cosmic Inflation
Other contributors to inflation theory are P. Steinhardt and A. Albrecht. Inflationary theory suggests why magnetic monopoles have not been discovered. Monopoles may have formed in the Big Bang and then dispersed during the inflationary period, decreasing their

density to such an extent as to make them undetectable.

Guth, A., *The Inflationary Universe*, NY: Perseus, 1997.
Musser, G., *The Complete Idiot's Guide to String Theory*, NY: Alpha, 2008.

1981, Quantum Computers
Other important names associated with quantum computing include C. Bennett, G. Brassard, and P. Shor.

Clegg, B., *The God Effect*, NY: St. Martins, 2006.
Kaku, M., *Visions*, NY: Oxford University Press, 1999.
Kurzweil, R., *The Singularity is Near*, NY: Viking, 2005.
Nielsen M., Chuang, I., *Quantum Computation and Quantum Information*, NY: Cambridge University Press, 2000.

1982, Quasicrystals
In 2007, researchers published evidence in Science of a Penrose-like tiling in medieval Islamic art, five centuries before their discovery in the West. R. Ammann independently discovered these kinds of tilings at approximately the same time as Penrose.

Pickover, C., *The Math Book*, NY: Sterling, 2009.
Gardner, M., *Penrose Tiles to Trapdoor Ciphers*, Freeman, 1988.
Lu, P., Steinhardt, P., *Science*, **315**: 1106; 2007.
Penrose, R., Bull. of the Inst. Math. Applic., **10**: 266;1974.
Senechal, M., *Math. Intell.*, **26**: 10; 2004.

1984, Theory of Everything
Note that in modern quantum theories, forces result from the exchange of particles. For example, an exchange of photons between two electrons generates the electromagnetic force. At the Big Bang, it was thought that all four forces were really just one force, and only as the universe cooled did the four forces become distinct.

Greene, B., *The Elegant Universe*, NY: Norton, 2003.
Kaku, M., *Visions*, NY: Oxford University Press, 1999.
Lederman, L., Teresi, D., *The God Particle* (Boston, MA: Mariner, 2006).

1985, Buckyballs
Graphene, resembling a hexagonal grid of carbon atoms, may have many potential applications, partly because electrons travel more quickly through this material than silicon under certain conditions.

Technology Review, *tinyurl.com/nae4o7*.

1987, Quantum Immortality
B. Marchal has also shaped the theory of quantum immortality.

1987, Self-Organized Criticality
B. Mandelbrot has also focused attention on many examples of power-law correlations in his study of fractals. This complex behavior may be studied as a whole system rather than focusing on the behavior of individual components.

Frette, V., et al., *Nature*, **379**: 49; 1996.
Grumbacher, S., et al., *Am. J. Phys.*, **61**: 329; 1993.
Jensen, H., *Self-Organized Criticality*, NY: Cambridge, 1998.

1988, Wormhole Time Machine
A feedback loop of virtual particles might circulate through the wormhole and destroy it before it can be used as a time machine.

1992, Chronology Protection Conjecture
Silverberg, R., *Up the Line*, NY: Del Ray, 1978.

1992, Quantum Teleportation
See also the work of G. Brassard, L. Smolin, W. Wootters, D. DiVincenzo. C. Crépeau, R. Jozsa, A. Peres, and others. In 2010, scientists teleported information between photons over a free-space distance of nearly ten miles.

1993, Stephen Hawking on *Star Trek*
Smolin, L., *Three Roads to Quantum Gravity*, NY: Basic Books, 2001.

1995, Bose-Einstein Condensate
W. Ketterle created a condensate made of sodium-23, allowing him to observe quantum mechanical interference between two different condensates.

1998, Dark Energy
Overbye, D., *http://tinyurl.com/y99ls7r*.
Tyson, N., *The Best American Science Writing 2004*, D. Sobel, ed., NY: Ecco, 2004.
Tyson, N., Goldsmith, D., *Origins*, NY: Norton, 2005.

2007, HAARP
The effects of HAARP on the atmosphere would only last from seconds to a few minutes before completely dissipating. Other ionosphere research facilities exist around the world, but HAARP is unique due to its range of facilities. The HAARP antennas are 72 feet high.

Streep, A., *tinyurl.com/4n4xuf*.

2008, Blackest Black
Yang, Z., et al., *Nano Lett.* 8: 446; 2008.

2009, Large Hadron Collider
The LHC is a synchrotron accelerator.

Bryson, B., *tinyurl.com/yfh46jm*.

36 Billion, Cosmological Big Rip
Today, physicists are not sure why the dark energy should exist, and some have considered explanations including unseen dimensions or parallel universes that affect our own. In the Big Rip scenario, dark energy is also referred to as *phantom energy*. In another model, 10^{-27} seconds before the Big Rip, the universe contracts, generating countless small, separate universes.

Caldwell, R., Kamionkowski, M., Weinberg, N., *Phys. Rev. Lett.*, **91**: 071301; 2003.

100 Billion, Cosmic Isolation
Krauss, L., Scherrer, R., *Scient. Amer.*, **298**: 47; 2008.
Verschuur, G., *Interstellar Matters*, NY: Springer, 1989.

100 Trillion, Universe Fades
According to physicist Andrei Linde, the process of eternal inflation implies an eternal "self-reproduction" of the universe, with quantum fluctuations leading to production of separate universes. He writes, "the existence of this process implies that the universe will never disappear as a whole. Some of its parts may collapse, the life in our part of the universe may perish, but there will always be some other parts of the universe where life will appear again and again. . . ."

Adams, F., Laughlin, G., *The Five Ages of the Universe*, NY, Free Press, 2000.

>100 Trillion, Quantum Resurrection
Banks, T., *tinyurl.com/yjxapye*.

Index

Photo Credits

Because several of the old and rare illustrations shown in this book were difficult to acquire in a clean and legible form, I have sometimes taken the liberty to apply image-processing techniques to remove dirt and scratches, enhance faded portions, and occasionally add a slight coloration to a black-and-white figure in order to highlight details or to make an image more compelling to look at. I hope that historical purists will forgive these slight artistic touches and understand that my goal was to create an attractive book that is aesthetically interesting and alluring to a wide audience. My love for the incredible depth and diversity of topics in physics and history should be evident through the photographs and drawings.

Images © Clifford A. Pickover: pages 111, 399, 469

NASA Images: p. 19 NASA/WMAP Science Team; p. 21 NASA; p. 75 NASA; p. 81 NASA/JPL/Space Science Institute and D. Wilson, P. Dyches, and C. Porco; p, 107 NASA/JPL; p. 113 NASA/JPL-Caltech/T. Pyle (SSC/Caltech); p. 119 NASA; p. 133 NASA and The Hubble Heritage Team (AURA/STScI); p. 141 NASA/JPL-Caltech/T. Pyle (SSC); p. 185 HiRISE, MRO, LPL (U. Arizona), NASA; p. 189 NASA/JPL-Caltech/R. Hurt (SSC); p. 191 NASA/JPL-Caltech; p. 223 NASA; p. 267 NASA/JPL/University of Texas Center for Space Research; p. 281 NASA/JPL; p. 289 NASA; p. 301 NASA, ESA and The Hubble Heritage Team (STScI/AURA); p. 307 NASA; p. 315 Bob Cahalan, NASA GSFC; p. 319 NASA, the Hubble Heritage Team, and A. Riess (STScI); p. 359 R. Sahai and J. Trauger (JPL), WFPC2 Science Team, NASA, ESA; p. 367 NASA/Swift Science Team/Stefan Immler; p. 369 NASA; p. 391 NASA/Wikimedia; p. 427 Apollo 16 Crew, NASA; p. 443 NASA/JPL-Caltech; p. 457 NASA; p. 459 NASA, Y. Grosdidier (U. Montreal) et al., WFPC2, HST; p. 473 NASA, WMAP Science Team; p. 489 NASA; p. 513 NASA, ESA, and the Hubble Heritage Team (STScI/AURA)-ESA/Hubble Collaboration; p. 515 NASA, ESA, and A. Feild (STScI)

Used under license from Shutterstock.com: p. 5 © Eugene Ivanov; p. 17 © Catmando; p. 18 © Maksim Nikalayenka; p. 27 © Sean Gladwell; p. 31 © Roman Sigaev; p. 33 © zebra0209; p. 42 © Zaichenko Olga; p. 43 © Andreas Meyer; p. 49 © lebanmax; p. 56 © Hintau Aliaksei; p. 57 © Tischenko Irina; p. 61 © William Attard McCarthy; p. 65 © Lagui; p. 67 © Graham Prentice; p. 69 © Chepko Danil Vitalevich; p. 76 © Mike Norton; p. 79 © Christos Georghiou; p. 86 © Roman Shcherbakov; p. 93 © Tobik; p. 95 © Graham Taylor; p. 97 © Steve Mann; p. 101 © Rich Carey; p. 105 © Harper; p. 109 © YAKOBCHUK VASYL; p. 115 © Awe Inspiring Images; p. 117 © Tatiana Popova; p. 129 © Patrick Hermans; p. 135 © Andrea Danti; p. 145 © STILLFX; p. 149 © Tischenko Irina; p. 151 © Ulf Buschmann; p. 153 © ynse; p. 155 © jon le-bon; p. 158 (lower) © Sebastian Kaulitzki; p. 159 © Solvod; p. 160 © Bill Kennedy; p. 163 © Martin KubÃ?Â¡t; p. 164 © Brian Weed; p. 165 © Norman Chan; p. 167 © Noel Powell, Schaumburg; p. 169 © Oleg Kozlov; p. 171 © Ronald Sumners; p. 173 © Kenneth V. Pilon; p. 175 © Mana Photo; p. 176 © S1001; p. 177 © Teodor Ostojic; p. 179 © anotherlook; p. 180 © Kletr; p. 187 © Awe Inspiring Images; p. 192 © Elena Elisseeva; p. 193 © Sebastian Kaulitzki; p. 195 © coppiright; p. 197 © bezmaski; p. 199 © Jim Barber; p. 203 © iofoto; p. 209 © Diego Barucco; p. 211 © Tischenko Irina; p. 212 © Mark Lorch; p. 213 © kml; p. 215 © Ellas Design; p. 217 © Gorgev; p. 220 © Jose Gil; p. 225 © Shchipkova Elena; p. 227 © John R. McNair; p. 229 © Johnathan Esper; p. 230 © Joseph Calev; p. 231 © Mark William Penny; p. 233 © Dmitri Melnik; p. 234 © Konstantins Visnevskis; p. 239 © Stephen McSweeny; p. 241 © nadiya_sergey; p. 245 © Yellowj; p. 247 © Allison Achauer; p. 249 © ErickN; p. 251 © nikkytok; p. 253 © Tatiana Belova; p. 257 © Scott T Slattery; p. 261 © Jessmine; p. 263 © WitR; p. 265 © Chad McDermott; p. 271 © Maridav; p. 273 © Lee Torrens; p. 279 © Jhaz Photography; p. 283 © beboy; p. 285 © Marcio Jose Bastos Silva; p. 287 © ex0rzist; p. 293 © patrick hoff; p. 299 © FloridaStock; p. 305 © Alex; p. 309 © Perry Correll; p. 313 © Chepe Nicoli; p. 320 © immelstorm; p. 321 © Mopic; p. 323 © 1236997115; p. 327 © Mark R.; p. 329 © Ivan Cholakov Gostock-dot-net; p. 335 © Norman Pogson; p. 339 © Alegria; p. 341 © EML; p. 344 © garloon; p. 347 © Eugene Ivanov; p. 361 © Geoffrey Kuchera; p. 375 © -baltik-; p. 377 © Olga Miltsova; p. 381 © Carolina K. Smith, M.D.; p. 385 © Kesu; p. 387 © Ken Freeman; p. 405 © JustASC; p. 407 © Vladimir Wrangel; p. 409 © photoBeard; p. 410 © Dwight Smith; p. 411 © Otmar Smit; p. 413 © 2happy; p. 419 © Sandy MacKenzie; p. 425 © Sandro V. Maduell; p. 433 © Joggie Botma; p. 435 © red-feniks; p. 440 © f. AnRo brook; p. 445 © Huguette Roe; p. 453 © argus; p. 455 © Todd Taulman; p. 460 © Bruce Rolff; p. 461 © Tonis Pan; p. 463 © Alperium; p. 464 © ErickN; p. 465 © Martin Bech; p. 470 © Petrunovskyi; p. 471 © Andrejs Pidjass; p. 476 (upper) © Tischenko Irina; p. 481 © sgame; p. 483 © Linda Bucklin; p. 484 © Elena Elisseeva; p. 485 © Evgeny Vasenev; p. 487 © edobric; p. 490 © Mopic; p. 491 © Tischenko Irina; p. 493 © Markus Gann; p. 494 © Anton Prado PHOTO; p. 505 © danilo ducak; p. 507 © Olaru Radian-Alexandru; p. 511 © sdecoret; p. 516 © iDesign; p. 517 © Jurgen Ziewe

Other Images: p. 28 ©istockphoto.com/akaplummer; p. 29 ©istockphoto.com/SMWalker; p. 35 Wikimedia/Ryan Somma; p. 39 Stan Sherer; p. 51 Wikimedia/Björn Appel; p. 53 Rien van de Weijgaert, www.astro.rug.nl/~weygaert; p. 55 John R. Bentley; p. 71 Sage Ross/Wikimedia; p. 77 © Allegheny Observatory, University of Pittsburgh; pp. 82-83 Wikimedia/Beltsville Agricultural Research Center, US Dept. of Agriculture/E. Erbe, C. Pooley; p. 85 N. C. Eddingsaas & K. S. Suslick, UIUC; p. 87 Shelby Temple; p. 89 US Air Force/J. Strang; p. 91 Softeis/Wikimedia; p. 99 Joan O'Connell Hedman; p. 127 Jan Herold/Wikimedia; p. 131 Bert Hickman, Stoneridge Engineering, www.capturedlightning.com; p. 134 Teja Krašek, http://tejakrasek.tripod.com; p. 137 © Bettmann/CORBIS; p. 147 Wikimedia; p. 161 Mark Warren, specialtyglassworks.com; p. 168 Wikimedia; p. 181 J. E. Westcott, US Army photographer; p. 186 Wikimedia; p. 201 Wikimedia; p. 205 © Justin Smith/Wikimedia/CC-By-SA-3.0; p. 219 Stéphane Magnenat/Wikimedia; p. 221 Hannes Grobe/AWI/Wikimedia; p. 235 John Olsen, openclipart.org; p. 237 US Navy; p. 243 ©istockphoto.com/iceninephoto; p. 259 Mila Zinkova; p. 269 ©istockphoto.com/isle1603; p. 275 Heinrich Pniok/Wikimedia; p. 277 ©istockphoto.com/PointandClick; p. 297 U.S. Army/Spc. Michael J. MacLeod; p. 303 Wikimedia; p. 311, 317 Courtesy of Brookhaven National Laboratory; p. 331 Wikimedia; p. 333 Frank Behnsen/Wikimedia; p. 337 Lawrence Berkeley National Laboratory/Wikimedia; p. 349 ©istockphoto.com/STEVECOLEccs; p. 353 Randy Wong/Sandia National Laboratories; p. 357 NARA; p. 363, 365 Courtesy of Brookhaven National Laboratory; p. 371 Idaho National Laboratory; p. 373 K. S. Suslick and K. J. Kolbeck, University of Illinois; p. 379 Alfred Leitner/Wikimedia; p. 382 Wikimedia; p. 383 Ed Westcott/US Army/Wikimedia; p. 389 NARA/Wikimedia; p. 395 U.S. Navy photo by Ensign John Gay/Wikimedia; p 397 Heike Löchel/Wikimedia; p. 403 Courtesy of Umar Mohideen, University of California, Riverside, CA; p. 415 Courtesy the American Institute of Physics/M. Rezeq et al., J. Chem. Phys.; p. 417 NIST; p. 421 Fermilab National Accelerator; p. 423 Courtesy Princeton Plasma Physics Laboratory; p. 431 The Air Force Research Laboratory's Directed Energy Directorate; p. 437 Courtesy of Brookhaven National Laboratory; p. 439 National Archive/U.S. Air Force; p. 441 Daniel White; p. 447 CERN; p. 449, 451 Courtesy of Brookhaven National Laboratory; p. 467 Keith Drake/NSF; p. 475 J. Jost/NIST; p. 476 (lower) courtesy of Edmund Harriss; p. 477 Wikimedia; p. 479 Courtesy of Brookhaven National Laboratory; p. 495 Official White House Photostream; p. 497 NIST/JILA/CU-Boulder/Wikimedia; p. 499 Lawrence Berkeley National Lab; p. 501 CERN; p. 503 OAR/ERL/NOAA's National Severe Storms Laboratory (NSSL); p. 504 HAARP/US Air Force/Office of Naval Research; p. 509 CERN